The
GREAT BETRAYAL

Also by Horace Freeland Judson

The Techniques of Reading

Heroin Addiction in England

The Eighth Day of Creation: Makers of the Revolution in Biology
(Expanded Edition, 1996)

The Search for Solutions

The GREAT BETRAYAL

Fraud in Science

*

HORACE FREELAND JUDSON

Harcourt, Inc.

Orlando Austin New York San Diego Toronto London

www.HarcourtBooks.com

Library of Congress Cataloging-in-Publication Data
Judson, Horace Freeland.
The great betrayal: fraud in science/Horace Freeland Judson.—1st ed.
p. cm.
Includes bibliographical references and index.
ISBN 0-15-100877-9
1. Fraud in science. I. Title.
Q175.37.J84 2004
344'.095—dc22 2004005906

Text set in Adobe Garamond
Designed by Ivan Holmes

Printed in the United States of America

First edition
A C E G I K J H F D B

for Olivia

Ach! Furchtbar ist Gewissen ohne Wahrheit.

—Sophocles, *Antigonæ,* Friedrich Hölderlin, translator

CONTENTS

PREFACE

On 22 March 1991, the front page of *The New York Times* carried the headline BIOLOGIST WHO DISPUTED A STUDY PAID DEARLY. The story, by Philip Hilts out of the *Times*'s Washington bureau, was a miniature profile of a scientist named Margot O'Toole, who had raised questions about a paper originating in two laboratories at the Massachusetts Institute of Technology and published five years earlier in the journal *Cell.* A senior author of the paper was David Baltimore, who when the *Times*'s account appeared was president of Rockefeller University, in New York. A co-author was Thereza Imanishi-Kari, who back when *Cell* ran the paper had headed the small laboratory at MIT where O'Toole was on a one-year fellowship. Baltimore had brushed O'Toole off as a "a disgruntled postdoctoral fellow." She had been driven from science; she and her husband had had to sell their house; she had gone to work answering the telephone at a moving company her brother owned. Now, however, a draft of a report about the dispute over the paper had leaked from the Office of Scientific Integrity, at the National Institutes of Health. The draft declared that O'Toole had been right about the paper, that she was a heroine who had "maintained her commitment to scientific integrity," and that data originating with Imanishi-Kari had been faked.

The case was certainly pungent. It was new to me, but took its place in my awareness of a rising number of scientific frauds that had been uncovered in recent years. I turned to the inside page. What immediately caught my eye was the name of a friend, Mark Ptashne, a molecular biologist at Harvard who, the account said, had taken up O'Toole's case, had got her a job back in science at Genetics Institute, a biotechnology company I knew he had founded, and who now spoke out on her behalf.

I was at Stanford University that year as a senior visiting scholar in history of science. I telephoned Ptashne: "You're getting your fifteen minutes of fame?!" He said, "No, wait! This is a serious and interesting case!"—and proceeded to tell me much more about it, about Baltimore's involvement, and about O'Toole.

That is how this book began. I got in touch with Walter Stewart and Ned Feder, two scientists at the NIH who were following the Baltimore–Imanishi-Kari affair closely; I went into the matter with O'Toole and asked to interview Baltimore, whom I had first met twenty years earlier when working on *The Eighth Day of Creation,* a history of molecular biology. I talked with a number of other scientists, among them Gerald Edelman, an immunologist of erudition, dominating intelligence, and philosophical bent, who was at Rockefeller University and had no love of Baltimore. Back at Stanford, sometime in January I got a call from Martin Kessler, editor in chief and publisher of Basic Books, in New York, a firm he had founded. I knew something of Kessler's house: big books on serious subjects, worthy, sometimes ponderous. He explained that he was Edelman's publisher and had learned from him that I was planning a book about fraud in science. He had been looking for just such a book; what I described to him, he said, was exactly what he had in mind. I had a contract, though, with my publishers, Simon and Schuster, for two other books, but one of these I had decided not to write, and the other, all but completed, they did not really want to publish. I told Kessler that I was negotiating to kill that contract.

By early summer, I was free to write a prospectus, and soon signed with Kessler. But then he moved to the Free Press. Before he could take my book with him, he died. His successor at Basic Books was promoted from the marketing rather than the editorial side; he was not interested in my project and wanted the advance back. The book was in limbo. I kept on with it. But I am cursed with indolence, accidie, the deadliest of sins, and had personal and professional preoccupations, for I had moved to George Washington University, bringing with me a large grant, to found there the Center for History of Recent Science. At about that time, though, I met Michael Bessie, an editor and publisher of enormous experience and wisdom, semi-retired. He in turn put me in touch with Jane Isay, editor in chief at Harcourt. The times had altered, the landscape of fraud had widened. She pressed me to restructure the book entirely. At first I resisted: a lot of interesting material would be sacrificed. But she was right. The result you have in your hand.

Many debts have accumulated over those too-many years. It is an honor and a great pleasure to thank Jane Isay and Michael and Cornelia Bessie. Drummond Rennie and C. K. (Tina) Gunsalus know more about scientific fraud, its epidemiology, cure, and prevention, than anyone else, and have been remarkably generous with information and encouragement; Tina read the whole and had excellent suggestions. Margot O'Toole read Chapter 5, on the Baltimore affair, and made essential points. Paul Ginsparg, a central figure in Chapter 8, on open-access publishing, read that chapter and commented sharply, which is just what one wants. Joe Cecil, at the Federal Judicial Center, directed me to the sources for the discussion in Chapter 9 of science versus law, and read that portion, showing me how to amplify it; Scott Kieff read that whole chapter and had trenchant observations. George Thomsen taught me the curious relevance of a fact about tax law. Robert Pollack read a lot of the manuscript in various drafts, most recently Chapter 9 and the Epilogue; he is an intellectual sparring partner of quick brilliance from whom I always come away lightly bruised and greatly exhilarated. Philip Kitcher read portions of the earlier version, to deft and bracing effect. Fellows and colleagues at the Center for History of Recent Science were critical, collegial readers. Noel Swerdlow saved me from an ignorant error about Claudius Ptolemy, who therefore does not appear. I thank my agent, Frank Curtis, for his brisk good sense and long patience.

Many others have been encouraging along the way: I must mention, with deep gratitude, Thomas Blake, Richard Cohen, Anne Fitzpatrick, Bill Glen and the others of the Heretics' Club, Lillian Hoddeson, and for critical reading as a historian and tenderness and laughter as a friend, Angela Matysiak.

Of course, always, I have been urged on by my children Grace Judson, Olivia Judson, and Nicholas Judson. Olivia and Nicholas are scientists, as I am not, and have been incisive frank readers.

The GREAT BETRAYAL

At Olympia, in the Peloponnesus, for centuries the site of the ancient Olympic Games, stood a row of images of Zeus, called the *Zanes,* cast in bronze and paid for by the fines levied on competitors who had cheated. The Zanes were lined up on pedestals along the north of the Altis, the consecrated precinct of the Olympian Zeus, and led directly to the gate of the private passage by which athletes entered the stadium: they were the last objects that met the eyes of those about to compete. When German archaeologists excavated Olympia, in the eighteen-seventies, they found sixteen pedestals in place.

PROLOGUE

About the light that fraud and other misconduct shine on the enterprise that the sciences at the opening of the new century have become

Consider the roles that defects have played in scientific research.

When early in the seventeenth century the anatomist William Harvey first attempted to analyze the motions of the heart and blood, he made crucial observations of animals or men that were in some relevant way defective. In healthy, normal mammals, the sequence of the beat of the four chambers of the heart, and the relationship of heartbeat to pulse, are too quick, continuous, subtle, and slimy to be distinguished by touch or hearing from outside or even by watching what happens in the chest cavity of a living beast freshly cut open—a dog or hog upon the table or a deer brought down in the chase. Harvey watched the hearts as the animals died. "The movements then become slower and rarer, the pauses longer, by which it is made much more easy to perceive and unravel what the motions really are"—so his Latin translates. Then he could see clearly that in the heartbeat the chambers move sequentially, with the two auricles (now called atria) contracting first, then the two ventricles. He showed that what we perceive from outside the chest as the beat is caused not by an active expansion of the heart, like a bellows pulling in air, to fill itself—in his day the received explanation—but by the muscular contraction, the tightening of the structure, expelling blood. The heart is a pump. A celebrated painting shows Harvey with King Charles I—he was personal physician to the king—explicating to his royal patient and patron the motion of the heart. He also contemplated the hearts of embryonic animals, chicks barely formed in the egg, for example. One mordant observation was of a patient who presented with a large, pulsating swelling at the base of the right side of his neck. This, Harvey wrote, was an aneurysm—an erosion of the wall—of the subclavian artery, which supplies the right arm, looping downwards from the collarbone, descending into the armpit. "This swelling was visibly distended as it received the charge of blood brought to it by the artery, with each stroke of the heart." (He noted, in a laconic parenthesis, "quod secto post mortem cadavere deprehensum erat," that post-mortem dissection confirmed the nature of the defect.) The pulse was thus dependent on the heartbeat, and unlike the heartbeat was passive and an expansion. Such observations of the abnormal opened to Harvey's understanding the normal physiology of blood flow. They became essential components of his demonstration that the blood circulates.

Examples multiply. In the nineteenth century, physicians encountering cases of goiter, or of cretinism, or of stunted growth, patients with galloping heartbeat and eyes so bulging that the lids would not close over them, instances of dwarfism, gigantism, acromegaly—this last a disease of adults where feet, hands, and facial bones begin to grow monstrously large—took the first steps to understanding the functions of thyroid and pituitary. Study of defects, in the clinic, engendered the science of endocrinology. Again, much of what neurobiologists know about the working of our brains comes from a century of attempts to use dysfunctions—the consequences of brain damage from accident, stroke, gunshot, or surgery in previously normal people—to analyze the neural basis of perception, thought, speech, and emotion.

Geneticists, from the start of the twentieth century to the present day, have constructed their science on patterns of transmission of variant characters—mutations—down the generations. Most of these are genetic defects: the lack of a pigment in the eye of a fly, the lack of a component in the blood-clotting mechanism in hemophilia, the change of a single unit of the molecular sequence of hemoglobin in sickle-cell anemia, and so on. Until the invention of the most recent techniques, the genetic map of fruit fly, mouse, or man has been almost entirely the map not exactly of genes but of gene defects. A leading medical geneticist has called this mapping "the morbid anatomy of the genome."

A cancer is a disturbance of the processes of cells of higher organisms, in particular of controls on their growth. Cancers will not be understood until these processes and controls are understood, and they are baffling—complex, fast, interacting, and difficult to trace out. In the nineteen-seventies, just as molecular biologists were putting together the armamentarium of new techniques that made the development and differentiation of normal cells appear solvable at last, President Richard Nixon declared a war on cancer. Biologists grumbled that money was being allocated to targetted research when fundamental, pure research was needed. They quickly learnt to write their grant applications to use cancer-research funds to ask those fundamental questions. Legitimately, too: because cancers result from faults in the normal processes of cells, they also offer a window into those processes.

In the physical sciences, to be sure, defects less often open the way to understanding the course of more usual phenomena. Yet quakes have told us of the structure of the interior of the earth, while astronomers scan the skies for new supernovas—and the prospect of a chain of comet fragments striking Jupiter, cataclysmically perturbing the planet's outer layers, entranced planetary scientists and made the front pages of the world's press for weeks on end.

The conspicuous recent example of the place of defects in biological research is, of course, acquired immune-deficiency syndrome, AIDS. The immune system is intricate, second only to the nervous system, recalcitrant to scientists' understanding. The human immunodeficiency virus sits in cells that command a central intersection of the pathways by which the system functions. Once more, as with cancer in relation to cellular biology, the attack on AIDS requires fundamental, general research into the system of which it is a perturbation. Once again, the defect opens a window.

In sum: in the sciences, defects can give access to the processes they disrupt, processes otherwise inordinately large in scale, complex, quick, inaccessible—or where deliberate intervention is impossible or unethical. Now take a step back. All those characteristics describe the enterprise of the sciences as they have developed in the latter half of the twentieth century. But fraud and related forms of misconduct are assuredly defects in the processes of science. The premise of this book is that scrutiny of the nature of fraud and other misconduct will reach to the heartbeat and pulse of what the sciences are and what scientists do at this, the start of the new millennium.

We do not know the incidence of scientific fraud. Those who have been looking into the problems most closely and long are most likely to be convinced that misconduct is widespread in scientific research. Nobody has yet demonstrated a way to measure it reliably. We are left with what's been called "anecdata."

Behind the problem of incidence lies that of definition. What exactly would we want to measure? The standard definition goes under the nickname FF&P, for fabrication, falsification, and plagiarism. Fabrication is faking data entirely; biologists call it "dry labbing." Falsification is manipulating data that have been obtained—picking only the favorable results, trimming off data points that seem to threaten the paper's conclusions, presenting readings barely above background levels as fully significant, combining the best parts of two experiments into what then appears to be one, and so on through all the permutations the imp of the perverse can whisper. Plagiarism is broader than mere copying, it is the misappropriation of intellectual property, the purloining of another's ideas, methods, results, and perhaps the expression of them, and publishing them as one's own. Judging from the cases that come to light, theft of intellectual property is epidemic in the sciences, and often the conflicts that result are devastating to the scientist seeking redress. Until recently, the standard definition went further, adding to fabrication, falsification, and plagiarism a catch-all clause—in one version, "or other serious deviation from accepted practices in proposing, carrying out, or reporting results."

To some, including most lawyers, the definition is far too loose. Many scientists defend it, primarily because it's brief, memorable, and not constricting. Many have objected to the "other serious deviations" clause as puzzlingly vague and therefore threatening, and although some recognize that it represents the reality that science is a collective, community process, the bureaucratic process is under way to delete it from federal rules, though institutions remain free to set more inclusive definitions. Rulings in recent cases have left the definition of scientific misconduct more confused than ever. Is there, for example, a boundary between fraud and sloppiness? To prove scientific fraud must one show intent?

Central to the problem of misconduct is the response of institutions when charges erupt. Again and again the actions of senior scientists and administrators have been the very model of how not to respond. They have tried to smother the fire. Such flawed responses are altogether typical of misconduct cases. Such behavior has deep roots in the group psychology of laboratories and the institutions that harbor them. A salient part of response is the treatment of someone bold enough to charge misconduct—the whistle-blower. Of course, in cases handled well one may never hear of the whistle-blower. Yet all too often the cases that go public have not been handled well: almost without exception, and we see the same notoriously in industry, government, the churches, the military, whistle-blowers even when their charges are justified are treated badly, often savagely, their careers ruined and lives twisted up.

Then we reach underlying, more general questions: whether in fact and to what extent science really is self-correcting, even when a purported finding may have been constructed fraudulently, and whether in fact the scientific community can be autonomous, self-policing, free from scrutiny by government bureaucrats and legislatures, even when these are the source of the funding. The institutions of self-correction overlap to a great degree with those of self-governance. Some are formal. These run—but the list is not exhaustive—from the organization of scientists' education and apprenticeships, and the varieties of social structures of laboratories including the roles of junior scientists, to the governmental agencies and private institutions that grant funds, the practices of journals in which scientists publish, the various professional societies of scientists in one field or another, and up to the National Academy of Sciences.

Closest to conduct as well as to substance are the specialized mechanisms by which scientists judge each other's work—peer review and refereeing. Peer review, *sensu stricto,* is the system whereby appointed, anonymous groups rate the quality of applications in their field for grants of funds, attempting to put the meritorious applications in order of worthiness. Proposed work is thus assessed prospectively. Retrospective assessment comes with the refereeing, again almost always anonymous, of papers submitted for publication in the journals. At each end, scientists evaluate scientists. Peer review and refereeing have been self-governance at its most disinterested and pure—in principle, at least, and for decades in practice. They are institutions that grew up after the second world war. For most scientists, science without them seems inconceivable.

Half a century on, though, peer review and refereeing are moribund. They have become dispirited, often ineffectual, and in some respects corrupt, infested with politics, rife with temptation to plagiarize. Refereeing faces particular problems, generated by the immense proliferation of papers and journals and by the increasing intensity of competition among scientists. Only since the mid–nineteen-nineties has a healthy alternative developed, with the growth of electronic publication. This growth is taking several forms and is accelerating. Its effects are already being felt. They justify that overworked adjective "revolutionary."

Other social structures seem informal. Yet they can be the more powerful for that; and they mediate the lives and interactions of scientists in all settings, to the most formal. These include the networks of judgements by which scientists rate other members of their communities, the hierarchies of respect and esteem—or of distrust. Most encompassing of these, though, are the lineages of science.

For science has lineages. To know with whom you trained, from whom you learned, will tell others your specialty, your likely quality, your attitudes towards work, towards collegiality and competition, towards the integrity of science generally. The power of lineages is manifest at every juncture. Above all, lineages establish the contexts of daily practices and unspoken examples—always more binding than precept—by which younger workers are inducted into the sciences. Niels Bohr, Thomas Hunt Morgan, Ernest Rutherford, Max Delbrück, Lawrence Bragg, Linus Pauling, Arthur Kornberg, André Lwoff: great lineages are known by their great mentors. But with growth and ever-narrower specialization, recent years have seen a thinning-out of lineages, the attenuation of mentoring.

Deeper yet lies the question whether in controversies over misconduct, however bitter, the interests of science, of the scientific community, of individual scientists are framable in our usual terms of individual guilt or innocence, determined in adversarial proceedings. For those brought up in the Anglo-American traditions of jurisprudence, alternatives are almost inconceivable. We haven't got names to know them by. Yet other legal cultures do employ alternative systems, systems that may be adaptable to the needs of the sciences, while in the last fifteen years change has been stirring in certain areas of American civil law—indeed, in just those areas where cases involve complex technical and scientific issues.

Deepest lies the inevitable coming to an end of the exponential growth of the sciences. This is the doctrine of Malthus applied to the greatest set of institutions, the greatest enterprise of intellect and spirit we shall ever know. Decades under way, and conditioning, threatening, wracking every aspect of the sciences is the relentless transition to the steady state.

1

A CULTURE OF FRAUD

So consider fraud. Emerging from the end of a century—gilded age? robber barons?—we find our sense of ourselves and our society in certain respects altered, and that not pleasantly: we are bombarded, saturated, harried by fraud. Day upon day, week upon week, the cases multiply, in social institutions of the most varied kinds—in finance and industry, or in the professions, or in the churches, or in sports, the media, the sciences. We can no longer suppose these are isolated instances within largely self-governing, self-correcting systems. We can no longer escape considering the shapes and contexts of fraud.

Where to begin? By the nineteen-nineties, we had forgotten history's lessons. Six decades and more after the great crash of 1929, hardly anyone was still alive who remembered the roaring twenties—and the financial scandals and frauds that had come to light once the bubble burst. Who now recognizes the names Shenandoah Corporation or

Blue Ridge Corporation, speculative investment trusts organized in 1928 and 1929 by Goldman Sachs, or the utilities holding company that Samuel Insull built and fraudulently ran, which at its height at the end of that decade owned more than five hundred power plants in the United States, with stock valued by the market at more than three billion dollars—then? Who now recalls Howard Hopson, or Ivar Kreuger, the Swedish match king, who speculated on a vast and fraudulent scale with other people's money, or the Union Industrial Bank of Flint, Michigan, whose officers conspired to embezzle millions to play the market? Yet these are but a few of the most flagrant of all those brought down and brought to light in the crash.

Walter Bagehot, the nineteenth-century British editor who built *The Economist,* wrote in 1873: "Every great crisis reveals the excessive speculations of many houses which no one before suspected." Oblivious to the past, we got the roaring nineties, which we all now recognize as an era of astonishing exuberance in enterprise and markets in the United States and worldwide. Fortunes beyond avarice, beyond need or the most extravagant luxury or even the possibility of their spending, counted in the tens and hundreds of millions, some in the billions, were gathered up by captains of industry and finance, of law and accounting, while millions of little people were towed happily along in their wake.

Then, instant notoriety—names we shall not forget in our lifetimes. One of the most exuberant enterprises of them all, an energy-trading company based in Texas called Enron, a Wall Street darling rated the fifth-largest corporation in the United States, was forced to acknowledge accounting practices that had kept huge liabilities off its balance sheets, and to restate its profits from 1997 through 2000—lowering its book value by a billion and a quarter dollars. On 2 December 2001, Enron went bankrupt. It was by far the largest corporate bankruptcy in world history. A brash upstart, politically well connected but hardly a member of the club, Enron might have seemed an aberration—a bad apple, however bloated.

Then came Global Crossing, a telecommunications company that had built a world-wide network of fibre-optic lines, which went bankrupt at the end of January 2002 when it was found to have inflated its revenues by still other dodgy accounting practices. Then on 4 April 2002 the story broke that John Rigas, founder of Adelphia Communications, a cable-television company, and two sons were charged with taking home upwards of 2.3 billion dollars in bank loans guaranteed by their corporation, without recording the sums on the books. Very bad judgements by deranged individuals, perhaps—an acute narcissism had led them to believe themselves immune to discovery, above the law. Such, anyway, is a standard psychiatric diagnosis of such actors. Plausibly applied, too, when at the end of May Dennis Kozlowski, head of Tyco International, a manufacturing conglomerate with a quarter-million employees and a stock that had risen fifteen-fold since 1992 when he became chairman, was indicted for evading New York City sales taxes on more than a million dollars worth of artworks. How petty this seemed. The artworks were as ill-judged as the evasion.

Then came WorldCom. An aggressively expansive telecommunications company beloved of Wall Street, in the spring of 2002 WorldCom was obliged to confess that it had manipulated its accounts to create 3.8 billion dollars in illusory earnings over a fifteen-month period. In April, Xerox, long thought a model of probity, paid a fine of ten million, admitting that over the previous five years it had fraudulently overstated its profits by 1.4 billion dollars. The net of corruption entangled others, perhaps chief among them Qwest Communications, the United States' fourth-largest local telephone network, which had been involved with Enron, with WorldCom, with Global Crossing in tricky schemes to inflate revenues.

Day upon day, week upon week, the scandals, the revelations spread. We remember the highlights. The accounting firm Arthur Andersen had been auditors of Enron, WorldCom, Global Crossing, and other companies that had hidden their liabilities or spiked their revenues; Andersen, long established and one of the big five accounting firms, was convicted of obstruction of justice, and driven out of business. The stink wafted through the corridors of brokerage houses. Merrill Lynch, though admitting no wrong-doing, paid penalties of one hundred million dollars to end New York State's investigation of whether its stock recommendations were tilted to favor companies whose investment-banking business Merrill coveted. On October 1, Morgan Stanley was censured and agreed to pay half a million to settle charges brought by the Securities and Exchange Commission.

More and more came out about the big ones. Mid-July of 2002, WorldCom filed for bankruptcy. In August, Scott Sullivan, once finance chief there, was indicted on charges of securities fraud and making false statements to the SEC; he pleaded not guilty, but in September two WorldCom executives of the next rank pleaded guilty to like charges. On September 13, *The New York Times* reported 2 TOP TYCO EXECUTIVES CHARGED WITH $600 MILLION FRAUD SCHEME. Not so petty, after all. Kozlowski's companion in the dock was Tyco's former chief financial officer, Mark Swartz. On September 24, the Rigas father and sons with two other Adelphia executives were indicted. On October 11, Troy Normand and Betty Vinson, two more former WorldCom officials, pleaded guilty to fraud and conspiracy charges—and by then the amount was up to seven billion dollars in expenses they had schemed to conceal. On November 1, *The New York Times* reported that "Andrew S. Fastow, the former chief financial officer of Enron, was indicted by a grand jury in Houston yesterday on 78 counts of fraud, money laundering, conspiracy and obstruction of justice." By mid-November, Qwest's overstatements were reported to have been upwards of a billion and a half. By year's end, in 2002 according to a count made by *The Economist* some two hundred and fifty American corporations had had to restate their accounts, compared to ninety-two in 1997.

Next to fall sick was the United States Olympic Committee. On New Year's Eve of 2002, *The New York Times* ran a story by Jere Longman, brief and on an inside page, saying that the committee was investigating its chief executive, Lloyd Ward, for abusing his position to help his brother's business get a contract. What developed from that was pure farce. The American Olympic movement was still living down the scandal of the award of the 2002 winter games to Salt Lake City, where it had turned out that Utah businessmen had given more than a million dollars in cash and other gifts to members of the International Olympic Committee. Now, seemingly every day, news stories reported a titanic—well, insensate—struggle between Ward and the United States' committee's president, Marty Mankamyer. Ward was the committee's fourth chief executive since 1999; Mankamyer was its third president since 2000, having taken over in 2002 after her predecessor resigned for falsifying her résumé. On January 13, the committee announced that although Ward had admitted the conflict of interest, its ethics panel had cleared him with a mild censure and the board had kept him on. On January 15, a member of the committee resigned, protesting the leniency. The next day another did the same. The next day, Friday, three more quit. On Monday, Mankamyer started a new ethics investigation. Whereupon on Tuesday seven officers of the committee demanded that Mankamyer resign. Custard pies flew in press and boardroom, and at the end of the week it emerged that the committee's ethics chairman, Kenneth M. Duberstein, who had cleared Ward, is a lobbyist for General Motors—on whose board Ward sits. The following Tuesday and Wednesday—by now we're up to January 28 and 29—the Commerce Committee of the United States Senate held hearings on the affair. Then Bill Briggs, a sports writer for *The Denver Post,* reported that a real-estate agent in Colorado Springs, where the committee has its offices, claimed that when Ward had bought land for a house, Mankamyer, who was a broker for a different firm, had "pressured her to surrender much of the commission she earned." On Tuesday, February 4, facing a no-confidence vote of her committee, Mankamyer resigned. The farce wasn't over. At the end of that week, the committee stripped Ward of a bonus of one hundred and eighty thousand dollars. The following Friday, the chairman of John Hancock Financial Services, a corporate sponsor that has given more than a hundred million dollars to the Olympics, said that the committee had acknowledged to him that it had "filed inconsistent and opaque" tax returns with the Internal Revenue Service. On Wednesday, February 26, *The New York Times*'s chronicler, Richard Sandomir, reported that "Ward . . . escaped an effort last night to oust him during a stormy 90-minute conference call held by the organization's executive committee." On March 1, Ward resigned. On Monday, 3 March 2003, the *Times* ran a seven-hundred-word wrap-up of the affair by a reporter named Jayson Blair. Keep Jayson Blair in mind.

The epidemic spread to Europe. Royal Ahold, an international food-service and retailing conglomerate based in the Netherlands, had grown to be the world's third-largest grocer by expensive acquisitions, chiefly in the United States: 2.9 billion dollars for Stop & Shop in 1996; 2.6 billion for Giant Food, a supermarket chain, in 1998; and in May of 2000, 3.6 billion for US Foodservice. On 24 February

2003, Ahold disclosed accounting irregularities at US Foodservice that had inflated earnings there and also at the parent company by more than five hundred million dollars. That same day, Cees van der Hoeven, the chief executive officer who had driven the expansion, resigned. He was soon followed by eight others at the parent company, while two top executives at US Foodservice were fired. By early May, the amount of the irregularities reached 880 million dollars; other American divisions were also infected; a headline in the *Financial Times*—the pink-paper, London-based alternative to *The Wall Street Journal*—warned that Europe faced "an Enron-style accounting debacle."

Nor was health care immune. On 19 March 2003, the Securities and Exchange Commission charged that HealthSouth, the United States' largest chain of for-profit rehabilitation hospitals, at the direction of its founder and chief executive, Richard Scrushy, had boosted profits by upwards of 1.4 billion notional dollars since 1999 and had inflated the company's assets by 800 million to cover up the fraud. The commission gave abundant detail of the occasions when Scrushy had ordered staff to falsify profit figures to maintain the stock price, even while he was selling shares to the value, then, of two hundred and fifty million dollars. The investigation quickly expanded to Medicare fraud, to other businesses Scrushy was connected with, to HealthSouth's investment bankers, UBS Warburg, and its accountants, Ernst and Young. One, then another of Scrushy's senior associates pleaded guilty—nine in all by the end of April. Scrushy was fired, later indicted for fraud and insider stock trading. He pleaded not guilty. His trial was set for mid-August 2004.

Revelations marched on. In the fall of 2003, the mutual-fund industry in the United States came under scrutiny. A whistle-blower charged that some funds were allowing big clients to execute trades after hours, a banned practice that could win great profits in anticipation of the opening prices the next day—short-changing small investors. Then on January 12 the Securities and Exchange Commission announced, after a nine-month examination of fifteen of the largest brokerage firms, that they had found that a number of mutual funds were paying off brokerage houses to recommend their funds to investors. The payoffs took several forms, including cash and steering their own stock trades to compliant brokers; the incentives would not have been illegal if fully disclosed to investors, but many were not.

That winter, a case erupted in Italy that commentators instantly compared, as usual, to Enron. Parmalat, a corporation based in Parma, Italy, had been built by Calisto Tanzi and his associates into a global giant in the dairy and grocery industry; though publicly traded, fifty-one per cent of its stock was retained by the Tanzi family. On 24 December 2003, Parmalat went bankrupt. Italian prosecutors said that the firm had created dozens of offshore companies to invent assets to cover up liabilities that might reach as much as eleven billion dollars. Four days later, an Italian judge ordered Tanzi's arrest. The scandal spread to the Netherlands, Brazil, the United States, to Italian and foreign banks, to Parmalat's former and current auditors, Grant Thornton and Deloitte and Touche, and to its outside law firm. On 18 January 2004, an Italian lawyer representing investors said that the New York branch of Bank of America held seven billion euros, or 8.7 billion dollars, of Parmalat's money, but the bank denied that such an account existed. By mid-March, Parmalat's debts were put at 18.3 billion dollars. Prosecutors pressed for fast-track indictments against Tanzi and twenty-eight other people, and against Bank of America and the auditing firms. On March 24, a judge in Milan turned down that request, saying the evidence was not strong enough to justify skipping normal preliminary hearings—which in Italy can last several years. At the end of May, prosecutors renewed the request.

In the United States in the first months of 2004, earlier scandals moved forward. Enron: On January 14, Andrew Fastow, the former chief financial officer, indicted on October 31 on seventy-eight counts, pleaded guilty to two of them, agreeing to coöperate with prosecutors. On February 19, Jeffrey Skilling, the former chief executive, was indicted on thirty-five counts of fraud, insider trading, and conspiracy. His trial was expected to begin the next winter. On July 8, in Houston, Kenneth L. Lay, the former chairman, was marched to the federal courthouse with his hands cuffed behind his back, to face a sixty-five-page indictment that charged him with eleven counts including wire fraud, securities fraud, and making false statements to banks. He pleaded not guilty.

Adelphia and the Rigas family: On March 1, John Rigas, founder of Adelphia, his sons Michael and Timothy, and Michael Mulcahey, the former assistant treasurer, went on trial in Federal District Court in Manhattan, charged with conspiracy, bank fraud, securities fraud, and wire fraud. The prosecution claimed that the Rigases hid 2.3 billion dollars in debt and stole one hundred million before the company sought bankruptcy protection in June of 2002. On July 8, in New York, after a trial of eighteen weeks, the jury found John Rigas and his son Timothy guilty on one count of conspiracy, fifteen counts of securities fraud, and two of bank fraud. The other brother on trial, Michael, was acquitted on some charges; the jury was unable to agree on the other counts against him, and the next day the judge declared a mistrial on those, setting him free. The fourth defendant, Michael Mulcahey, Adelphia's former assistant treasurer, was acquitted on all counts.

WorldCom: On March 3, Bernard Ebbers, the former chief, appeared in handcuffs in Federal District Court in Manhattan and was charged with securities fraud, conspiracy to commit fraud, and participation in filing false corporate records with the Securities and Exchange Commission. He pleaded not guilty and was released on ten million dollars bail. At the end of May, prosecutors added six more charges of securities fraud. His trial was scheduled to begin the next November. Meanwhile, the corporation, of course under new managers and struggling to emerge from bankruptcy, changed its name to MCI; in mid-March it restated its financial results from 2000 through 2002 and got Oklahoma—with Tulsa one of its regional headquarters—to drop felony charges, promising to create sixteen hundred jobs there and to help in prosecuting former executives, including Ebbers.

Tyco International: The case against Dennis Kozlowski and Mark Swartz had centered on the charge that they had taken six hundred million dollars by stealing from the company and manipulating its stock. After a six-month trial on thirty-two separate counts, eight thousand pages of evidence, a twelve-day deliberation, tabloid newspaper stories fingering one juror as a holdout and a threatening letter to that juror, on April 2 the judge declared a mistrial. The Manhattan district attorney's office promptly announced it would move to retry the case. On May 6, in a blaze of prosecutorial and defense rhetoric, Mark Belnick, the former general counsel of Tyco, went on trial for grand larceny and securities fraud on a scale of tens of millions of dollars.

Meanwhile, in sordid counterpoint, years-long, systematic corruption had come to light in the Roman Catholic church: priests were accused of sexually abusing children. The pattern of this corruption, of the institution's response, and of its disclosure is instructive about the shapes and contexts of fraud. The scandal had been bubbling to the surface here and there for decades. Yet church authorities could reassure the faithful that cases were isolated, involving troubled individuals—all the while, as it turned out, moving child-molesting priests from one diocese to another and paying off families to keep quiet.

That broke in January of 2002. An especially egregious case of a priest moved from parish to parish over many years in the archdiocese of Boston, and an especially stubborn resistance to its acknowledgment by the archbishop, Cardinal Bernard Law, prompted men who as boys had been victims of this priest and of others, many others, to come forward in shame and outrage. Victims have leveled accusations against sixty-five priests in that archdiocese alone. The stain blossomed. Accusations arose in dioceses across the nation: St Louis, Chicago, Milwaukee, Richmond, Tulsa, Kalamazoo, Louisville and Lexington—in Kentucky one hundred and fifty priests have been accused, a record so far—Albany, Bridgeport, San Diego, Pittsburgh, New York, day upon day, week upon week. Echoes came from overseas, as bishops and laity confronted accusations in England, Ireland, Germany. Cynics observed, of course, that jokes about priests and choir boys go back centuries, to Rabelais and further. Yet the scale has been stupefying: in a two-year span *The New York Times* alone carried stories or references to sexual abuse by Roman Catholic priests in fully ten thousand articles. Within the year, Law's resignation was forced on the Vatican. Lawsuits continued to multiply in Boston and across the nation.

Through the storm, two features have been inescapable. Disturbing and offensive has been the resistance, vacillation, and backtracking of so many in the church hierarchy, from the Vatican downwards, to acknowledging the problem and dealing with the priests, the victims, the angry laity—and with the ramifications. And for our purpose in anatomizing fraud, the Roman Catholic church, with its authoritarian, hierarchical organization and its dogmas about sexuality and celibacy, makes glaringly obvious that the problem of priests molesting children is built in. As the evidence about the Boston archdiocese grew yet nastier, one Roman Catholic priest in Massachusetts tried to contemplate the incomprehensible. "I don't think it was protecting the hierarchy from scandal. I think it was protecting the leadership from scandal," Robert Bullock said in an interview on public radio. Asked about the culture of the church, he defined the problem: "It is clericalism. Clericalism is a state of secrecy, of immunity, of privilege, of lack of accountability."

Indeed. The evidence, massive in scale, bewildering in variety, forces the recognition that much fraud is structural, and that it arises in all sorts of institutional cultures that are characterized by secrecy, privilege, lack of accountability. Significantly, it arises in institutions and professions that claim to be self-governing, self-regulating. Accounting firms are supposed to observe a code of standards that has been elaborated by professional committees. Their compliance is monitored by periodic audits carried out by, yes, other accounting firms—and predictably, even in the full glare of the Andersen collapse, as Congress considered new federal supervision they used the influence bought by great campaign contributions to lobby to weaken proposed legislation and then to keep tough critics of the profession off the regulatory agencies. The securities industry and the stock exchanges were self-governing through the nineteen-twenties. But in the early thirties Congress gave the Federal Reserve Board new powers, established the Securities and Exchange Commission, and erected walls prohibiting banks from acting as brokers and brokerage houses from banking—and predictably during the nineteen-nineties the great brokerage houses and banks lobbied successfully to start tearing down those walls. The law, medicine—there as elsewhere the ideal of self-regulating institutions is ardently defended, but the reality is that it does not survive scrutiny.

The sciences are another of the great institutions that have claimed to be self-governing, self-correcting. And since the second world war, after all, the sciences have enjoyed the biggest, longest bull market of all time.

In the United States before the early nineteen-forties, except for research in agriculture and geology little government money went into science and what funding there was came from private sources. The sciences were tiny then, to a degree we find hard to imagine. In 1943 and 1944, most of the theoretical and experimental physicists in the country together with their graduate students, and including many refugees from Europe, could be gathered up and put to work in a former boys' boarding school in an isolated small town atop a mesa in New Mexico called Los Alamos. In the later nineteen-forties, when the types of research that came to be called molecular biology were emerging, those concerned with them numbered fewer than three dozen, world-wide. But at the end of the second world war a new policy took shape—radically new, though now it seems tame.

Prophet of the new was Vannevar Bush. An electrical engineer who at the Massachusetts Institute of Technology in the interwar decades had pushed that school decisively towards basic research, from 1941 onwards Bush had been director of the new Office of Scientific Research and Development. He reported directly to President Franklin D. Roosevelt, and oversaw the exploitation of civilian science and technology for war—including, for example, the improvement of radar, the mass production of sulfa drugs and penicillin, the building of the atomic bomb. Although these triumphs were for the most part technological, he held that basic research is technology's feedstock, and became convinced that after the war the federal government would have to continue to organize and pay for scientific research and the training of new scientists. He arranged for Roosevelt, in November of 1944, to direct him to produce recommendations for the government's role in science after the war. In July of 1945, Bush delivered to Harry Truman his report: *Science—The Endless Frontier.*

The timing was right, the audience ready. The war's massive spending, bringing the nation out of the Great Depression, had demonstrated among much else the effectiveness of science-driven technology. Bush's presentation was masterly—clear, reasonable, compelling. He framed the argument for basic science in terms of practical payout: for "the war against disease," for new products and industries to drive full employment, for military preparedness including the rapid development of new weapons. (The atomic bomb was unmentionable, still being tested, Hiroshima still a month off.) "But without scientific progress no amount of achievement in other directions can insure our health, prosperity, and security as a nation in the modern world." Even before the war, private funding of science had been dwindling. Further, "We can no longer count on ravaged Europe as a source of fundamental knowledge." Thus, governmental support was necessary and right. The United States, due to the war's preëmption of manpower, faced a deficit of some hundred and fifty thousand science and technology graduates with bachelor's degrees, he wrote, producing in ten years a shortfall of perhaps seventeen thousand advanced degrees. Thus, the government must provide scholarships and fellowships to find and train the essential talent. To run the enterprise, Bush proposed a single, civilian National Research Foundation. Congress was to appropriate; the foundation, carrying out no research itself, was to distribute the money to "the centers of basic research which are principally the colleges, universities, and research institutes." All this was unprecedented.

The uniquely compelling part of Vannevar Bush's proposal, though, was its comprehensive ideology. Basic research would surely produce practical payout—but when and from where could not be predicted. Thus, for example:

> Discoveries pertinent to medical progress have often come from remote and unexpected sources, and it is certain that this will be true in the future. It is wholly probable that progress in the treatment of cardiovascular disease, renal disease, cancer, and similar refractory diseases will be made as the result of fundamental discoveries in subjects unrelated to those diseases, and perhaps entirely unexpected by the investigator. Further progress requires that the entire front of medicine and the underlying sciences of chemistry, physics, anatomy, biochemistry, physiology, pharmacology, bacteriology, pathology, parasitology, etc., be broadly developed.
>
> *Progress in the war against disease results from discoveries in remote and unexpected fields of medicine and the underlying sciences.*

The italics were his gift to the scientific community.

A fundamental principle followed:

> Scientific progress on a broad front results from the free play of free intellects, working on subjects of their own choice, in the manner dictated by their curiosity for exploration of the unknown.

*

In the event, administration of federal support of the sciences did not get centralized. Back in May of 1930, President Herbert Hoover had signed into law an act establishing a National Institute of Health. In 1937, Congress legislated a separate National Cancer Institute, and gave it authority to award extramural research grants. Though starved of funding during the Depression, by the mid–nineteen-forties the institute was flourishing; in 1948, Congress added further institutes, reorganized the lot, and made the name plural: the National Institutes of Health. Meanwhile, the armed forces had also been reorganized, and centralized under the Department of Defense, which retained control of weapons research. Only in 1950 did Congress create the National Science Foundation, and even then in its first years funding was niggardly.

Vannevar Bush foresaw growth. He could hardly have imagined its extent. Yes, the money has fluctuated, in some years has even been cut back. The trend for half a century has been up. The trend for sixty years has been up. The budget for the National Institutes of Health in 1995 was 67.1 million dollars; 2004 was set at 27.9 billion dollars, while the National Science Foundation got 5.5 billion.

A quick sampling of the statistics of growth: In 1953, total federal outlays for research and development, basic and applied science and engineering, was 2.783 billion dollars, while by 2002 it was 28 times larger, 78.185 billion. Industrial research and development grew faster, more than 80-fold, from 2.247 billion dollars—slightly less than the federal figure—to 180.769 billion in 2002. But those figures must be adjusted for inflation. The National Science Foundation offers them in constant 1996 dollars. In those terms, federal research and development expenditures in 1953 were the equivalent of 5.273 billion dollars, rising to 21.505 billion 1996 dollars in 2002, a four-fold increase. Applying the same deflator to outlays by industry, the increase was 26-fold.

Such statistics don't allow for the increasing cost, in constant terms, of the technologies of most of the sciences. Other measures may be as significant. Between 1980 and 2000, the average annual growth rate of the total American civilian work force was 1.1 per cent, while the number of workers in science and engineering grew at an average annual rate of 4.9 per cent. The Bureau of Labor Statistics, in the Department of Labor, predicts that between 2000 and 2010 the total work force will grow by 15.5 per cent, while employment in science and engineering will grow by 47 per cent. By still another index, 1988 saw 178,000 scientific and engineering articles published in the United States, out of a world total of 363,000, while in 2001, 201,000 articles were published in the United States, a 12 per cent increase, yet the rise of research in Europe, Japan, and elsewhere in Asia brought the world total to 530,000, a jump of 46 per cent in thirteen years—but reducing the United States' share from 38 to 30 per cent.

Consider the growth another way. Of all scientists who have ever lived, eight out of ten, perhaps more, are alive today. Of work in the sciences that has ever been done and published and woven into the fabric of reliable knowledge, the vast bulk has been done by people who are alive today. The momentum of the sciences, their reach towards fundamental questions, their interest, has never been greater; the influence of sciences on everyday life is pervasive and increasing. The sciences are the dominant form of intellectual as well as practical activity of the last sixty years, and will continue to be so. These astonishing statements have become truisms.

Leave scientists free to choose their problems and basic research will pay out richly, though unpredictably and in the long run: Vannevar Bush's optimistic argument is still the ideology to which scientists hold. Yet the extraordinary growth of the enterprise has been transforming the ways they work. We can state one aspect of the change in general, structural form: as the number of scientists and the complexity of their subjects increase, the networks of relationships among scientists in any particular field grow larger and more intricate. The increase in complexity is steep: ten strangers in a room require a total of forty-five introductions, but twenty strangers shake hands a hundred and ninety times. With such complexity the competitive landscape changes, even as the passing on of the traditions of ethical behavior becomes attenuated: greater pressures, weaker constraints.

Most radically transformed are the ways scientists communicate their work while it is in progress—to each other and each to himself. Scientists used to write letters. They used to fill their daybooks with speculations and inner dialogues. Newton and Darwin were unusual only in the enormous volume of their note-writing and (in Darwin's case) correspondence. But those practices have largely been superseded. Scientists today communicate at meetings, by telephone, and ever more by electronic mail, to the neglect of means that leave a permanent record. The communication that *is* committed to paper is more ephemeral. Even the traditional, rigorous standards for the keeping of notebooks in the laboratory, the observatory, and the field are relaxing. Where are the bound ledgers of times past? At the extreme, lots of data are now entered directly into computers, leaving no paper.

As in other institutions, so in the sciences, such exuberant growth and its accompanying transformations are the background and context of fraud.

The grandees of the scientific establishment regularly proclaim that scientific fraud is vanishingly rare and that perpetrators are isolated individuals who act out of a twisted psychopathology. As a corollary, they insist that science is self-correcting. Typical is the lofty assertion by Roald Hoffman, a professor of chemistry at Cornell University, at the annual meeting of the American Chemical Society in 1996. Fraud in science "is not a real problem," he said. "This is because of the psychology of the perpetrators of fraud . . . the psychopathology of fraud." And, he said, "there are extraordinarily efficient self-corrective features in the system of science." The grandees make these claims as a matter of faith. They could not be so dogmatic if they had considered what evidence there is that might back up general conclusions, positive or negative, about the nature and incidence of scientific fraud. Their claims about science are unscientific.

Shamelessly so, for fraud arouses profound anxieties and the proclamations spring from obvious self-interest. The grandees are torn by opposed necessities; they are impaled on a dilemma. On the one horn, throughout the world recent science is paid for—exceedingly richly, almost entirely—by governments. Only the methods of allocation vary somewhat from nation to nation. Where private money contributes big, as do the Howard Hughes Medical Institute in the United States (endowment twelve and a half billion dollars at the end of 2003) or the Wellcome Trust in Britain (which in the single year 2001 gave out grants totalling some six hundred million pounds, or nearly a billion dollars), the donors operate in a daylight of scrutiny and accountability. For funding to be maintained and regularly increased, scientists must court, placate, never upset the sources. Thus funding of science has an inescapable political aspect—which means that the independence and self-government of the sciences are always potentially at risk, sometimes actually. (One small sign of this is the use of earmarks, that is, allocations within research appropriations of support to projects specified by the legislators rather than by the normal scientific granting agencies. Earmarks reached an unprecedented number in the United States in 2002.) This threat is the other horn of the dilemma.

In sum, the scientific communities believe that public funding is their right, but so is freedom from public control. The grandees, the guardians, must defend the honor of science: fraud, if it proved to be more than rare and individual, perhaps even to be built into the ways institutions of science work, would engender distrust among legislators and the public and so threaten both the money stream and self-governance.

If the twentieth century has not been the century of fraud, it was surely the century of its exposure. Work of some of the greatest scientists has come into question. Isaac Newton, Gregor Mendel, Charles Darwin, Louis Pasteur, Sigmund Freud, Robert Millikan, other prominent figures have been attacked and defended, generating a considerable literature and, indeed, a standard list of classic cases and controversies. Among them they display a full range of features. They suggest how understanding of fraud—more generally, of norms of scientific behavior—has evolved over three centuries.

The recent list reads differently. A quarter-century of rising concern about fraud in science began unobtrusively enough in 1974 with William Summerlin and the case of the painted mice. The dishonor roll swelled at the end of that decade: John Long, Elias Alsabti, Vijay Soman, and Philip Felig. Then in 1981 the case of John Darsee broke—crucial for its extent and for its implications for other scientists involved. More followed; names that scientists will not soon forget include Robert McCaa, Stephen Breuning, Robert Slutsky. The most notorious have been the matters of David Baltimore and Thereza Imanishi-Kari, and of Robert Gallo. New cases continue to be found out. The Office of Research Integrity, in the Department of Health and Human Services, made fourteen determinations of scientific misconduct in 2001, comprising ten cases of fabrication and/or falsification, three of plagiarism in combination with fabrication or falsification, and one of plagiarism only. In 2002, forty-one new cases were opened, the most since 1995, and a number of these were still under review at the end of the year; of cases closed, misconduct was found in thirteen, all fabrication and/or falsification. In 2003, the figures were much the same, with twelve findings of misconduct posted.

That fraud is almost always the work of isolated, twisted individuals seems inherently unlikely. Yet to understand fraud we must see it in relationship to the rest of science—and, to begin with, to the standards of behavior that scientists set themselves, as scientists, and that the communities of scientists enforce or believe they should enforce. These are called variously the ethos of science, or the norms of science. In less lofty language: Why not cheat?

Although scientists would be embarrassed to use the term, when the controversies over cases and remedies have raged, at root they have been about the norms of science. The problem of norms has been central to the sociology of science from its first emergence as a distinct line of inquiry. The emergence began with Max Weber's celebrated lecture at the University of Munich in 1918, "Science as a Vocation." Celebrated, but seldom read.

One quick definition of sociology is the study of institutions—meaning not buildings but the social structures—that we have erected in one sphere of human action or another, and that we absorb and that shape our behavior within that sphere. For the sociologist, durable institutions are like Plato's axe. Plato had this axe, you see, the essence of axe, a fine bronze blade and a strong handle of boxwood from the Caucasus, and he bequeathed it to his intellectual epigones. At some later date, the handle splintered and was replaced, but this time made of oak. Passed on down the generations of philosophers, it made its way to Asia Minor, where in the Dark Ages of the Christian west the Arabs were the custodians of the philosophical and scientific wisdom of Greece. The blade corroded, an attack of bronze disease, otherwise cuprous chloride, a simple compound that will eat its way through an old-fashioned English penny in twenty-four hours. So the blade was replaced—this time, though, with fine Damascus steel. Still later, Plato's axe made its way to the New World, where the handle, giving trouble again, was replaced—with hickory. The axe in its twenty-five hundred years has gone through other repairs, and now resides in the basement of the Metropolitan Museum in New York, still sharp, still elegant, the essence of axe—where I saw it just the other day. (The curator, alas, doesn't realize what they've got there.)

The question is, of course, in what sense is this eloquent object Plato's axe? Continuity and change: the parable of Plato's axe epitomizes the problem of the sociologist's study of institutions. Including, for example, the institutions of the law: In what sense is the jury system in American criminal or civil law today the system prescribed in the sixth and seventh amendments to the Constitution of the United States? Or in what sense are the institutions of recent science, science today, the same as those of 1918, or of the nineteen-forties, before the bull market began?

For Max Weber, science as a vocation meant more than science as a choice of a career. Science he saw as a lifelong dedication, a total commitment—fully in the sense that one speaks of a religious vocation. His concern was to contrast this vocation to others, the vocations of the politician, of the artist, of the religious believer. To begin with, scientific work is unique in being always subject to change, to being superseded. "[S]cience has a fate that profoundly distinguishes it from artistic work," he said. "In science, each of us knows that what he has accomplished will be antiquated in ten, twenty, fifty years. That is the fate to which science is subjected; it is the very *meaning* of scientific work." He went on, "Every scientific 'fulfillment' raises new 'questions'; it *asks* to be surpassed and outdated." And he said, "We cannot work without hoping that others will advance further than we have."

Secondly, for millennia and today at an ever-accelerating pace, it is science that has driven "rationalization and intellectualization and, above all, . . . the 'disenchantment of the world'," the removal of the supernatural from our view of the way things are.

In sum, Weber's norms of science require that total commitment, that acceptance of change and of being superseded, and above all that dedication to rationality and demystification. For him, the culminating conclusion to be drawn was that science and religion are radically incompatible. He wrote that "the decisive characteristic of the positively religious man" is what he termed *"das intellektuelle Opfer,"* the "intellectual sacrifice"—the surrender of judgement to the prophet or to the church. He respected those who genuinely can make that sacrifice; he viewed with contempt and pity those who cannot "bear the fate of the times," who need "to furnish their souls with, so to speak, guaranteed genuine antiques."

> In doing so, they happen to remember that religion has belonged among such antiques, and of all things religion is what they do not possess. By way of substitute, however, they play at decorating a sort of domestic chapel with small sacred images from all over the world, or they produce surrogates through all sorts of psychic experiences to which they ascribe the dignity of mystic holiness, which they peddle in the book market. This is plain humbug or self-deception.

And, he said, "[T]he tension between the value spheres of 'science' and the sphere of 'the holy' is unbridgeable." Science for Max Weber was a high and ascetic calling.

This seems remote from what we now mean by the norms of science. Yet consider: perhaps it is not. As noted, Weber distinguished science from other possible callings, commitments, attitudes of mind—and the difference that summed up all the others was that refusal to make the intellectual sacrifice. Norms have generated a considerable literature since then, though the canonical line is slim enough. Later sociologists have been concerned with distinctions within the scientific enterprise—to put it roughly, with the attitudes of mind, enforced by the community, that facilitate good science rather than bad. For the individual, the norms promote the achievement of new knowledge, while for the community they require protection of the integrity of the scientific record and the scientific process. This is a phrase that will recur.

The next statement, the starting point for most discussions of the norms of science, came from Robert Merton, elegant progenitor of the modern sociology of science. Merton's sociological lineage traced back primarily not to Weber but to Talcott Parsons, whose abiding concern was the dynamics of social systems and their construction, and who asserted "that all social action is normatively oriented, and that the value orientations embodied in these norms must to a degree be common to the actors in an institutionally integrated interactive system. It is this circumstance which makes the problem of conformity and deviance a major axis of the analysis of social systems." (Indeed, the use of scientific fraud as a window into the normal processes of the sciences is a particular variant of a noble pedigree in sociology.)

In 1942, in the article "Science and technology in a democratic order" (reprinted several times over the years as "The normative structure of science"), Merton wrote that he was responding to "incipient and actual attacks upon the integrity of science." The goal of science "is the extension of certified knowledge," that is, "empirically confirmed and logically consistent statements of regularities (which are, in effect, predictions)." He continued, in part:

> The ethos of science is that affectively toned complex of values and norms which is held to be binding on the man of science. The norms are expressed in the form of prescriptions, proscriptions, preferences, and permissions. They are legitimized in terms of institutional values.

Merton celebrated "four sets of institutional imperatives—universalism, communism, disinterestedness, organized skepticism." Universalism, briefly stated—but his discussions were qualified and often subtle—asserts that science evaluates work by "preestablished impersonal criteria," and knows no boundaries of nation, ethnicity, or political ideology. He went on, "The acceptance or rejection of claims entering the lists of science is not to depend on the personal or social attributes of their protagonist." In short, "Objectivity precludes particularism." In the phrase of another prince among observers of science, Joseph Needham, modern science is ecumenical.

Communism, in the sense of common ownership, Merton called "a second integral element of the scientific ethos." Because "the substantive findings of science are a product of social collaboration," he said, "the equity of the individual producer is severely limited." Scientists own their results jointly. "Property rights in science are whittled down to a bare minimum by the rationale of the scientific ethic." The social system of science returns rewards only on intellectual property that is given away. The rewards proper to scientists, he mentioned here and elaborated elsewhere, are conferred by the scientific community, by one's peers. These rewards are two: to be recognized for originality, which necessarily puts high value on priority, and to be held in esteem by other scientists. The two are obviously interdependent.

Disinterestedness, the third of Merton's norms, follows from the second, communism, but pushes it further: the scientist is to behave as though his possible personal stake in his results is overridden by the power of the community. Merton's discussion of disinterestedness says most clearly how he believed the norms of science are supposed to work. He held that it is "a distinctive pattern of institutional control of a wide range of motives which characterizes the behavior of scientists." He went on, "For once the institution enjoins disinterested activity, it is to the interest of scientists to conform on pain of sanctions and, insofar as the norm has been internalized, on pain of psychological conflict."

Organized skepticism, his fourth norm, is the one that harks back most obviously to Weber and the refusal to make the intellectual sacrifice. Merton, too, recognized the inherent, fundamental conflict between the scientific ethos and "other attitudes toward these same data which have been crystallized and often ritualized by other institutions." But the central importance of organized skepticism is as a mechanism: it enforces the other norms, notably disinterestedness.

Some great scientists have behaved as though acknowledging such norms. In 1969, Max Delbrück was awarded the Nobel prize in physiology or medicine, along with Salvador Luria and Alfred Hershey. (That was the year the prize for literature went to Samuel Beckett, whose work Delbrück found to resonate with the deepest paradoxes of the scientist's endeavors. But Beckett did not come to Stockholm, which Delbrück always regretted.) In the lecture he gave upon receiving the prize, Delbrück compared the scientist and the artist, reflecting upon the vocation of the scientist in terms that recall Weber. Speaking, with grave courtesy, in Swedish, he said, in part:

> The books of the great scientists are gathering dust on the shelves of learned libraries. And rightly so. The scientist addresses an infinitesimal audience of fellow composers. His message is not devoid of universality but its universality is disembodied and anonymous. While the artist's communication is linked forever with its original form, that of the scientist is modified, amplified, fused with the ideas and results of others, and melts into the stream of knowledge and ideas which forms our culture. The scientist has in common with the artist only this: that he can find no better retreat from the world and also no stronger link with the world than his work.

Merton's essay has been of enormous and abiding influence. Yet today the Mertonian norms seem sadly naïve, idealistic, old-fashioned.

For some, indeed, the term "Mertonian norms" has come to be used in derision. Can these four concepts genuinely be the norms of practicing scientists? Today it requires an active suspension of disbelief to think that they could have been thought so even during the second world war and before Vannevar Bush, the beginning of massive governmental funding for the sciences, and their high-exponential growth. In the last twenty years, every one of these has come into question. Universality? Common ownership of the results? Disinterestedness and organized skepticism? In language that now evokes a sorrowful shake of the head, Merton wrote of the effects of these as though describing the real world, noting what he saw as "the virtual absence of fraud in the annals of science, which appears exceptional when compared with the record of other spheres of activity." He went on:

> By implication, scientists are recruited from the ranks of those who exhibit an unusual degree of moral integrity. There is, in fact, no satisfactory evidence that such is the case; a more plausible explanation may be found in certain distinctive characteristics of science itself. Involving as it does the verifiability of results, scientific research is under the exacting scrutiny of fellow experts. Otherwise put—and doubtless the observation can be interpreted as lese majesty—the activities of scientists are subject to rigorous policing, to a degree perhaps unparalleled in any other field of activity.

Though I risk lese-majesty to say so, science doesn't work that way. Outsiders speak of "the scientific method." An honest scientist will tell you, "You get there how you can." To be sure, many particular rules obtain, concerned with, for example, the design of control experiments, rigor in statistical inference, or in clinical trials the right use of double blinding. But as previously noted, many kinds of observations, sometimes called natural experiments, cannot deliberately be repeated. The new supernova or pulsar, the earthquake, the brain damage in a car crash which turns out to provide clues to neurological functioning—such phenomena are evidently out of reach, being too remote, too unpredictable, too large in scale, or ethically prohibited.

At the other end of the spectrum lies clinical research, studies of the effectiveness of pharmaceutical products or medical procedures, or of the long-term consequences of such matters as smoking or diet. This takes many forms, ranging in scale from reports of a few dozen patients through to vast projects enrolling thousands of subjects, perhaps carried out at many locations and perhaps over a period of many years. Often the collection of clinical data is hard to monitor. Usually the results are not of the sort that gets built into the next step, as in experimental science: clean or bent, these findings affect only the treatment of later patients. But even in the broad middle of the research spectrum, where "verifiability of results" would seem to reign, rarely are experiments replicated, the findings reported in a paper directly tested.

In practice, verification is more complex, tenuous, and indirect. What happens, rather, is that when you publish an interesting new result, your closest competitor on reading your paper slaps his forehead and exclaims, "Why didn't I think of that!" But then he realizes, "If that's true, why then the next step is *X*, and I'd better get back to the lab pronto." Confirmation is indirect: if the new finding works, it can be built into the growing edifice.

A more complex version of that process has been called triangulation. In much science, confirmation comes about by the use of independent but neighboring and convergent approaches and results—hence, triangulation. At the top level of theorizing, the mid–nineteenth-century philosopher and historian of science William

Whewell called this by the satisfying phrase "the consilience of inductions." (The most celebrated and compelling example of consilience, though perhaps unfamiliar to the general reader, comes from the history of physics: the determination of Avogadro's number. As recently as a hundred years ago, though we may find it hard to believe, whether atoms actually exist was still for some eminent physicists a matter of dispute. Long before, in 1811, the Italian physicist Amedeo Avogadro had offered the conjecture that in equivalent amounts of different gases held under the same conditions of temperature and pressure the number of smallest particles—atoms or molecules—will be the same. The idea seemed untestable. But a few years after 1900 that number emerged from a variety of unrelated methods, from relativity to crystallography. It is enormously large: 6.02×10^{23}, or the number 602 followed by twenty-one zeroes. But all the different determinations agreed quite closely. Atoms must be real.) Triangulation can yield new findings. It allows the scientist building on another's claimed result to receive credit. It provides cross-checks that will undermine a claim or make it more robust.

Here at the middle of the spectrum, three types of findings don't fit this pattern of indirect confirmation. They don't fit Merton's scheme, either. First, the brute fact is that many published papers, tens of thousands every year, are ignored. They are not woven into the fabric; they are cited rarely or not at all except perhaps by their authors. Such vanishing papers are not part of live science.

Secondly, and in contrast, a paper that offers an advance in technique, a useful new method or device, will be repeated in laboratories everywhere, as scientists rush to learn how. Such papers are among the most frequently cited of all, showing up as endnotes in the "methods" section of every paper where relevant. Indeed, again and again the most important technical advances have won Nobel prizes. Examples:

William Lawrence Bragg and his father got the prize in physics in 1915, for x-ray crystallography. Theodor Svedberg, chemistry, 1926, for the ultracentrifuge. Arne Tiselius, chemistry, 1948, for electrophoresis; and Archer John Porter Martin and Richard Laurence Millington Synge, chemistry, 1952, for paper chromatography—these all being methods for separating and identifying extremely small quantities of nearly identical large molecules. Frederick Sanger, chemistry, 1958, for sequencing of the amino acids in protein chains, and a second Nobel in chemistry, 1980, this one with Walter Gilbert, for sequencing DNA. David Baltimore and Howard Temin (with Renato Dulbecco), the prize in physiology or medicine, 1975, for demonstrating the existence of an enzyme we now call reverse transcriptase, which reads genetic sequences from RNA back to DNA. Daniel Nathans and Hamilton Smith (with Werner Arber), physiology or medicine, 1978, for restriction enzymes, which cut up DNA. Reverse transcriptase and restriction enzymes are fundamental molecular tools in genetic engineering. César Milstein and Georges Köhler, physiology or medicine, 1984, for monoclonal antibodies. Ernst Ruska, physics, 1986, with two younger scientists, fifty years after he invented the electron microscope. (Longevity pays, Nobel prizes not being awarded posthumously.) Kary Mullis, a California surfer, chemistry, 1993, for the polymerase chain-reaction, which allows tiny samples of DNA to be selected and duplicated a million-fold in a few hours.

Some of these were discoveries in a conventional sense, to be sure, but all are rightly considered inventions. The Nobel committees reward such works because they open up new fields, allowing scientists to pose types of questions that were previously beyond reach. Thus in a conversation a while ago in his office at the Salk Institute, in La Jolla, California, Francis Crick remarked that for the last quarter-century the new molecular biology of development and differentiation has not been driven by big discoveries or big theories: "I think what really stands out is the rate at which they invent new techniques."

Special cases aside, though, in an experimental science when an investigator does repeat someone else's published work, that's often a sign that the work has come into question. (Thus, that the problem of fraud is felt to be pervasive is indicated by the practice I have been told has arisen at some pharmaceutical companies of duplicating every published report they might apply, before relying on it.) A scientist wanting to check another's work by direct repetition—even where this would be possible and relevant—encounters formidable obstacles as the enterprise is organized today. In a report in *The New England Journal of Medicine* in 1987, a committee of scientists and a philosopher analyzed a flagrant case of fraud that had just been uncovered at the University of California in San Diego. They pointed to the problem. "Replication, once an important element in science, is no longer an effective deterrent to fraud because the modern biomedical research system is structured to prevent replication—not to ensure it. It appears to be impossible to obtain funding for studies that are largely duplicative," they wrote. Academic credit—and publication, though they didn't mention this—"tends to be given only for new findings. Furthermore, attempted replication will detect only incorrect results: it will not detect correct results based on fraudulent data."

For three reasons, then, significant laboratory results don't normally get verified. Two of them are institutional: repeating other scientists' work is not an enterprise that attracts funding, and journals rarely publish negative results. The third reason lies in the practical problems of laboratory research: some experiments can't be rerun. In many areas of science, experimental conditions are notoriously difficult to duplicate exactly. In biology, typically, the setup of an experiment depends heavily on the suppliers of materials, reagents, animals; on the skills of the technician; and on the unstated procedures—called tacit knowledge—which will vary from laboratory to laboratory. In immunology, for example, strains of laboratory mice, cell lines, preparations of immune sera or monoclonal antibodies, and so on, often cannot be reproduced in every detail with confidence. Classical molecular biologists have long distrusted immunology for just those generic difficulties. Another aspect of the norms of science, an important one, comes into play: What makes an experiment, even an entire field, acceptably rigorous? Yet a great deal of research can be like that—for a reason having nothing to do with chicanery. Significant, original work must often be out beyond the frontier, stretching the limits of what the available methods can reliably produce.

In the last decades of the twentieth century, the idea of the norms of science came under attack from another direction, from new fashions in the sociology of science. The subversion began with Thomas Kuhn and his book *The Structure of Scientific Revolutions.* Kuhn was a physicist turned sociologist. He put forth the claim that science changes not by steady progress but rather by periods of "normal science" interrupted spasmodically by revolutions. This claim had several components. The most conspicuous was that scientific revolutions occur because the ideas, methods, and ruling preconceptions of a period of normal science accumulate anomalies they cannot explain—and eventually break down under the weight of these and are replaced *en bloc* by new ones. Kuhn called such revolution in ruling preconceptions by a new term, "paradigm shift." (The word was taken over from classical grammar, where a paradigm is a model that systematizes correct usage, as in the first declension singular of Latin nouns, *puella, puellae, puellae, puellam, puella.* Kuhn's term has come into mushy popular use like other borrowings from the sciences, "epicenter," for example, or "quantum jump.") Close examination of the history of scientific revolutions has long ago shown that Kuhn's scheme doesn't work. But its minor component, the idea of normal science, which he notoriously dismissed as "puzzle solving," has had lasting influence. It brought the scientific enterprise down with a thump, from a high Weberian calling to an everyday occupation like many others—however lit by occasional flashes of excitement.

Kuhn was a precursor of more divisive changes. Sociologists' attempts to analyze the workings of science became rent by bitter disputes over the extent to which scientific theories, or even what most people take as scientific facts, can claim a reliable grounding in a reality independent of presuppositions and prejudices of the scientists themselves. A camarilla of sociologists of science flourished in the nineteen-eighties and nineties, who declared it their program to show that all sciences, their theories or hypotheses, methods and apparatus, and the facts themselves, are constructed by the agreement of scientists—in the strong, or extreme, program, by nothing other than a consensus of scientists achieved by negotiation—and that no grounding of these in some putative objective reality can be proven. Scientists' agreement, their consensus about what is the case, according to such doctrines is not about truth but about social control, hierarchies, power relations in a larger, capitalist context—and in this sense, for some, all science must appear fraudulent. Social constructionism, the doctrine is called. Its practitioners congregate in departments and journals that carry the label "science and technology studies." They talk of breaking "the Mertonian hegemony." Wrote one leading activist in 1995, "Marxism is at the root of most of what is innovative about science studies in relationship to traditional history, philosophy and sociology of science." Baldly put, the doctrine is compounded of elements of sociology, often indeed of a Marxist tint, and aspects of a movement fashionable in academic literary studies called post-modernism or critical theory. Curiously, perhaps not coincidentally, social constructionism reached its apogee just in the period when cases of scientific fraud arose and multiplied.

To be sure, the controversies have renewed historians' attention to such things as laboratory practice, the ways that the instruments, materials, and methods at the bench shape scientists' questions (you do what you hope you have the means to do) and their results, the handing on to apprentices of tacit knowledge and methods (that is, how to make experiments go), the political and hierarchical relations in and among laboratories and other scientific institutions, and so on. Due attention to such factors is a gain—though they had already got the attention to a large extent, and without the doctrine, for they are indeed necessary components of the scientific process. But the strong program asserts that they are also sufficient. This claim, this challenge that seemed so radical, agitated what in the mid–nineteen-nineties were called the science wars, an internecine, generational, academic-political conflict in the United States and Canada, in Britain and Europe, within academic history and sociology of science. Practicing scientists, or those among them who paid constructionism any heed, were puzzled, alarmed, sometimes outraged: for a while, the science wars raged virulently between these sociologists and their allies among historians, on the one hand, and on the other scientists and more traditional historians who styled themselves defenders of the dignity of science. The wars were waged with polemic heavy and light, an entertaining spectator sport.

Fraudulent science, though, may well be socially constructed. After all, it can hardly claim to be grounded in a reality. So the threatening question cannot be dismissed: Is fraud in any significant sense intrinsic to the institutions of science? If indeed we are to see fraud in relationship to legitimate science, we need a basis for the contrast. We need data. We need case studies.

2

WHAT'S IT LIKE?

A TYPOLOGY OF SCIENTIFIC FRAUD

Scientific inquiries are more exposed than most others to the inroads of pretenders; and I feel that I shall deserve the thanks of all who really value truth, by stating some of the methods of deceiving practised by unworthy claimants for its honours, whilst the mere circumstance of their arts being known may deter future offenders.

—*Charles Babbage, 1830*

Cases abound. In present-day sciences, the incidence and character of fraud change across the spectrum of specialties. The phenomena also have a temporal dimension; they have shifted over the centuries. Surprisingly, with exceptions, little has been done to analyze these cases, classic or recent—the circumstances, the charges, the defenses—for shared or persistent features and thus to begin to construct a typology of fraud.

First to attempt a typology was Charles Babbage. He is celebrated today as the nineteenth-century English physicist who spent a considerable part of his professional life attempting to build a computer—a mechanical one, possibly to be steam driven—which he called the "difference engine," and which was to automate and make more accurate the calculation of such things as tables of logarithms and certain kinds of sophisticated calculus. For twelve years, he was

Lucasian Professor of Mathematics in the University of Cambridge, which is the chair of theoretical physics; Isaac Newton had been the second to sit there, and its most recent occupant is Stephen Hawking. Babbage's interests were wide, running from astronomy and meteorology to rationalization of factory production and even to the use of tree rings to investigate climate in past centuries. Statistics and probability were central among his concerns from cryptanalysis to linguistics to actuarial science. He was a founder of the Astronomical Society, and of the Statistical Society of London. Even as an undergraduate he had become convinced that in mathematics England lagged far behind other countries, and had campaigned for improvement. In the eighteen-twenties, he was a leader in a movement for reform, modernization, and increased rigor in education at Cambridge, including new scientific subjects and chairs.

Babbage remained concerned with the quality of English science. An aspect of that, he saw, is fraud, on which his observations constitute the *locus classicus.* In 1830, he published *Reflections on the Decline of Science in England, and on Some of Its Causes.* In language by turns earnest and disdainful, he deplored the ignorance of science among the aristocracy and the governing classes; the absence of governmental funding, particularly important for the pure sciences; the lack of incentives for young men without private means to pursue the sciences; the failure to honor successful scientists. The book's chief subject was the state of the Royal Society, founded under Charles II in 1660–1662 as an assembly of leading naturalists and practical men to promote knowledge, but now corrupted by indolence, ignorance, favoritism, and patronage. Babbage waxed sardonic about everything from the nomination of new members to the awarding of the society's several annual medals. He proposed reforms—among them the founding of an Order of Merit, to counterbalance the fact that knighthoods and new peerages honored military men. (In 1740, Frederick the Great, of Prussia, had established the first such order, *Pour le Mèrite.* Other European countries had adopted the idea; but Britain established the Order of Merit only in 1902. Members, appointed by the monarch and limited to forty at any time, have included novelists, artists, philosophers, and, indeed, a number of scientists. Francis Crick is a present-day member.)

In his fifth chapter (of six) Babbage wrote of "the art of making observations and experiments." He began with a warning against becoming obsessed with the appearance of precision. "The extreme accuracy required in some of our modern inquiries has, in some respects, had an unfortunate influence, by favouring the opinion, that no experiments are valuable, unless the measures are most minute, and the accordance amongst them most perfect." The warning was prescient; the problem is serious, now as then. He then addressed "the frauds of observers," offering a classification that prefigures present-day definitions.

> There are several species of impositions that have been practiced in science, which are but little known, except to the initiated, and which it may perhaps be possible to render quite intelligible to ordinary understandings. These may be classed under the headings of hoaxing, forging, trimming, and cooking.

He looked on outright hoaxes with ironical amusement, giving an extended example of a report from Naples, forty years earlier, of a new family and species of mollusc, described and pictured "with great minuteness" including its structure, its shell, and its mode and speed of movement. The discoverer had named it after himself. The new creature got picked up, diagrams and all, by a French encyclopedia.

"The fact, however, is, that no such animal exists." The discoverer had found on the Sicilian seashore three bits of bone from a known species of mollusc, and had then, Babbage said, "described and figured these bones most accurately, and drew the whole of the rest of his description from the stores of his own imagination. Such frauds are far from justifiable; the only excuse which has been made for them is, when they have been practised on scientific academies which had reached the period of dotage."

He went on:

> *Forging* differs from hoaxing, inasmuch as in the latter the deceit is intended to last for a time, and then be discovered, to the ridicule of those who have credited it; whereas the forger is one who, wishing to acquire a reputation for science, records observations which he has never made. . . . Fortunately instances of the occurrence of forging are rare.
>
> *Trimming* consists in clipping off little bits here and there from those observations which differ most in excess from the mean, and in sticking them on to those which are too small; a species of "equitable adjustment," as a radical would term it [i.e., in British politics of the day, one who agitated for more equal distribution of wealth], which cannot be admitted in science.
>
> This fraud is perhaps not so injurious (except to the character of the trimmer) as cooking, which the next paragraph will teach. The reason of this is, that the *average* given by the observations of the trimmer is the same, whether they are trimmed or untrimmed. His object is to gain a reputation for extreme accuracy in making observations; but from respect for truth, or from a prudent foresight, he does not distort the position of the fact he gets from nature, and it is usually difficult to detect him. He has more sense or less adventure than the Cook.

Babbage drew most of his examples from fields most familiar to him, mathematics, physics, and especially astronomy (hence "observers"). His types flourish in biology, of course, and in the social sciences. He wrote *"Of Cooking"* as though solemnly offering recipes, "receipts."

> One of its numerous processes is to make multitudes of observations, and out of these to select only those which agree, or very nearly agree. If a hundred observations are made, the cook must be very unlucky if he cannot pick out fifteen or twenty which will do for serving up.
>
> Another approved receipt, when the observations to be used will not come within the limit of accuracy, which it has been resolved they shall possess, is to calculate them by two different formulæ. The difference in the constants employed in those formulæ has sometimes a most happy effect in promoting unity amongst discordant measures. . . .
>
> It sometimes happens that the constant quantities in formulæ given by the highest authorities, although they differ amongst themselves, yet they will not suit the materials. This is precisely the point in which the skill of the artist is shown; and an accomplished cook will carry himself triumphantly through it.

Yet cooking was not without risks.

> There are some few reflections which I would venture to suggest to those who cook, although they may perhaps not receive the attention which, in my opinion, they deserve, from not coming from the pen of an adept. . . .
>
> In all these, and in numerous other cases, it would most probably happen that the cook would procure a temporary reputation for unrivalled accuracy at the expense of his permanent

> fame. It might also have the effect of rendering even all his crude observations [we would say, raw data] of no value; for that part of the scientific world whose opinion is of most weight, is generally so unreasonable, as to neglect altogether the observations of those in whom they have, on any occasion, discovered traces of the artist. In fact, the character of an observer, as of a woman, if doubted is destroyed.

Babbage's forging is, of course, what we now call fabrication. His trimming and cooking are varieties of falsification; the distinction is useful. His sardonic dry tone is inimitable.

The most prominent of the classic frauds illustrate Babbage's typology in all its variety. Eight frauds and a hoax demand our attention. The hoax is Piltdown Man. The other cases implicate some revered names: Isaac Newton, Gregor Mendel, Charles Darwin, Louis Pasteur, Robert Millikan, Ernst Haeckel, Sigmund Freud, Cyril Burt.

We look at the classic cases with a double vision, though many commentators seem not to realize how this may affect their judgement. We cannot help assessing them by today's standards—not that today's standards are necessarily higher than those of the past, except perhaps in the view of the more smug of the leaders and defenders, and certainly not that the standards are more often lived up to, but rather that they are more sophisticated about what goes into research, careful or sloppy, honest or fraudulent. Simultaneously, though, we must see the classic cases in their own terms, in the scientific community of their time and the wider milieu in which they took place. Such is of course the central task of the historian. But to think the thoughts of the past is never more difficult than in the sciences.

Must it be said that the raising of questions does not prove guilt? Yet universal and conspicuous in accounts of these classic cases has been the dismay evinced by those who themselves raise the questions, bring out the evidence. Scientists, often, and not far below the crust, are sweetly idealistic. They have heroes. More than that, for many scientists science itself is a heroine—the French have got the gender right, *La Science*—whose innocence and pristine reputation must be defended. When great scientists are attacked, the prosecutors themselves leap to the defense of the accused. Mendel. *Mendel? The monk in the garden?* The problem is that some of his data are too perfect. Yes, well, "Mendel was the first to count segregants at all. It is rather too much to expect that he would be aware of the precautions now known to be necessary for completely objective data." Darwin. *Darwin?* The problem is that some of the photographs in *The Expression of the Emotions in Man and Animals* were doctored. Ah, but, "In many ways publication of *Expression* marked the birth of empirical photography. It could not conform to rules about scientific photography, because it was part of the creation of those rules." So strong is the compulsion to save the great men, to protect the reputation of science herself, that defense turns nastily *ad hominem.* Pasteur. *Pasteur? The hero of generations of French schoolchildren, the inspiration of countless careers in science?* The problem is that his entries in his private notebooks contradict, significantly, his public claims for how he had achieved his most celebrated victories, including immunization against rabies. No, no, "complete rubbish," his accuser is "trying to pull a great man down" and is himself "guilty of unethical and unsavory conduct when he burrows through Pasteur's notebooks for scraps of supposed wrongdoing."

Amidst weighty accusations and earnest analyses, Piltdown Man emerges as a fraud, yes, but a comical one, a monumental hoax of exactly the sort Babbage would have enjoyed. At a meeting of the Geological Society of London in December of 1912, Arthur Smith

Woodward and Charles Dawson announced that they had found, in a gravel pit at Piltdown, in Sussex, an early-stone-age human jawbone and fragments of skull, lying with stone and bone tools. Smith Woodward was a paleontologist, the keeper of geology at the British Museum. Dawson was a solicitor and amateur scientist. From the start, Piltdown man was distinctly odd, but the English press and public were eager to welcome him. Half a century earlier, in August of 1856, a blast at a quarry in the valley of the Neander River, in Germany, had uncovered a skull and a lot of bones that were undoubtedly human, yet radically different from any present-day human. Neanderthal man launched human paleontology. Then in the first decade of the twentieth century paleolithic tools, cave paintings, and human bones were turning up all over France and Germany. Now England had a Man of her own—and a candidate for the "missing link" that popularizers of evolutionary biology were then calling for.

Dawson died unexpectedly in 1916; Smith Woodward lived to 1944. Posthumously, in 1948, he published a popular book with the sublime title *The Earliest Englishman.* It contained, among other delights, the observation, about a large bone tool that had been found in that gravel pit in 1914, that he and Dawson "were surprised to find that the damaged end had been shaped by man and looked rather like the end of a cricket bat."

With a cranium that appeared human and an apelike jawbone, Piltdown Man never fit the pattern of human evolution that was emerging from other discoveries. In 1925, Raymond Dart and Robert Broom announced the find in Africa of another early human ancestor, whom they christened *Australopithecus,* and which had a small, apelike cranium but jaws and teeth that had evolved towards the human. Piltdown Man was a puzzle, inexplicable, to be set aside for lack of evidence—for lack above all of connectedness. Then in

1953, Joseph Sydney Weiner, an anthropologist from Oxford, began to wonder whether the pieces of Piltdown were really from a single creature. Chemical tests he arranged showed that the skull and jawbone did not belong together and that all the pieces were modern, stained with chemicals to fake age. The rest of the finds from the gravel pit were easily shown to have been planted. Piltdown Man was the Piltdown Hoax.

Who was the hoaxer? Dawson was the obvious possibility. Pierre Teilhard de Chardin, a Jesuit priest, paleontologist of sorts, and mystic, was on the scene and implicated at least as a dupe. Nothing could be proved. In the fall of 1978, a new suspect emerged. He was William Sollas, who had been professor of geology at Oxford at the time of the original discovery and until his death, in 1937. Sollas was accused by his successor in the chair, J. A. Douglas, who had died earlier in 1978—and who had left a tape recording, made ten years earlier, after his retirement. The tape was played publicly for the first time at a symposium of paleontologists. Douglas on tape made two sorts of points. The first was motive: Sollas had thought Smith Woodward a pretentious and ignorant fool, and in early scientific debates had several times held him up to public scorn. The second was means: Sollas had had access to the right sorts of bones, and Douglas remembered particular occasions when Sollas had, for example, taken apes' teeth from an Oxford department or had obtained a packet of one of the chemicals with which the Piltdown skull had been stained.

Here was one instance, though not answering to the usual formula, where science was self-correcting. But note the mode of correction—and that it was smoothly under way well before the discovery of the hoax. Piltdown Man, in the forty years from his birth to his death, was anomalous. He didn't fit. Even while accepted as genuine, as the human family tree was traced out Piltdown became increasingly isolated, irrelevant. The community left him encapsulated and got on with debating the tree.

Newton and Mendel provide variants of trimming, or falsification. The case against Newton is clear-cut and not denied, although it took a quarter of a millennium to be exposed fully. He adjusted his calculations of the velocity of sound, and of the precession of the equinoxes, and his early work on the orbit of the moon, to make them correlate more closely with his theories. In effect, he worked not from observation to theory but in reverse, at least in the final stages—from the top down, from what he was convinced was the truth the data must bear out. Newton's biographer Richard Westfall, though overawed by Newton's incomparable intellect and achievements, has confronted the evidence:

> Having proposed exact correlation as the criterion of truth, he took care to see that exact correlation was presented, whether or not it was properly achieved. Not the least part of the *Principia*'s persuasiveness was its deliberate pretence to a degree of precision quite beyond its legitimate claim. If the *Principia* established the quantitative pattern of modern science, it equally suggested a less sublime truth—that no one can manipulate the fudge factor so effectively as the master mathematician himself.

Deliberate pretense to extreme precision: just the syndrome Newton's successor Babbage described. We shall see it again, along with the arrogance of scientists of lesser genius who make the data conform to what they think they know to be so.

The case against Mendel is the earliest example—and certainly the most contentious and tantalizing—of one of the most frequent types of problems, data that seem statistically improbable. In 1936, Ronald Aylmer Fisher, an irascible Englishman who was one of the founders of population genetics, broached problems about Gregor Mendel's experiments with garden peas that had established the elementary foundation of genetics. Mendel was an Augustinian monk, a teacher of science, and had carried out his experiments beginning in 1857 in a small garden plot in front of the monastery at Brünn, modern Brno, in the then-Austrian province of Moravia. He read a report of his results to meetings of the Brünn natural-history society on 8 February and 8 March 1865. A year later, the paper appeared in print in the journal of the society, which was fairly widely distributed but in this instance, as we all know, almost universally ignored. At the turn of the century, and well after his death, several naturalists rediscovered Mendel's rules of heredity and his paper, bringing him recognition.

Fisher's article has come down in history as charging that Mendel fudged some of his data. What Fisher wrote was more complex than that, and full of respect, of praise. Comparing Mendel to later workers, he wrote, "The theoretical consequences of his system he had thought out thoroughly, and in this respect his thought is considerably in advance of that of the first generation of geneticists which followed his rediscovery." (Certainly so, as the simplest comparison of his paper with those of his rediscoverers will show.) In summary, he saluted Mendel's researches as "conclusive in their results, faultlessly lucid in their presentation, and vital to the understanding not of one problem of current interest, but of many."

Mendel had offered his data in meticulous detail, but his findings can be sketched simply enough to make Fisher's charge comprehensible. He chose pea plants that exhibited seven different *pairs* of characters, or traits: for example, flowers that were spread along the stems versus those that were bunched at the tips, unripe pods that were green versus those that were yellow, and so on. He had found that the inheritance of any one of these pairs, green pods versus yellow pods, say, was independent of any of the other pairs. He found further that of each pair one of the traits was what he called "dominating," the other "recessive." That is, plants that he knew to be hybrid exhibited only one of the alternatives in each pair. All flowers were spread, all pods green, and so on. But the recessive character, though it didn't show in the hybrids, was not extinguished. When he crossed hybrid with hybrid, the recessive trait reappeared in a quarter of the next generation. Those plants showing the recessive character continued to breed true in successive generations, for example, always yellow pods. But repeated crosses of plants showing the dominant trait proved, by their progeny, that one-third were pure-breeding dominant, two-thirds again hybrid. Thus, the 3:1 overall ratio became a 1:2:1 ratio. These patterns are the essence of what Mendel found. He worked through the various pairs of traits, and ran two-factor and three-factor crosses. The behavior was consistent.

Too consistent? In 1911, ten years after the rediscovery of Mendel's rules and his paper, Fisher, then a final-year undergraduate at Cambridge, noticed the problem. In a talk to the Cambridge University Eugenics Society, he said, in part:

> It is interesting that Mendel's original results all fall within the limits of probable error; if his experiments were repeated the odds against getting such good results is about 16 to 1. It may have been just luck; or it may be that the worthy German abbot, in his ignorance of probable error, unconsciously placed doubtful plants on the side which favoured his hypothesis.

In the mid–nineteenth century, when Mendel (never German, not yet abbot) was crossing his peas, the science of statistics was developing rapidly, in the analyses of such men as Babbage and Francis Galton, cousin to Darwin. "The test of significance of deviations from expectation," Fisher wrote in 1936, "had been familiar to mathematicians at least since the middle of the eighteenth century." Indeed, in two years at the University of Vienna Mendel had studied mathematics and physics. An important aspect of his originality lay in his statistical treatment of heredity. None the less, statistical treatment of data was still rudimentary and rare—and as Fisher also noted, Mendel's "confidence and lack of scepticism" are evident throughout his paper.

With the alert sensibility of a practicing scientist, Fisher was able to think through the experiments, the eight years of work and thousands of pea plants. He wrote, "There can, I believe, now be no doubt whatever that his report is to be taken entirely literally, and that his experiments were carried out in just the way and much in the order that they are recounted." Yet his scrutiny brought Fisher to think that in many places the data Mendel published were too good to be true.

The point is that in any real experiment the results will deviate from the ideal, purely by chance—sometimes considerably, and the more so the smaller the sample. Though Mendel reported thousands of plants in total, for many individual experiments the number counted was indeed small. Using an analytical tool called the chi-square test, Fisher tabulated the deviation of Mendel's actual data from theoretical expectation. This method indicated that Mendel's ratios of dominant to recessive characters were often suspiciously close to the ideal values predicted by the hypothesis. In one analysis, Fisher wrote, "so low a value could scarcely occur by chance once in 2000 trials. There can be no doubt that the data from the later years of the experiment have been biased strongly in the direction of agreement with expectation." The nub of the matter: "Fictitious data can seldom survive a careful scrutiny, and, since most men underestimate the frequency of large deviations arising by chance, such data may be expected generally to agree more closely with expectation than genuine data would."

Fisher tried all he could think of to explain the discrepancies. Sources of bias can be many. For one thing, careful reading of the paper makes clear that from the early stages Mendel had his theory worked out. Perhaps when sorting peas by color and by shape he tended unconsciously to put borderliners where they would bring the numbers closer to expectations. Yet such an explanation is no help for the pairs of traits in most of the experiments. Again, from the paper one can tell that Mendel did not always count all the plants of a particular cross: perhaps he tended to quit counting when he approached the ideal result, or perhaps he selected—this would have been hard to avoid—the most robust plants to count. Yet in certain experiments all the plants had to be counted. Writhing on the spike on which he had impaled himself, Fisher advanced a novel notion: the phantom gardener. "Although no explanation can be expected to be satisfactory, it remains a possibility among others that Mendel was deceived by some assistant who knew too well what was expected."

The mark of sainthood in the paradise of science is the "-(i)an" standard, as in Copernican, Galilean, Newtonian, Darwinian, Einsteinian—and Mendel marches among that number. In the decades since Fisher, and with mounting ingenuity in books and scores of articles, scientists and historians have attempted to wriggle out of the anomalies. The hopeful defenders, often distinguished geneticists, make up a roster of frustration and bafflement.

Perhaps the simplest explanation was offered by Sewall Wright, an American geneticist of Fisher's generation, in 1966 in a brief note to a reprinting of Fisher's paper. He recalculated Fisher's tests and came out with "substantially the same result." But he suggested that Fisher did not allow enough for the cumulative effect "of a slight subconscious tendency to favor the expected result in making tallies." He went on, "Mendel was the first to count segregants at all. It is rather too much to expect that he would be aware of the precautions now known to be necessary for completely objective data." Making repeated counts is surprisingly difficult: Wright quoted an experiment by a noted statistician "in which 15 trained observers obtained extraordinary differences in sorting and counting the same 532 kernels of corn. Checking of counts one does not like, but not of others, can lead to systematic bias toward agreement." For most of Mendel's data, even tiny shifts in the numbers—"less than 2 such misentries in 1000"—could lead to what Fisher took to be large discrepancies. Wright conceded that several of Mendel's experiments showed bias greater than that. "Taking everything into account, I am confident, however, that there was no deliberate attempt at falsification."

A thorough treatment was published fifty years after Fisher by Anthony W. F. Edwards, an English geneticist. He examined all previous analyses and explanations and concluded "that only Fisher (1936) could withstand detailed criticism." He went on, "For many years I supposed that Fisher's analysis was going to be able to be faulted . . . , but a complete review of the whole problem has now persuaded me that his 'abominable discovery' must stand." And, "As to the precise method by which the data came to be adjusted I would rather not speculate, though it does seem to me that any criticism of Mendel himself is quite unwarranted. Even if he were personally responsible for the biased counts his actions should not be judged by today's standards of data recording." Edwards gave Fisher's first words the last word. "Perhaps after all the undergraduate Fisher came close to the correct solution 75 years ago: 'It may be that the worthy German abbot, in his ignorance of probable error, unconsciously placed doubtful plants on the side which favoured his hypothesis.'"

The problem remains awkward. The most recent combatants have not tried to excuse Mendel's supposed psychology but have attacked Fisher's own statistical methods, and by extension Edwards's. The argument has become intricate and abstruse. Several geneticists and statistical theorists have taken part, with papers flying, notably in *The Journal of Heredity.* The chi-square test, on which Fisher relied, is indeed standard for all sorts of experimental work. However, the latest generation of scientists rebutting Fisher assert that the method doesn't apply to Mendel's data in the way Fisher supposed. The genetics of garden peas has turned out to be more complex, in many details, than Mendel could have known or that Fisher could have taken into account. Perhaps the most nearly definitive of these re-analyses was performed by a German botanist, Franz Weiling, and published in 1986. It is recondite both genetically and statistically. Alain Corcos and Floyd Monaghan, geneticists at Michigan State University, published in 1993 an excellent book-length guide to Mendel's great paper. In an appendix, "Where is the bias in Mendel's experiments?," they summed up by writing that Weiling "demonstrated that the conclusions drawn by Fisher were wrong because the assumptions on which his use of the chi-square test was based had not been met in Mendel's experiments. It looks as though Mendel has been vindicated."

In a recent conversation, Corcos said that this judgement is the present consensus. The World Wide Web boasts several active Mendel web sites which will be the battlefields for any further controversy.

His actions should not be judged by today's standards. Newton, Mendel, Darwin, Pasteur: again and again, considering the classic cases of scientific fraud, we encounter that defense. Implicit in the cases of Newton and Pasteur, however, is a motive not noted by Babbage: the scientist's conviction that he already knows the answer. From this arises a second and rather different defense. *They were right.* Their scientific conclusions are unimpeachable. Yet *therefore*—what? Are we really to conclude that in such instances the fudging or faking doesn't matter, must be condoned? The argument is all too plausible. As we shall see, it is all too frequently offered in exculpation of present-day misconduct. A variant defense: *They are great scientists.* But, after all, does that not make the offense more rank?

Fisher on Mendel is a first instance of a recurrent phenomenon. In any taxonomy of fraud, classical or recent, statistical anomalies are among the most frequent signs of trouble. Certainly in the present day, scientists of all stripes have become acutely aware that at their peril do they ignore the relation of their data to standard statistical tests.

So consider the tale that Heinrich Matthaei told me—showed me—in a conversation in Göttingen in 1976. Recounting this tale in the context of this book, I must emphasize that it is in no way an instance of fraud in science: rather, it is a quirky illustration of scientists' understanding of the kind of problem Fisher had brought to everyone's attention. German, tall, lean, and precise, Matthaei had a wistful air of grievance, a man who had had a share in a crucial discovery but not in the consequent Nobel prize. Back in the late 1950s, the leading problem of molecular biology was to break the genetic code—that is, to determine which sequences of chemical units along a strand of DNA specify which amino acids in the protein chain. (DNA and RNA carry sequences of four types of chemical subunits, called bases, while the amino acids twenty, so to specify one of them requires a three-base set, called a codon.) Matthaei, young and unknown then, had been collaborating with an equally unknown American biochemist,

Marshall Nirenberg, when he got the first codon: the triplet of bases that specifies the amino acid phenylalanine.

Matthaei brought out the notebooks. They were of the classical type (a rare and endangered species), tall, bound ledgers with pale gray, stiff covers and heavy pages, the entries in ink. "I would say I was the hard, accurate worker, neat worker," Matthaei asserted. He opened a notebook. "Fairly neat," he said in an apologetic tone. He turned pages. "Of course, my, because we were in a hurry, my notes don't all look very neat—" They were very neat. He and Nirenberg had been working with a few feet of bench space at what was then the National Institute of Arthritis and Metabolic Diseases; Nirenberg was a junior research scientist, Matthaei still more junior, a NATO postdoctoral fellow. They planned all sorts of experiments—Nirenberg lived by the shotgun approach, try everything and run with what works—and Matthaei had set out the full protocols in one notebook. When he performed an experiment he entered the results in a second, parallel ledger. From papers recently published by others, they had taught themselves how to put together in a test tube a system that had no whole cells but contained all the cellular components necessary for protein synthesis. They wanted to supply the cell-free system with some sort of RNA strands of known sequences that would prompt synthesis of strands of amino acids whose sequence they could then determine, breaking the code. They had dreamed up a variety of tricks to try. One was to use artificial RNA strands, made up entirely of just one of the bases, which ought to produce a string of just one amino acid. Senior colleagues Leon Heppel and Maxine Singer happened to have some of these RNAs already prepared, stored in a freezer.

"Marshall was not a precise experimenter. He couldn't pipette within five per cent"—that is, he could not transfer fluids accurately from one container to another, drawing them up into a hollow glass rod graduated to serve also as a measuring tool. "I was very precise." Matthaei turned to the pages where in May of 1960 he had entered the sequence of experiments that had got that first codon, and spoke of the mounting excitement of working all week, and then all night, the night of Saturday, May 28, and the sense of triumph by dawn. Nirenberg was in Berkeley that week, visiting another lab.

Matthaei leafed back to earlier pages. One day several months before the breakthrough experiment, he said, Nirenberg came in and looked at some of Matthaei's recent results along a different trail. "And he said, 'These are too close to the expectation. They'll never believe it. They're too good.' And Marshall took out a pen and crossed out some of the data entries and wrote in new numbers that were farther away." Matthaei found the page. "He said, 'Let's see how it would look like this.'" Sure enough, several of the immaculate entries had been scribbled over with a ball-point pen and new numbers put in, in a sloppier hand. Even though the original numbers were still perfectly legible under the scribbling, Matthaei had rewritten them on pieces of the ledger paper the size of postage stamps and attached them, with transparent tape, as tiny flaps. They were very neat. The original data were not lost. That particular work never generated a paper anyway, he said. Matthaei looked again at the page, still bemused.

A peccadillo—it is hardly more than that—of Charles Darwin's affords a crisp instance of that defense so regularly offered: *His actions should not be judged by today's standards.* In November of 1872, thirteen years after the first edition of *On the Origin of Species,* he published *The Expression of the Emotions in Man and Animals,* the earliest treatment of the evolution of behavioral traits. This was one of the first scientific books to use photographs, and these have become famous. They were of people's faces, expressing what he considered to be elemental and universal emotions, grief, joy or high spirits, anger, contempt, disgust, surprise, fear, horror, shame. Although the photographs were all of Europeans, in the text he gathered up accounts that related the same facial expressions to such emotional states from human groups worldwide. (From animals, too, this being a study of evolution—including such poignant examples as that elephants, when captured and immobilized or when grievously wounded, weep.) His interest in the emotions remained lively: he annotated heavily the margins of his copy of the book and drafted new materials in his notebooks. In 1889, after his death, his son Francis Darwin included some of these in a second edition of *The Expression of the Emotions.* Then in 1998, reprinting the book as a third edition with full scholarly treatment, Paul Ekman, a social psychologist and Darwinist at the University of California at San Francisco, inserted the rest of Darwin's new materials and added running commentaries on the present state of the phenomena and issues Darwin had raised. He also found additional photographs that Darwin had mentioned but not published, as well as the originals of many that were in the first edition.

Ekman addressed the fact that some of those famous photographs had been altered. Darwin had had difficulty assembling satisfactory pictures, especially because photographic plates were not fast enough to capture fleeting expressions, such as a baby's change from fear to tears. He had made no particular secret of the fact that some photographs were posed, others modified—but Ekman found from the Darwin archives and correspondence that the alterations were more extensive than had been known. He appended an essay on this problem by Phillip Prodger, an English art historian. For example, a physiologist in mid-century Paris, Guillaume-Benjamin Duchenne de Boulogne, working at La Salpêtrière, a hospital for the indigent, had devised a way to stimulate groups of facial muscles with electrodes to force various expressions. Duchenne had published a book about the muscles with photographs of patients submitted to such treatment. Darwin used eight of these. Of several, he had engravings made that retained the stimulated expressions but eliminated the electrodes. Photographs from other sources Darwin directed to be retouched, for example to add lines on the forehead. The London photographer Oscar Rejlander proved especially skillful at securing the expressions Darwin wanted. The first and most celebrated photograph in *The Expression of the Emotions* is of a weeping baby. Although based on a photograph, it turns out actually to be a drawing Rejlander made to look like a photograph. Rejlander had also sometimes posed for his own camera, Prodger found, while in one picture his wife produced for Darwin a most convincing sneer.

The dilemma is delicate. "Darwin aspired to produce an objective study of expression, using photography as evidence for the minutiae of expressive behaviour," Prodger wrote. He went on:

> Darwin and Rejlander have been widely criticized for the compromises they made in preparing the illustrations. Although much of this criticism may be justified by contemporary standards, it is wise to remember the historical context in which the book was produced. The distinction between *evidence* and *illustration* is blurred because there was little precedent for the acceptance of photographs as scientific data. Rules about photographic objectivity did not yet exist, in part because photographers frequently found it necessary to manipulate their work to enhance the visual appeal and clarity of their images. In many ways publication of *Expression* marked the birth of empirical

> photography. It could not conform to rules about scientific photography, because it was part of the creation of those rules. Before it was published, photographs were judged on how real they looked, not on how scrupulously they had been produced. Later, as scientists began to use photography as evidence of events invisible to the unassisted eye, viewers began to demand proof that photographs were accurate. . . . *Expression* was produced on the cusp of this change of attitude.

Fueled by cases where fraud is charged, debates have simmered recently on the relationship between a scientist's notebooks and published work. Some find a legal puzzle here, to do with who pays for the research and actually owns the notebooks. Some find a philosophical puzzle, to do with epistemology, or the study of how we know what we claim to know. The puzzlement becomes acute when a scientist's notebooks are discovered to disagree, even to contradict, the published work.

The accomplishments of Louis Pasteur were many, resounding, the pride of every Frenchman, every French schoolchild. He led the campaign for the germ theory of disease; he marshalled and publicized the evidence against the belief that life originates by spontaneous generation. We pasteurize our milk (so never mind that his research into fermentation was unable to improve French beer). With his loyal team of comrades in science, Pasteur found vaccines against chicken cholera, a scourge of French farmers (no relation to the cholera of humans), and against anthrax. His legendary triumph came in 1885, against rabies. He was the heroic model of the scientist, stern and reticent, working long and tirelessly, relentlessly methodical, brilliantly intuitive—himself the epitome of his oft-quoted motto: *Chance favors only the mind that is prepared.* Pasteur to the

French has been the scientific method made flesh. With all that, he was a masterful polemicist in an era when science was combat, and a showman setting up public challenges to his work and meeting them audaciously. At a time when nationalist feelings burned high he was the champion of French science in both substance and style, particularly in competition with Germans such as Robert Koch in Berlin and Justus von Liebig in Giessen. In 1887, eight years before Pasteur's death, Auguste Lutaud said it: "In France, one can be an anarchist, a communist or a nihilist, but not an anti-Pastorian. A simple question of science has been made into a question of patriotism."

Imagine therefore the consternation and obloquy in the fall of 1995, when Gerald Geison published *The Private Science of Louis Pasteur.* Geison, who died in the summer of 2001, was a historian of science at Princeton. Over a span of twelve years, he had immersed himself in Pasteur's laboratory notebooks, one hundred and two of them, some ten thousand pages obsessively detailed in crabbed hand and cryptic style. He was the first historian or biographer—or scientist, for that matter—able to do so. Pasteur had been an intensely secretive man. In his lifetime he had allowed no one, not even his closest collaborators, to look into the notebooks. By his command to his family, after his death they had been closed to all. In 1964, Pasteur's last surviving direct male descendant gave the notebooks to the Bibliothèque Nationale as part of an immense, uncatalogued archive. Only in 1971 did they become available.

Geison asserted among much else that at the two crucial, acclaimed demonstrations of his work—the vaccination of sheep against anthrax and the inoculation of the boy Joseph Meister against rabies—Pasteur had significantly misled scientists and the public about the vaccines he used, his methods of preparing them, and his previous tests of their safety and efficacy.

Eminent scientists were angered. Not only the French were upset. Max Perutz, for example, reviewed Geison for *The New York Review of Books.* (Perutz was a very senior crystallographer, who got his Nobel prize for elucidating the molecular structure of hemoglobin. Though he didn't say so in the review, Pasteur had been a boyhood hero.) He wrote, in incandescent rage and among much else, that Geison was setting himself up as judge over a scientist whose work he failed to understand. "Geison," Perutz wrote, "rather than Pasteur, seems to me guilty of unethical and unsavory conduct when he burrows through Pasteur's notebooks for scraps of supposed wrongdoing and then inflates them out of all proportion." Succinctly, Perutz in a telephone conversation told me the book was "complete rubbish" and said that Geison was "trying to find things to pull a great man down."

Such opinions are oblivious to the facts of what Geison intended and accomplished. History, not the exposure of fraud, was his central concern. Recent historians had already been adding darker shades to Pasteur's portrait. Geison wrote, and said to me as well, that he had felt consternation as he realized what the notebooks told. Yet he came to believe that the great myth of Pasteur, constructed during his lifetime and carefully perpetuated for a century, although it had served France and the scientific enterprise well in many ways—as emblem for patriots, role model for science, inspiration for the young—must now be replaced by a more complex, more subtle history, yet to be written. From the tension between Pasteur's conduct, his mixed motives, and his remarkable, undeniable accomplishments he emerges as a far more interesting scientist. "The resulting transformation in the story of Pasteur may also serve useful functions for our time and beyond." The new version, Geison thought, deepens our understanding of the ways the scientific process really works.

The anthrax incident is the more minor of the two. It took place in 1881. The disease kills sheep. Losses from anthrax had reached an estimated twenty to thirty million francs a year. Pasteur had a general theory—well, a roughly formulated idea—of immunity, and with it a theory of vaccines whereby living cultures of a pathogenic organism were to be progressively attenuated by prolonged exposure to the oxygen in air. Competitors were skeptical of all this, yet a year earlier he had tried the method successfully with his vaccine against chicken cholera. In February of 1881, he published a paper announcing that he had made a vaccine against anthrax by the method, described in detail. Other scientists were trying different approaches, including one of Pasteur's collaborators, Charles Chamberland, who attenuated *his* cultures with an antiseptic, potassium bichromate.

Pasteur set himself a public challenge. At a farm near Paris, he and his colleagues established a herd of fifty sheep. On May 5 and again on May 17, they injected twenty-five of these with vaccine Pasteur provided. The other sheep got nothing. On May 31, all fifty were injected with a culture of fully virulent anthrax. Pasteur announced that those vaccinated would all survive, the others all die of anthrax. Two days later, on Thursday, 2 June 1881, a crowd of more than two hundred officials, politicians, journalists, and farmers—Geison's narrative is vivid—gathered in the farmyard. Pasteur and his colleagues, escorted by officials, arrived at two o'clock. Applause, congratulations: twenty-three of the twenty-five unprotected sheep were dead, the other two dying, while of the twenty-five that had been vaccinated twenty-four were healthy, one pregnant ewe was dying. "Pasteur had aroused great excitement by predicting such decisive results in what was, after all, the world's first public trial of a laboratory vaccine."

Then and in all Pasteur's accounts of the trial, in the weeks, months, years that followed, he implied, though he never quite said so, that the vaccine he provided had indeed been prepared by the method of attenuation in air. Geison found in the notebooks, however, that in point of fact Pasteur had used vaccines, the first strongly attenuated, the second less so, that he had prepared by Chamberland's method, with potassium bichromate. Here and throughout, Geison reproduced photographically the relevant pages of the notebooks.

"The conclusion is unavoidable: Pasteur deliberately deceived the public, including especially those scientists most familiar with his published work"—and these included his longtime close colleagues—"about the nature of the vaccine actually used at Pouilly-le-Fort." Yet did that matter? Here Geison was inclined to be benign. Tenacity was a dominant trait of Pasteur's. "The same laboratory notebook that reveals the secret of Pouilly-le-Fort also shows that Pasteur had begun to achieve increasingly secure results with his oxygen-attenuated vaccines even as the Pouilly-le-Fort trial was underway." Within a month of his triumph, he tested these new vaccines on a flock of seventy-five sheep. "The results of these and subsequent trials were overwhelmingly positive."

Geison took the analysis one step further, to open to our understanding the full context, the intense, sometimes vicious rivalries in the professional world in which Pasteur was working, the vigor of the opposition his ideas faced both in France and especially from his arch-rival Koch in Berlin. In summation:

> Pasteur knew his enemies well. In the end, it is mainly a measure of the importance attached to originality in modern science—and of the competitive environment in which Pasteur lived, moved, and had his being—that a significant and undeniable element of deception should have entered into the most celebrated public experiment by one of the greatest heroes in the history of science.

This is not pulling a great man down. And among much else it recognizes the double vision with which we see the conduct of scientists in the past.

Four years later, on 26 October 1885, before the Académie des Sciences, Pasteur announced that the previous July, employing vaccines attenuated by a new method, he had successfully treated a boy, Joseph Meister, who had been badly bitten by a rabid dog, and that even as he spoke he was working with a second and even more severely bitten patient, a young man named Jean-Baptiste Jupille. Within days, Pasteur was a national and international hero. Within months, centers for treatment by his method had sprung up in many countries. Within a year, the method had been applied to more than two thousand bite victims in the Paris region alone.

Two characteristics of rabies seriously complicate judgement of the success of vaccine trials. The incubation period before symptoms and signs appear can be long, sometimes many months. Even when untreated, the vast majority of animals or humans bitten by rabid dogs—perhaps eighty-five per cent—never do develop the disease.

In the paper of October 26, Pasteur asserted that before treating Meister he had already immunized fifty dogs against rabies without a single failure. He went on (in Geison's translation), "My set of fifty dogs, to be sure, had not been bitten before they were made refractory"—that is, immune—"to rabies; but that objection had no share in my preoccupations, for I had already, in the course of other experiments, rendered a large number of dogs refractory after they had been bitten."

Back to Pasteur's notebooks. These reveal a number of discrepancies, which build to a conclusion at variance from the legend. In the four years since the anthrax demonstration, Pasteur had tried a great variety of ways to produce strains of rabies that were maximally virulent and others that were suitably attenuated. He sacrificed hecatombs of rabbits and dogs and quite a few monkeys, before finally settling on a method of attenuation by air-drying the spinal cords of rabbits that had died from the most virulent strain. The drying took several days; the longer the time the greater the attenuation. This was the method, as he reported to the Académie, by which he had prepared the vaccines used on Meister. He—or rather, a qualified physician under his supervision—gave the lad a series of thirteen injections over as many days, under the pinched-up skin of the belly, half a cubic centimetre each time of material made from cords that had been dried for fifteen days, the weakest, fourteen days, twelve days, eleven days, all the way to the thirteenth and last injection, a fully virulent preparation. His audience naturally supposed that this was the method he had used with that pack of fifty dogs and with that "large number of dogs refractory after they had been bitten." Geison:

> The first remarkable conclusion to emerge from a close study of Pasteur's laboratory books is that this "large number" was in fact less than twenty. More important, in the course of producing immunity in these bitten dogs—no more than sixteen, by my count—Pasteur failed to save ten dogs treated at the same time and by the same methods.

The trials with bitten dogs began on April 13 and ended May 22. "More than that, the success rate in these dogs treated after sustaining rabid bites was essentially no different from the survival rate of otherwise similar dogs that were simply left alone after their bites." Still further, in his experimental trials of rabies vaccines Pasteur used few or no controls. Thus far, the analysis demonstrates, in short, that Pasteur had no statistically justifiable basis for claiming he had created a successful vaccine. No one but Pasteur himself could have known this. None of the data was published.

The first problem is the methods Pasteur had used with the twenty-six dogs that had been bitten. The notebooks show that the sequence of injections was the reverse of that used two months later with Meister. Following the theory of immunity Pasteur entertained that spring, he began with a vaccine prepared from the most virulent rabbit cords, moving to progressively more attenuated materials. Geison's italics: *"As only Pasteur's laboratory notebooks reveal, not a single one of those twenty-six dogs, including of course the sixteen that did develop immunity to rabies, was treated by the method later applied to young Meister."*

The second problem concerns the trials with dogs that had not been bitten. These began only on May 28, when Pasteur gave a cohort of ten animals a first injection. He started ten more on June 3, yet another ten on June 25, a last group of ten two days after that, June 27. The total was forty. All of these dogs Pasteur did indeed treat by the method soon to be used with Meister, that is, beginning with the most attenuated, least virulent preparation and moving each day progressively through more and more virulent vaccines. On July 6, when Pasteur began treating Meister, all forty dogs were healthy. None, though, had ever been bitten by a rabid animal. To be sure, the virus preparations might have caused rather than prevented rabies—but the earliest inoculations had been done less than six weeks previously, the most recent just ten days earlier. "None of the dogs had survived as long as thirty days since their last (and highly lethal) injection." The immunity of these dogs was not proven, nor was the safety of the method. And Meister was symptom free. "Against this background, it should come as no great surprise that Pasteur never did publicly disclose the state of his animal experiments on the 'Meister method' as they stood at the point at which he decided to treat the boy. Nor, indeed, have they been revealed in print until now."

All this established, Geison laid italics aside and investigated what was for him the important questions that led him to the notebooks in the first place. Historically, again, what was the context in which Pasteur worked? In his laboratory, what were his relations with his principal colleagues? What questions or doubts did critics at the time raise to the celebrated paper of 26 October 1885? Scientifically, what were his laboratory practices, in fact? And just how did Pasteur's theory of immunity change during that vital period? Geison's account runs clear and deep, his motive in writing complex. Pasteur's misrepresentations were a discovery along the way—unsettling, yet opening a unique and invaluable view into the man. The Pasteur who emerges is a shadowed and devious mind, obscure sometimes even to himself, driven by an extraordinary mix of ambition, brilliance, hunch, tacit understandings not formulated even in those private notes, and luck—in short, the volatile compound we call genius.

Notebooks again—but the case of Robert Millikan raises other questions and takes us deeper into the ways scientists really work, and therefore into the nuances of the problems of fraud. To follow these, we need details: the science of the notebooks, of his published claims, and of their setting in an international world of physics riven by rival theories, personalities, and polemic.

Millikan received the Nobel prize in physics in 1923 for a variety of accomplishments. The earliest of these, the foundation of his reputation then and now, was his measurement of the charge on the individual electron—a value called *e*. (This is obviously not the same *e*-for-energy as in the Einsteinian $e = mc^2$.) He got into the work in 1907 at the University of Chicago, published a first rough paper on it in 1909, and by 1913 had shown that *e* is the same for every electron and had determined its value with great precision. The value *e* is a fundamental physical constant. Indeed, in 1917 he called it *"the most fundamental of physical or chemical constants, but also the one of most supreme importance for the solution of the numerical problems of modern physics."* (He had a penchant for italics.)

Millikan's determination and his italics were driven by a larger and bitter controversy. A coterie of physicists, mostly European, some eminent, were tenaciously holding out, even that late, against the theory that matter is made up of atoms. Millikan directly opposed the view of one of these, Felix Ehrenhaft, which was that electricity should be thought of as analogous to a liquid, made up of particles, "subelectrons," whose charge varied continuously over a considerable range; in that view, *e* was no more than the average of values. For Millikan to fix the value of *e,* and with such precision, was a polemical masterstroke—and he knew it—central to final establishment of the atomic theory.

After trying several experimental approaches, Millikan and an assistant rigged up a chamber with a horizontal pair of brass plates between which a strong electrical field could be turned on or off. His aim was to observe the behavior of single, microscopically tiny drops of oil, electrically charged, between the plates. He observed each droplet through a short-focus telescope with a scale in the eyepiece. In more than five years' work, he improved the apparatus markedly. He kept temperature and air pressure rigorously constant; even the thin beam of light that illuminated the drops had the infrared, heat-producing rays filtered out. A drop would fall slowly through the air, but with the electrical field turned on it would fall yet more slowly, and as it evaporated slightly it would remain stationary, perhaps for ten to fifteen seconds, then begin to rise. He could change the charge on the drops with a beam of x rays. When the charge changed, the drop's behavior changed abruptly—dramatic evidence that charge came in discontinuous units. He timed each drop's motions. (The experiment is notoriously finicky to do, even as it is sometimes set up today as a demonstration in introductory physics courses.) Millikan carried out hundreds of these observations and recorded the results in notebooks. He calculated the charge on each drop. Thus he established that the charges were always integral multiples of a definite irreducible minimum value. This is *e*. Beginning in 1910, he published a series of papers on the oil-drop experiments.

In 1978, Gerald Holton, who is simultaneously a physicist and a historian of science at Harvard, published a collection of studies of *The Scientific Imagination.* His second chapter was about the Millikan-Ehrenhaft dispute. Holton knew the work, understood the philosophical animus of the continental physicists, and had visited the California Institute of Technology to study Millikan's notebooks, in the archive. (The place was founded in 1891 as Throop Polytechnic Institute, and got its present name in 1920. Millikan had moved there in 1921, after government service during the first world war, and was, his biographer says, "in effect the president of the school." He built Caltech in its formative decades, and is venerated there.)

Holton's study—early, seventeen years before Geison's of Pasteur—set a high standard for subsequent attempts to analyze and interpret scientists' work through the relationship between their notebooks and their published papers. His primary concern was to contrast two of what he called themata, "the thematic concept of the continuum rather than the thematic concept of atomism." What he found in the notebooks, though, was that Millikan had recorded many oil-drop observations that he threw out when calculating *e*. This was true of the whole run of papers on the determination of *e*. In 1913, by which time Millikan's apparatus and technique were sophisticated, he published in *Physical Review* the definitive paper in the sequence. It reported a series of observations of fifty-eight different oil drops. He wrote—the italics as usual are his—*"It is to be re-*

marked, too, that this is not a selected group of drops but represents all of the drops experimented on during 60 consecutive days, during which time the apparatus was taken down several times and set up anew." In his autobiography, Millikan repeated this sentence and added, "These drops represent all those studied for 60 consecutive days, no single one being omitted."

The trouble is that the notebooks show that lots of drops during the sixty days were omitted. Even as he jotted down the readings, Millikan made rough calculations and noted his judgement of each run. Many pages, Holton reported, carried notes, such as, "This is almost exactly right & the best one I ever had!!!," or "Publish this Beautiful one," or "Exactly right," or "Perfect Publish." Many other pages bore comments like, "Very low Something wrong," or "Agreement poor. Will not work out," or "This drop flickered as tho unsymmetrical," or "No. Something wrong with the therm." Holton reproduced photographically two consecutive pages from this notebook. The first is labelled "Friday March 15, 1912," and shows columns of data and of calculations; at the lower left, "Beauty," and "Publish this surely beautiful!!" The second page is "Friday March 15, 1912 Second Observation 5:00 pm." The columns are shorter, the calculations few, the run terminated at "5:35 pm." At the right, "Error high will not use," and—added later, judging from the writing as well as the date—"Can work this up & probably is ok but point is not important. Will work if have time Aug. 22." Holton called this "a failed run—*or, effectively, no run at all.*"

The fifty-eight drops Millikan reported in the paper of 1913 were selected from about one hundred and forty, by Holton's count. In Babbage's terms, Millikan proved an accomplished cook, selecting the observations to report. The aim, once more, was to achieve great precision. But Holton's judgement was benign. He called the notebooks "private science." He was interested primarily in how Millikan had discovered—had so luckily stumbled upon—the right experimental setup; in how his thoughts and imagination seemed to work moment to moment as, in communion with the experimental apparatus, he made and evaluated the observations; in how the conclusions were buttressed by different work being done elsewhere, notably by Ernest Rutherford, in England, calculating the charge on the alpha particle; in how Ehrenhaft indeed might have used those suppressed notebook pages to reach different conclusions. Behind all, Holton saw his own explanatory scheme, the controlling power of the thematic preconceptions, atomism as against the continuum. Of one of Millikan's brisk annotations—"$e = 4.98$ which means that this could not have been an oil drop"—Holton wrote:

> This remark illustrates again that the results of Millikan and of Ehrenhaft were quite sensitive to the treatment of data—and, before that, to the decision about what is the relevant or even crucial aspect of the experimental design, which data are discordant or suspicious, and which may be dismissed on plausibility grounds. As is generally true prior to the absorption of research results into canonical knowledge, the selection of the relevant portion of experience from the in principle infinite ground is guided by a hypothesis, one that in turn is stabilized chiefly by success in handling that "relevant" portion, and by the thematic presupposition which helps focus attention on it.

Others have interpreted Holton's Millikan differently. Taken in isolation, talk like Holton's of "selection of the relevant portion of experience from the in principle infinite ground" delights social constructionists. Constructionism is detestable to Holton, who joined vigorously in the polemics around it.

More generally, responses to the exposure of Millikan's cooking have been polarized. Defenders say that Millikan was operating in the context of a juggernaut new research program—quantum physics and its insistence that energy comes in discrete packets—and that he was a superb experimentalist, with the great bench scientist's gift of an intuitive feel for what was working or not working. Critics say that Millikan was unduly biassed by his preconceptions, wantonly threw out data he didn't like including a high proportion of the runs for the definitive paper—and that then, never mind genius, he lied about it blatantly and repeatedly. In a shrewd and extended analysis of the Millikan controversy, published in 1995 in the inaugural volume of the journal *Science and Engineering Ethics,* Ullica Segerstråle, at the Illinois Institute of Technology, wrote, "What is interesting about this case is that, in the hands of different authors, the Millikan case has become an example of *either* scientific misconduct *or* of good scientific judgment."

Most careful of Millikan's defenders has been Allan Franklin, a physicist and philosopher-historian of science at Cornell University. Writing in 1984 in *The American Journal of Physics,* Franklin reanalyzed Millikan's notebook entries with different questions in mind. Here come the nuances.

By Franklin's count, the notes for the paper of 1913 showed 177 drops altogether. He divided and subdivided the runs.

Considering only the drops that Millikan omitted, Franklin looked first at all that were observed before the date of the earliest observation Millikan actually used. These were 68. In turn, the first 15 of these, Franklin determined, were done while the apparatus was being tuned up. That left 53. Franklin then calculated *e*—as Millikan oddly had not—for those alone. The result was perhaps surprising to Millikan's detractors: they yield a value "which is itself more precise than any previous measure." Franklin supposed that by the time Millikan reached the first drop that he did publish, he was satisfied that the apparatus and his mastery of it were functioning aright. In effect, drop 69 became drop number 1.

A trial period—experimenting with the experiment—is universal in original experimental work, and such renumbering is routine.

At this point, then, 107 drops remained. But of these Millikan published 58 and left out 49. Franklin subdivided yet again. Of the 49 left out, Millikan ignored 22 but did calculate 27. Franklin took the 27 drop by drop. Of them, 21 gave values close to what Millikan expected, but none the less had been excluded—19 because in one way or another they did not meet the conditions of the experiment, 2 "because of no apparent reason, presumably because they were not needed."

The remaining 6 to be accounted for were indeed left out because they were too far from what was expected—yet even of these, 5 were not more than two per cent out. By this point in the argument, Franklin had rehabilitated Millikan from a lot of cooking to a very little trimming—strictly in Babbage's sense, apparently to "gain a reputation for extreme accuracy in making observations." The sixth drop did yield a value forty per cent out—though even this result may have been due to some falling-off from the standardized experimental conditions.

Alas, others who present themselves as Millikan's champions are more crude than careful. "Millikan did not simply throw away drops he didn't like. That would have been fraud by any scientist's standard." Thus David Goodstein, a physicist at Caltech, in a pompous, sloppy, anodyne essay in the literary quarterly *The American Scholar* in 1991. Goodstein, too, had examined Millikan's notebooks, wrote indeed that he had used them, especially Millikan's comments on the quality of various observations, in teaching physics to undergraduates. "To discard a drop, he had to find some mistake that would invalidate that datum," Goodstein explained. He sketched a defense, including the point—yes, offered in exculpation—that "in Millikan's case, any mistake would seem like confirmation of Ehrenhaft's contention." He concluded sagely, "Millikan was not committing fraud. He was exercising scientific judgment."

Comment is otiose. Except, what do we make of Goodstein's blithe condoning of the possibility that excluded drops would have helped Ehrenhaft? Fortunately, the burden of Franklin's analysis was that this is not so.

Holton has said explicitly that Millikan's decision not to report unused runs "is not acceptable in the present ethos of science." Yet for Holton, Millikan's sin, if sin it was, reduces to no more than that. He returned to the case in the spring of 1996, in the elite intellectual journal *Daedalus,* where he wrote, in part:

> Every experimenter to this day, particularly when working with newly invented equipment, must have a keen sense about whether external circumstances—in this case voltage fluctuations, temperature changes, turbulences in the chamber—may be interfering with the presuppositions on which the experiment is built. Galileo had analogous problems with his new telescope. Today, our strategies for dealing with discordant data are very different, and in the light of our current, much harsher rules, it is all too tempting in retrospect to accuse Millikan of mischief.

By now, we are familiar with this defense of past scientists.

Beyond these specific views of Millikan's case, two issues about scientific practice and laboratory notebooks remain that we must distinguish. First, what are we to make of this notion of private science, introduced by Gerald Holton, elaborated (and used in his book's title) by Gerald Geison? It has come up in recent cases, too. The close reading of great scientists' laboratory notebooks has become a specialty in its own right. Stillman Drake and a number of Italian scholars have sweated the papers of Galileo. Innumerable historians have mined away at the notebooks of Newton and of Darwin. In this industry, Frederic Lawrence Holmes, at Yale, was acknowledged master. Holmes began in 1974 with a book devoted to the fine grain of the research of the revered nineteenth-century French physiologist Claude Bernard, and went on to similarly tightly focussed studies of the relationship between notebooks and publications in the lives of the eighteenth-century French chemist Antoine Lavoisier, in 1985, and then of the Nobel prize–winning biochemist of metabolism Hans Krebs, in two volumes, 1991 and 1993. Holmes conducted scores of interviews with Krebs, discussing passage after passage of the notebooks and the resultant papers.

Geison, scrupulously making clear what he was *not* saying about Pasteur, considered these notebook studies and others including Holton's and his own:

> Much remains to be done in this line of research. But in every case thus far in which records of "private science" have been closely investigated, one can detect discrepancies of one sort or another between these records and published accounts. Even the best scientists routinely dismiss uncongenial data as aberrations, arising from "bad runs," and therefore must omit or "suppress" them from the published record. Equivocal experiments are sometimes transformed into decisive results. The order in which experiments were performed is sometimes reversed. And the actual nature or direction of research is otherwise simplified, telescoped, and generally "tidied up." There is rarely anything

> sinister about such practices, rarely any intention to deceive, and their existence has long been recognized. As long ago as the seventeenth century, Francis Bacon noted that "never any knowledge was delivered in the same order it was invented," while Leibnitz expressed his wish that "authors would give us the history of their discoveries and the steps by which they have arrived at them." From time to time ever since, scientists and others, including the influential American sociologist of science Robert K. Merton, have drawn renewed attention to this "failure of the public record to record the actual course of scientific inquiry."

Geison wrote of Holmes's work that never does he "suggest that his historical actors engaged in deliberately deceptive practices."

> Instead, he maintains that Lavoisier, Bernard, and Krebs simply and wisely adopted the standard practices and rhetorical strategies that always intervene between private laboratory records and their effective and persuasive presentation in the public domain.
>
> Against this background, it should be clear that Pasteur was not committing "scientific fraud" whenever his laboratory notebooks reveal a course of research different from that recorded in his public works.

Well and good, yet this fails to cover certain potential practices. What if notebook entries themselves are fabricated or falsified? The charge has arisen in recent highly publicized cases. It is compounded by a fact of which many a scientist may be unaware: outrageously, some hold, under the regulations that govern federally financed research in the United States today, notebooks and other data don't belong to the individual scientist but are legally the property of the research institution through which the grant is made. In that case, can notebooks today be called private science in any meaningful sense?

Still further, implicit in some defenses of Mendel, explicit in many of Millikan (not Holton's), is the notion that misconduct underlying your immediate publications can be justified if later research, your own or others', proves your faked or fudged results were right. The assertion that present fraud, even if it leaves the supposed conclusions of a paper without foundation, can be excused *post facto* if those conclusions are later confirmed, has also appeared in controversy over recent cases. It is a seductive idea. It marks one of the deepest divides among practicing scientists today.

Now three classic cases of Babbage's forging, or outright fabrication—by Ernst Haeckel, by Sigmund Freud, and by Cyril Burt.

The German biologist Ernst Heinrich Haeckel has recently been exposed as a forger, whose fabrications misled embryologists and evolutionary biologists for decades. In 1874, Haeckel published a set of drawings of vertebrate embryos that he claimed to have made from direct observation of specimens, each at three successive stages of development. Fish, salamander, tortoise, chick, hog, calf, rabbit, man—remarkably, these drawings showed that at comparable stages of development the embryos of widely different creatures were almost identical. Haeckel produced his drawings in support of what he advanced as a universal law of evolution, that "ontogeny recapitulates phylogeny," which is to say that the embryos of higher vertebrates, such as mammals, birds, or humans, as they develop from the egg pass through the forms of their more primitive ancestors, only diverging late in development by adding on new stages. The notion of recapitulation and the catchy phrase enjoyed great influence and a long life. The drawings of embryos were picked up by George John Romanes, a popularizer of evolutionary theory after Darwin's death, and for decades were routinely reproduced in textbooks.

Those drawings, so widely known, so influential, were faked. To be sure, Haeckel's law itself has long been dismissed among evolutionary biologists. Almost since the drawings first appeared Darwinists, among them Darwin's friend Adam Sedgwick, had found them dubious. Indeed, Haeckel in his own day had been pushed into admitting that he had worked from memory with considerable freedom—but his admission was forgotten. In the mid–nineteen-nineties, the drawings caught the eye of Michael Richardson, an embryologist then at St George's Hospital Medical School, in London. They didn't fit what he knew about the rates at which vertebrate embryos develop their characteristic features. He and colleagues (from Australia, Canada, France, and the United States) ran their own comparative study, with photographs, and published the results in the summer of 1997. The embryos "often looked surprisingly different." Among much else, Haeckel had left out many details, such as the limb buds in human embryos. He had added others, such as a curl in the tail of a chick embryo that made it resemble a human's. He had drawn all the specimens as if to the same scale even when they differed ten-fold in size. Richardson wrote, "It looks like it's turning out to be one of the most famous fakes in biology," and added later, "The altered drawings support theories which the originals did not."

Many issues besmirch the standing of Sigmund Freud and psychoanalysis. A century after its first propounding, controversies rage interminably. Among them, Freud as fraudster comprises a small, recent set of charges—yet they dynamite the plinth beneath the heroic statue.

Since the earliest days, critics had raised problems: the claims of the master to his method's validity as real science; the question of whether his model of the mind in mental illness, his many successive models and those of his epigones, interpreters, and anti-disciples, ever bore any relation to reality; the essential circularity of reasoning, as in the counterattack to criticism that "your resistance demonstrates the truth of the method"; the sordid gossip of his poisonous relations with his followers, complemented by the ideology-ridden, cult-like behavior of psychoanalysts in good standing; the helpless fascination with Freudian ideas in some schools of literary criticism—right down to the pragmatic but not negligible question of whether the method produces cures. The nineteen-eighties and early nineties saw massive, cogent, systematic new attacks.

Among the most important were three. The first was mounted by Adolf Grünbaum, a professor of philosophy and of research in psychiatry at the University of Pittsburgh. In *The Foundations of Psychoanalysis: A Philosophical Critique,* published in 1984, Grünbaum analyzed the epistemological basis of Freud's clinical theory and its variants, demonstrating that they cannot be shown to be true. In 1991, Malcolm Macmillan, then professor of psychology at Monash University in Australia, published an immense, meticulous labor of erudition, *Freud Evaluated: The Completed Arc.* Working patiently through the historical development of psychoanalysis, he scrutinized the methodological claims and practices of Freud and followers. Six hundred pages on, he summed up, in part:

> Should we therefore conclude that psychoanalysis is a science? My evaluation shows that at none of the different stages through which it evolved could Freud's theory generate adequate explanations. From the very beginning, much of what passed as theory was description, and poor description at that.

Macmillan continued, "In every one of the later key developmental theses, Freud assumed what had to be explained."

A third critic was Frank Sulloway. Back in 1979, in *Freud, Biologist of the Mind: Beyond the Psychoanalytic Legend,* Sulloway had defended the young Freud, at length and persuasively, as a scientist, winning the gleeful praise of Freudians and a prize from the History of Science Society. But Sulloway is a relentless, even obsessive scholar. By the end of the eighties, he had repudiated that defense in journal articles and a new edition of the book.

The handbook to these controversies is *The Memory Wars: Freud's Legacy in Dispute,* which reprints a series of articles by Frederick Crews that appeared in 1993 and 1994 in *The New York Review of Books,* together with the bitter exchanges of letters to the editor that those articles provoked. Crews took down the entire Freudian enterprise, writing a meticulous review of a decade and more of the literature of attack and defense—but no usual review, for he is a joyous polemicist, wickedly ironical. In the course of this he first brought to the attention of a general intellectual audience the fact that Freud had faked and fudged the evidence.

Freud always insisted that his theories and methods were founded on his research with cases—his experience with particular, real patients. He did so from the start. In the early eighteen-nineties, in what was the false dawn of psychoanalysis, Freud developed the explanation that female hysterics were victims in childhood of actual sexual seduction by adults. In a talk in Vienna in the spring of 1896, to an audience of neurologists and psychiatrists, he claimed that the seduction theory had been confirmed by "some eighteen cases of hysteria," which he had treated with considerable success. As is part of the grand history of the movement, a year after that talk to the Viennese profession Freud abandoned the charge of actual seduction in childhood by an adult. He switched the blame to the hysterical women themselves, who he said as infants and small children had had fantasies of seducing their fathers—fantasies that, repressed and unrecognized, were causing their adult neurotic symptoms. The infant's wish to seduce the parent of the opposite sex, and the mechanism of repression, became the first and central tenet of psychoanalytical theory. The charge, then, brought in various forms by Grünbaum, Macmillan, Sulloway, and others, is that these and later revisions of Freud's theory and practice were arbitrary and without scientific basis.

All this, though, is standard stuff. What's new is the finding that Freud's cases are bogus from start to finish. For the pregnant years 1887 to 1904, the essential source for Freud's activities are his letters to an intimate friend and colleague, Wilhelm Fliess, a physician in Berlin. Only in 1950 was any of this correspondence published, in a German edition edited by three of Freud's followers including his daughter, Anna, herself a psychoanalyst. The editors censored the correspondence heavily, as became clear when a different editor published it in full in 1985. Misrepresenting cases and their outcomes began early. Thus, at the time of the talk before the Vienna physicians Freud admitted in a letter to Fliess that he had treated thirteen hysterical women, and not one successfully.

Next came the appalling treatment of Emma Eckstein. Fliess was more than Freud's confidant. He deeply—weirdly—influenced Freud's beliefs about the treatment of patients. Fliess had developed a theory that associated the genitals with the nose, in such a way that disorders of the nose could cause psychosexual problems, "nasal reflex neurosis"—which had to be treated by cauterizing and applying cocaine to the spongy bones of the nasal passages. Freud was a convert to Fliess's bizarre notion, and early in 1895 sent him an attractive young patient, Emma Eckstein. Fliess operated, removing some bone.

When Eckstein returned to Vienna, she developed a grave infection and a near-fatal hemorrhage. Freud diagnosed the cause: Eckstein was "bleeding for love" of him. Her life was saved when it was found that during the surgery Fliess had left behind half a metre of gauze. She was permanently disfigured.

In the Freudian chronology, these incidents were pre-analytic, embarrassing juvenilia. Their suppression for ninety years is the very model of cover-up. The case histories on which Freud based psychoanalysis, his theories and his bold claims of cures, were startlingly few. Frank Sulloway, assessing them in an article in 1991, could count only eight. Two examples will suffice.

The earliest case Freud published in 1905, though it dated back to the fall of 1900. This was the story of "Dora," to which he gave the title "Fragment of an Analysis of a Case of Hysteria." "Dora" ranked for long, among the orthodox, as "the classical analysis of the structure and genesis of hysteria"—so described as late as 1962 by Erik Erikson, an eminent psychoanalytic practitioner and theorist. We now understand "Dora" as a classic case of malpractice. One can hardly overstate the enormity of the facts. Briefly: "Dora" was Ida Bauer, aged eighteen, caught in an unpleasant net of adult relationships. Her father, though evincing the late stages of syphilis, had for some years been carrying on an affair with an attractive woman, identified as Frau K. Ida Bauer was fond of Frau K. She had long known of the affair. Herr K was a friend of the Bauers, and often gave Ida presents. His sexual interest in the girl was evident; she remembered with revulsion the time, when she was fourteen, that he forced a kiss on her. When she was sixteen, Herr K, during a country walk, made a sexual advance to which she reacted with disgust, slapping him and running away. She reported the incident to her father, who confronted Herr K, who denied it altogether. She insisted, vehemently and repeatedly, that they break off relations with the Ks. Her father took Herr K's side, and went on with the affair. She developed signs and symptoms including weeks-long attacks of coughing, a vaginal discharge, instances of bed-wetting, and thoughts of suicide. Her father sent her to Freud to be cured of these—and of her rebelliousness. She went to Freud altogether against her will.

Treatment was stormy and lasted only three months. None the less, in that time Freud was able boldly to interpret several of Ida Bauer's dreams and to elaborate a diagnosis. Briefly: He insisted to her that she was unconsciously in love with Herr K. Her bed-wetting and vaginal discharge he regarded as proof that she had been a masturbator—which he believed an unhealthy practice—although she denied it. The disgust she reported at the forced kiss masked the fact that she had actually been sexually, clitorally excited by it—which Freud thought was the only healthy response of a pubescent girl. The urgency of her insistence that they break with the Ks masked the fact that she was in homosexual love with her mother as well as with Frau K as a mother surrogate, and jealous of her father's affair. And so on. Her breaking off the analysis Freud reported to Fliess as a great pity, because he was on the brink of convincing her of the error of her ways.

We are offended by Freud's arrogant bullying of the patient, his exculpation of the pedophilic Herr K, and his toadying to her father. Those are not the point here. Entire books have been written exposing the case of "Dora." Macmillan dissected it in fifteen packed pages. The scientific fraud, disentangled as far as possible from the messy malpractice, lies as Macmillan showed in Freud's systematic, wonderfully agile rejection of Ida Bauer's knowledge of the facts and of her actual reactions and associations, always in favor of his own preconceived construction—which he then presented as the first triumph of his new science.

The greatest triumph and cure Freud reported was the case of the Wolf Man. He was Sergei Pankeev, from earliest childhood cripplingly depressed, anxiety ridden, obsessive, with a phobia about wolves that began with a nightmare at the age of four. Freud kept him intensively in analysis for five years, beginning in 1910. This was the period of his grand theory building. Freud settled upon the traumatic effects of the so-called primal scene, when an infant witnesses his parents copulating; the Oedipus complex and the fear of castration; and the tripartition of the mind into id, ego, and superego. As is general knowledge, Freud made castration anxiety and the Oedipus complex not merely the cause of neurosis but the indispensable, universal cause of the formation of each individual's superego, and thus the crucial factor in the repression and control of primitive impulses—and so in the preservation of civilization. The analysis of Pankeev was central to the foundation of all this, a great part of the putative evidence.

Freud published the case in 1918, as "From the History of an Infantile Neurosis." There he claimed to have cured Pankeev completely, freeing him of all his fears and obsessions. This was a lie. Further, those all-important memories from early childhood were not, in fact, offered by Pankeev but were imposed on him by Freud. The truth is that for nearly seventy years Pankeev was in and out of analysis with Freud and various of his followers, and in and out of institutions, his condition worsening, until his death. He was urged to keep silent, and intermittently paid a pension. In the nineteen-seventies, an Austrian journalist, Karin Obholzer, found him and interviewed him at length. Pankeev told her, in despair, "The whole thing looks like a catastrophe. I am in the same state as when I came to Freud, and Freud is no more."

Sulloway showed that every one of Freud's case histories is rendered worthless by gross fabrications and falsifications. Frederick

Crews summed up: "Freud's theories of personality and neurosis—derived as they were from misleading precedents, vacuous pseudo-physical metaphors, and a long concatenation of mistaken inferences that couldn't be subjected to empirical review—amounted to castles in the air."

Cyril Burt lies on the border between classic and modern frauds. In the typology of fraud, his work exhibits an assortment of significant traits. To begin with, it had great political importance, especially in the United States, where it has been used to support racist and elitist arguments; its exposure as riddled with fraud ignited controversies that persist today. That being so, it is astonishing that glaring flaws went unnoticed throughout a career that spanned more than half a century and loaded him with honors. Burt's publishing practices ought to have aroused suspicion—and warned us of the crucial role that we will see publication plays in the processes of science. And with Burt once again we encounter the telltale presence of statistical anomalies—in his case, statistical absurdities there for all to see. But who looked?

Burt was an English educational psychologist, who for the first half of the twentieth century was a massive influence in the study of intelligence—the gifted and the subnormal. His specialty was the testing of children's intelligence, and the genetics—the heritability—of intelligence. He was the leading expert in the statistical methods of such studies, and was skillful in presentation of the conclusions to general audiences. For decades, beginning at the time of the first world war, he was chief psychologist of the school system of greater London. From 1909 through 1975, he published upwards of four hundred papers, chapters, encyclopedia entries, books, popular articles, radio talks, and so on, ranging from technical analyses to inquiries into, say, the paranormal experiences claimed by Carl Jung.

(Oddly, Burt had a lifelong, highly sympathetic interest in psychical research.)

Particularly important for analyses of heritability are twin studies. Burt invented the field and dominated it. The method is to find pairs of monozygotic, or identical, twins—from the same egg—and pairs of dizygotic, or fraternal, twins. The idea is to test them for their levels of certain psychological traits, to see whether the identical twins are more similar to each other than are the fraternal. If so, then the argument is that that trait is to a measurable degree inherited. Especially interesting are identical twins who have been separated near birth and raised apart, in different families, for these when compared to identical twins reared together would seem to offer a way to distinguish the heritable from the environmental causes of similarities.

In Burt's work the trait at issue was, of course, intelligence. His post in London gave him access to children and to school records. He had begun collecting twins perhaps as early as 1913. He first published twin data in 1943, in a paper "Ability and income," which he said was based on 62 pairs of identical twins, of which 15 had been raised apart, and 156 pairs of non-identical twins. The correlation of I.Q.s between identical twins not separated was 0.86, that of identical twins that had been separated was 0.77, and that of fraternal twins was 0.54. With correlations stated in this form, 1 is a very large number—and so is 0. A correlation of 1.00 says that two events or phenomena, two types of data, are always linked: know one and you know the other. A correlation of 0.00 says that they are never linked, an equally strong statement. If two phenomena are altogether independent, going together half the time, the correlation is 0.5. The obvious import of Burt's figures was that intelligence, as measured by I.Q. scores, had a high genetic component. How and when the twins were found he never said. He did not publish the raw data.

In 1946, Burt was knighted. In 1947, he and a colleague, Godfrey Thomson, started a new journal, sponsored by the British Psychological Society, the *British Journal of Statistical Psychology.* No more than a hundred people ever subscribed to it. Thomson died in 1954, leaving Burt the sole editor. He published much of his later work there—papers that were long, dry, wordy, and not peer reviewed.

In 1955, Burt returned to twin studies, with a paper titled "Evidence for the concept of intelligence." The number of pairs was up, and he thanked an assistant, Jane Conway, for her help in finding them. The correlations reported were of the same order as in 1943. A year later, he published a long paper in his journal, listing a co-author named Margaret Howard. The correlations for separated twins were calculated to four decimal places. Details were scanty; again, he reported no raw data. The conclusion was that approximately eighty-seven per cent of variance in I.Q. scores was due to genetics, thirteen per cent due to environmental differences.

A year later still, in 1957, Burt gave the Bingham Memorial Lecture, an annual event of the British Psychological Society. His topic was "The inheritance of mental abilities." He offered no precise figures for the numbers of twin pairs in the several groups, though he said he had "over thirty" pairs of identical twins raised apart. The correlations, though, were the same as those in the 1955 paper. In his audience was Arthur Jensen, an American educational psychologist and an admirer of Burt's.

In 1958 and 1959, Burt's journal published two articles about twins, by Jane Conway. These stated that the number of identical pairs reared apart was now 42, double the number in his paper just three years earlier. Then in 1966 he published a paper of his own—not in his own journal—titled "The genetic determination of intelligence: a study of monozygotic twins reared together and apart." The pairs of identical twins separated were now 53; the numbers in the other groups were also different. None the less, many of the correlations he reported were exactly the same as in previous papers, to three decimal places.

At the time, various educational psychologists noted some anomalies, but little was made of them. Burt's studies, his statistical claims, his numbers of twin pairs in the different categories, his tables of correlations, were utilized as data by others carrying out similar twin studies. Yet his data and conclusions became increasingly controversial politically. Conservative commentators in Britain, such as the behaviorist Hans Eysenk, and in the United States, such as Arthur Jensen and Richard Herrnstein, used Burt's material and related work by others as evidence that differences in I.Q. between whites and blacks were largely innate, inherited—and therefore that education could not be relied on to bring blacks to the level of whites. In 1969, Jensen pursued this at 123-page length in an article, "How much can we boost I.Q. and scholastic achievement?" in the *Harvard Educational Review.* He relied in great part on Burt's twin studies, pooling them with others' including some of his own.

Richard Lewontin, at Harvard, was first to rise up against Jensen. Lewontin is a population geneticist of good scientific reputation, a political activist, and a Marxist. In the March 1970 issue of the *Bulletin of the Atomic Scientists* he published a scathing attack. Grounds for rejection were many. For one thing, separated twins were typically placed with families roughly the same in socioeconomic, ethnic, religious, or educational status, which suggested that the contribution of environment to intelligence was masked. But the heart of Lewontin's analysis then and in years of subsequent polemics has been the assertion that the data—anybody's data—from studies of heritability within particular groups do not permit meaningful comparisons between different groups. Some found the point technical, subtle, not commonsensical, but it is correct and vitiates the political claims.

On 10 October 1971, in his eighty-ninth year, Burt died, full of honors. The breaking down of his reputation had hardly begun.

Lewontin had written *contra* Jensen. The first major attack on Burt directly came from Leon Kamin, at Princeton, a sociologist and man of the left. In April of 1972, in the psychology department there, Kamin held a small conference about the heritability of intelligence. In September of that year, he elaborated the attack in a public lecture at the University of Pennsylvania, and appeared at other universities that fall and winter, culminating in an address before the Eastern Psychological Association in May of 1973. In October of 1974, he published a book-length analysis, *The Science and Politics of I.Q.*

Conflicts and contradictions among Burt's various reports; an almost total lack of information about how the data had been gathered and from what groups; innumerable small errors—but the overriding cause for suspicion was the statistically impossible identities, in the correlations Burt had reported over the years, even as the numbers of pairs of twins was doubling and more. The sizes of the sets of twins in all three categories changed from report to report. The correlations remained the same to three decimal places—twenty such coincidences out of a total of sixty correlations. Kamin reduced some of these to a table.

DATE OF PAPER	**1955**	**1958**	**1966**
Pairs of identical twins reared apart	21	over 30	53
Correlation of IQ for such twins	0.771	0.771	0.771
Pairs of identical twins reared together	83	not given	95
Correlation of IQ for such twins	0.944	0.944	0.944

Kamin wrote, in deadly understatement, "The numbers left behind by Professor Burt are simply not worthy of our current scientific attention."

One of those most shocked by these revelations was Arthur Jensen. He had come to know Burt well in the summers of 1970 and 1971. He had relied heavily on Burt's material in his own work. Jensen was at the University of California at Berkeley, politically volatile in the nineteen-sixties and seventies. He had been at the center of the raging American controversy over the relations of race and intelligence—the I.Q. wars. Students had demonstrated against him, packing his lectures and shouting him down; indeed, for a while he had been accompanied when on campus by a university bodyguard. But Jensen is an honest man, and went to London to examine the evidence himself. "The correlations are useless for hypothesis testing," he concluded.

Then on 24 October 1976, the British medical journalist Oliver Gillie, writing in *The Sunday Times* of London, opened the gathering evidence to the general public—and took the charges one step further—with a sensational front-page headline and story, "Crucial data was faked by eminent psychologist." Among much else, Gillie asserted that those astonishing correlations could only be explained by supposing that Burt had worked backwards, making up the data to fit. In all this, Gillie was for the most part making explicit the unavoidable inference from the findings by Kamin, Jensen, and others. Gillie's most startling accusation was that Burt's two assistants, Margaret Howard and Jane Conway, themselves authors or co-authors of papers on the twins, had never existed.

Controversy raged for months. In recent years, attempts have been made to rehabilitate Burt. These have failed.

These classics are tales of great men perhaps gone astray: this is largely why they have come to light and why we find them interesting. The bias is marked, though. What's lacking is an assortment of what one can call everyday fraud committed in those earlier times by lesser figures doing more routine work. Of course, the scientific enterprise was almost unimaginably smaller, even up to the second world war, yet surely some such cases occurred. Furthermore, the way the great classics are recounted provides too little sense of their laboratory or larger settings. From Newton to Burt the scientists are presented as loners. To be sure, Geison's account of Pasteur, though it emphasizes his secretiveness, does place him in his milieu, tellingly. Other exceptions, to some degree, are Darwin and Freud—diametrically different though they are.

Babbage's typology, which started us on our way towards the problem of anatomizing fraud, holds up well. He found motive in the wish for an appearance of great precision: although we cannot divine motive, certainly that appearance of over-precision emerges again and again—Newton, Millikan, Burt—as a bad sign. Piltdown of course was hoaxing, though uncovered, alas, too late to give the perpetrator his full measure of joy. Haeckel, Freud, and Burt were forgers, Newton a trimmer (and Mendel, too, if anything), Pasteur and Millikan accomplished cooks. In the course of centuries, notebooks remained evidentiary, standards evolved. Prodger defends Darwin's *Expression of the Emotions* as "produced on the cusp of this change of attitude" towards photographs as data rather than demonstration. Westfall on Newton and Holton on Millikan acknowledge that the acts they can't quite bring themselves to condemn outright would not be acceptable today.

What does not appear to have evolved is the judgement defense. If the cooking, systematic and considerable, of scientists of genius—Newton, Millikan—be excused as the proper exercise of scientific judgement, a most troublesome question arises for the everyday work of science. Judgement calls are indeed inescapable and necessary, moment after moment, in everyday science, but when do they veer into fraud? How are the standards of judgement learned? Do they vary from field to field? Are we really admitting, after all, that fraudulent behavior is OK if the conclusions are borne out?

This is a recurrent, nay, an almost universal theme. When a great scientist got the right answer, especially when it's the work that made him great, we revere him; but if it turns out later that he fudged to do it, we are deeply ambivalent. So put the question another way: Is there a difference between intuition and lies? Does examination of such cases bring us to a more complex appreciation of the nuances of scientific work?

3

PATTERNS OF COMPLICITY

RECENT CASES

One can only judge the rare acts that have come to light as psychopathic behavior originating in minds that made very bad judgments—ethics aside—minds which in at least this one regard may be considered deranged.

—*Philip Handler, then president of the National Academy of Sciences, testifying before Congress, 1981*

In the classic instances of fraud, the perpetrators acted alone. Or so the stories are told: they fit suspiciously well that archetypal image of the solitary genius making—faking?—the revolutionary discovery. Recent scientific frauds are different. In almost every case, to be sure, some one individual gets blamed, but these frauds cannot be presented even as anecdotes without an accounting of the relationships among many people within the laboratory and the larger institutional setting. The cases exhibit multiple, tangled complicities.

The complicities fall into certain standard, even predictable syndromes—clusters of telltale signs. Names for them? The dominant form is *the prodigy;* others are *the mentor seduced,* the *folie à deux,* and the *arrogance of power.* These are by no means limited to the sciences: indeed, we can best begin with a fraud that was not in the sciences at all but that displays in bitter detail the most common syndrome, the prodigy. This was the flagrant case of fabrication and plagiarism by a journalist named Jayson Blair that broke in May of 2003 and befouled the reputation of *The New York Times.*

The first acknowledgement of the debacle came on Friday, May 2, under the front-page headline TIMES REPORTER RESIGNS AFTER QUESTIONS ON ARTICLE. Jayson Blair had resigned the day before, after an article of his the previous Saturday, about the family in Texas of an American soldier missing in action in Iraq, had been found to incorporate passages from an account published earlier by the *San Antonio Express-News.* What original reporting Blair had put into the piece, if any, had not yet been determined. A terse "Editors' Note" accompanied the story, saying the *Times* was apologizing to the family and investigating Blair's work.

The *Times,* it must be said, reacted with what they do best, fast and massive multiple reportage. Nine days later, Sunday, May 11, the front page carried the headline CORRECTING THE RECORD: TIMES REPORTER WHO RESIGNED LEAVES LONG TRAIL OF DECEPTION. The byline to this report carried seven names; the article ran 7,397 words and was paired by another 7,023-word piece itemizing the fabrications and frequent plagiarism uncovered in Blair's published work back to late October of 2002. At that time, as a fledgling reporter, he had been seconded to the Washington bureau, initially to add to the paper's coverage of the suburban sniper attacks. The move was instigated by the two men at the top of the *Times's* news operation, Howell Raines, executive editor, and Gerald Boyd, managing editor. Raines had judged Blair "a hungry guy."

The self-exposé covered four full pages. Blair had been prolific, some seven hundred published articles in a career of less than four years, seventy-three published in the last six months, fifteen of them in April alone. After two sentences stating the fact of "frequent acts of journalistic fraud," the *Times*'s account went on:

> The reporter, Jayson Blair, 27, misled readers and Times colleagues with dispatches that purported to be from Maryland, Texas and other states, when often he was far away, in New York. He fabricated comments. He concocted scenes. He lifted material from other newspapers and wire services. He selected details from photographs to create the impression he had been somewhere or seen someone, when he had not.
>
> And he used these techniques to write falsely about emotionally charged moments in recent history, from the deadly sniper attacks in suburban Washington to the anguish of families grieving for loved ones killed in Iraq.

By the time the exposé was put to bed, the investigators had "uncovered new problems in at least 36" of those seventy-three articles; spot checks through Blair's work of the previous years had found "other apparent fabrications." He had been ingenious. "His tools of deceit were a cellphone and a laptop computer—which allowed him to blur his true whereabouts—as well as round-the-clock access to databases of news articles from which he stole." With the invasion of Iraq, his range expanded. He got stories published on interviews with wounded soldiers in the National Naval Medical Center, in Bethesda; with the families of missing men, in a northern suburb of Baltimore and in Cleveland; and even with the family, in West Virginia, of Pfc. Jessica Lynch, the soldier who had been captured (and was later rescued). The articles were often rich with visual detail and poignant quotations. They seemed to justify Raines's and Boyd's confidence. Yet repeatedly they raised problems. Front-page revelations about the handling of the suspects in the sniper case provoked prosecutors to call news conferences to denounce them. Time and again, articles turned out to be riddled with errors; he had a growing reputation for sloppiness. But his hustle was impressive: "Between the first coverage of the sniper attacks in late October and late April, Mr. Blair filed articles claiming to be from 20 cities in six states."

Then the *San Antonio Express-News* charged plagiarism. The frauds unravelled. Among lies by the score, Blair had not visited the hospital in Bethesda. He had not visited that family north of Baltimore. He had not shown up in Cleveland. He had exploited Jessica Lynch's family for five published pieces, but it turned out he had probably never gone to West Virginia. "Blair pulled details out of thin air." He wrote that Private Lynch's father "choked up as he stood on his porch here overlooking the tobacco fields and cattle pastures." But "The porch overlooks no such thing." He said the house was on a hilltop. "It is in a valley." Much more—and no member of the family recalled talking to him. A lot of the time, it turned out, he was not on the road at all but was sending his stories in by e-mail from New York. The investigators wrote, "In the final months the audacity of the deceptions grew by the week, suggesting the work of a troubled young man veering toward professional self-destruction."

The exposé was long on detail, much of it irresistibly titillating, even ludicrous. It was long on piety, institutional breast-beating. But on analysis it came up short.

The prodigy syndrome shows telltale signs. The perpetrator is charming and plausible. Blair first came to the *Times* in the summer of 1998 as an intern; he was a student at the University of Maryland and had previously interned at *The Boston Globe,* which is owned by the *Times*; he had "glowing recommendations and a remarkable work history." When he returned, a year later, bucking for a full staff position, he became noted for his affability. One colleague told the investigators, "He had charisma, enormous charisma." Another sign is apparent hard work, long hours, astonishing productivity—and the product seems impressive, important. Yet audacity becomes carelessness. Often the perpetrator becomes increasingly sloppy, even reckless: as the *Times*'s account said, Blair seemed to be courting self-destruction.

One need only ask how all this came to pass, to see the essential point that the syndrome has another side, in the characteristics of the professional setting. Typically the problems begin with a failure to check the background. All too often, inquiries after the fact discover misrepresentations or previous problems. Thus, the *Times*'s account, citing another journalistic source—other papers rushed in with investigations of their own—reported that while at *The Boston Globe,* "Mr. Blair apparently lied about having interviewed the mayor of Washington, Anthony Williams." When he came back to the *Times* in 1999, he gave the impression that he had completed his work at the University of Maryland, though it turned out that he still had more than a year to go.

Again, characteristically as the frauds are proceeding there are signs of trouble, yet these are ignored—sometimes with what looks like willful blindness. In Blair's case this pattern was extreme. Over the months and years, various members of the *Times*'s mid-level staff had come to distrust his work as flagrantly careless. Indeed, the *Times* said, in April of 2002 "Jonathan Landman, the metropolitan editor, dashed off a two-sentence e-mail message to newsroom administrators that read: 'We have to stop Jayson from writing for the Times. Right now.'" None the less, that fall Raines and Boyd, although they knew that Blair had had problems, promoted him and sent him to Washington—and nobody warned his new boss, Jim Roberts, the national editor.

Failure to supervise, failure to communicate, the *Times*'s exposé said. The problem goes deeper. Underlying all else, in this case of fraud as in most others, is the network of power relationships among the mid- and senior-level people who have been in touch with the actual perpetrator. The nature of those relationships is set from the top. For understanding the etiology of fraud, the case of Jayson Blair at *The New York Times* offers one advantage above all, the openness with which the paper reported the affair. The publisher of the *Times* is Arthur Sulzberger, Jr; he, Raines, and indeed the entire staff recognized that they had no alternative.

Three days after the publication of the four-page exposé, Raines called a meeting of the news staff. The next day, May 15, the paper reported it. The meeting was closed; the report was written by Jacques Steinberg, who deliberately did not attend. Complaints and anger had been bitter. One passage epitomizes the problems and Raines's responses.

> Mr. Raines made clear that he viewed the session as something more: a forum on his 20-month tenure as the newsroom's leader. During this period the paper has won eight Pulitzer Prizes—six for its coverage of the Sept. 11, 2001, terrorist attacks—but it has also been a time of dissension. A growing number of employees have expressed deepening concern about what is viewed as a top-down management style that, they say, could have contributed to Mr. Blair's ability to do what he did undetected for so long.
>
> "You viewed me as inaccessible and arrogant," Mr. Raines said, ticking off a list he had compiled from his own newsroom interviews in recent days. "You believe the newsroom is too hierarchical, that my ideas get acted on and others get ignored. I heard that you were convinced there's a star system that singles out my favorites for elevation.
>
> "Fear," he added, "is a problem to such extent, I was told, that editors are scared to bring me bad news."

Raines said he would not resign. Sulzberger, present at the meeting, said he would not accept the resignation if offered.

The Jayson Blair affair differed in one respect, anyway, from cases of fraud in science. Three weeks later, on June 5, Howell Raines and Gerald Boyd resigned.

Of scores of cases of scientific fraud in the past thirty years, here follow eighteen, some briefly, some in considerable detail. Nine are American. But the sciences these days are global, and so is fraud; the other nine are Indian, British, and German. As many more could be included, these I have chosen for multiple reasons. They come from across the spectrum of the sciences, though most are from the biological side. They illustrate various forms of fabrication, falsification, and plagiarism. And among them, these eighteen cases lay out the syndromes, the various patterns of the construction of fraud—and its exposure.

These cases had another consequence. Cumulatively, these are the ones that have made the general problem visible and urgent. They overflowed the bounds of the scientific community, attracting the attention of funding agencies, legislatures, and journalists. Several had long-lasting repercussions, notably the scandal of John Darsee—which also developed a curious connection to the David Baltimore affair. Recent fraud can be said to have come to public notice beginning in the spring of 1974, with the affair of William Summerlin, in whom we will recognize a prodigy, and his painted mice.

*

William T. Summerlin was chief of transplantation immunology at the Sloan-Kettering Institute for Cancer Research, a component of the Memorial Sloan-Kettering Cancer Center, which also includes a celebrated cancer hospital, on the upper East Side in Manhattan. He was charming and persuasive, and was doing important research. He had gone to Emory University, in Atlanta, as an undergraduate and then as a medical student, had interned in surgery in Texas, but then went to Stanford University for four years as a resident in dermatology. At Stanford he got into research on the biochemistry of skin. In the summer of 1971, he moved to the University of Minnesota to join the laboratory of Robert Good, a bold, powerful immunologist, often controversial for his theoretical ideas and his aggressive style.

Summerlin got surprising results. Notoriously, animals reject grafts of foreign tissue except when the donor is genetically identical to the recipient or very nearly so. In fact, modern immunology is in considerable part rooted in work done during and soon after the second world war which opened up the genetic basis of the rejection of grafts of skin, as with burn patients. But Summerlin reported experiments that demonstrated that rejection can be avoided if the graft materials are treated for several weeks by a form of tissue culture before transplanting. He worked with some human subjects, but chiefly with mice and rabbits, grafting corneas, certain glands, and chiefly skin—sometimes even across species barriers. Borne out and extended, such results would have transformed basic immunology as well as transplant surgery.

In January of 1973, Good moved to New York and took over the Sloan-Kettering Institute as president, top man for research. His boss was Lewis Thomas, known to laymen as an elegant essayist, who was president of the entire complex, the Memorial Sloan-Kettering Cancer Center, both institute and hospital. The institute's science and reputation had been slipping. Good was a high-pressure man in a high-pressure setting. In April, he brought Summerlin to the institute. During the next six months, though, questions about the research arose. Scientists elsewhere in the United States and abroad, and a colleague in Summerlin's own laboratory at Sloan-Kettering, John Ninnemann, could not reproduce his Minnesota experiments with mice and rabbits. Late in the year, Good insisted that he, Ninnemann, and Summerlin prepare a paper about the difficulty for the journal *Transplantation.* On 26 March 1974, at seven in the morning, Summerlin was to see Good about that paper. To show that the work was now going better, he brought with him a cart with cages holding eighteen white mice that had received grafts of skin, on their backs, from black mice. In the elevator, Summerlin pulled out a black felt-tip pen and darkened the graft areas of two of the mice.

The unravelling began almost instantly. After the meeting, Summerlin wheeled the mice back to the animal room. There, according to the report of a review committee, a month later, "A senior laboratory assistant, James Martin, in returning the cages to their places, noted that the appearance of the supposed black grafts on two white mice was unusual. On applying alcohol, he discovered black material that could be washed away." The finding raced up the chain of command, Martin to a senior research technician, then to a visiting research fellow, then to a colleague of Ninnemann's in Summerlin's laboratory, then by about eleven o'clock to Lloyd Old, executive vice president of the institute, and to Good. Though Summerlin insisted he had acted impulsively and regretted it deeply, by noon or shortly after Good suspended him. Within a week, Good appointed six Sloan-Kettering scientists to review Summerlin's work and possible misrepresentations. The committee moved fast and hard, making telephone calls, interviewing seventeen people, and securing notarized affidavits and written statements. They brought Summerlin before them for some eight hours.

More questions arose. From Summerlin's work in Minnesota, only one mouse successfully engrafted with skin from a different strain still survived—and now a blood test showed it to be a hybrid of the two strains, in which the graft would normally take. The committee found that data on much of the work was missing, vague, or organized too poorly to be analyzed. Summerlin had made conflicting statements on grant applications, for example asserting in one that in mice "foreign grafts are routinely accepted without rejection," but in another that the grafts went unrejected "in 50% of such animals which were followed for up to five months." He had submitted a paper to *The Journal of Experimental Medicine*—a distinguished journal for work in immunology, published by Rockefeller University—containing a table in which the absolute numbers of mice did not match the claimed percentages of acceptances of skin grafts. The paper had not yet appeared, so Good with Summerlin's agreement got it withdrawn.

Worst, though, were Summerlin's claims for corneal transplants. By a couple of different surgical techniques, carried out by ophthalmologist colleagues in Minnesota, portions of corneas had been transplanted into rabbits from a variety of sources—guinea pigs, chickens, humans, and other rabbits. In an abstract (a précis giving advance notice of a longer paper yet to be published) in May of 1973, Summerlin had said that the corneas prepared by tissue culture had survived and functioned up to six months, while control grafts were rejected in two weeks. Other abstracts made similar claims. The committee learned, though, that no grafts from guinea pigs or humans had lasted longer than forty days; that a few but not all grafts from chickens survived longer, though none more than six months; and that even rabbit-to-rabbit grafts were eventually rejected. Furthermore, Summerlin's Minnesota colleagues told the committee that they were not persuaded that the tissue culturing made a difference.

Nor does this sketch exhaust the problems that turned up. On

May 20, the committee reported to Lewis Thomas that Summerlin had extensively misrepresented his results. At a press conference four days later, the facts were made public. Summerlin was placed on medical leave, not to return.

The parallels with the scandal of Jayson Blair are exact, although, of course, Robert Good did not resign. In 1985, aged sixty-three, he became physician in chief of All Children's Hospital in St Petersburg, Florida, and research professor at the University of South Florida; he died in June of 2003.

In March of 1981, the investigative subcommittee of the Committee on Science and Technology of the House of Representatives held hearings. The subcommittee's chairman was Albert (not yet Al, not yet senator) Gore, Jr, Democrat of Tennessee. He had called the hearings in response to an alarming cluster of cases of scientific fraud that had emerged just in the previous year. These have since become part of the canon of misconduct—the dishonor roll.

The Long case: Beginning as a resident at Massachusetts General Hospital in Boston, hard by the Charles River, John Long worked closely with Paul Zamecnik, a prominent biochemist and molecular biologist, on Hodgkin's disease, a cancer. He rose to associate professor, with grants from the National Institutes of Health mounting up to $759,000. In 1979, he began a collaboration with David Baltimore right across the Charles at the Massachusetts Institute of Technology. But early in 1980 a junior colleague charged that Long had faked certain results; in a meeting with the chairman of his department, he admitted this and resigned. Scrutiny of the rest of his work found that of four independent lines of cultured Hodgkin's cells that he had established, three were actually derived from a South American monkey, while the fourth though human was from a patient who did not have the disease. By the time Long resigned, $305,000 of those NIH grants had been spent. The cost of his fraudulence in money and time wasted by others who pursued his leads or used his cell lines, the blasting of their careers, is incalculable.

The Alsabti case: Born in Iraq, claiming Jordanian citizenship, a medical degree, and a PhD, in the course of three years Elias A. K. Alsabti talked his way into research posts at universities and hospitals in Philadelphia, in Houston, in Roanoke at a hospital affiliated with the medical school of the University of Virginia, and lastly at a hospital associated with Boston University. In his mid-twenties, he had constructed a list of publications sixty papers long. Most of these, perhaps all, were plagiarized. He had lifted articles from obscure journals, changed their titles, put his own name to them, sometimes with names of nonexistent collaborators, and sent them off otherwise verbatim to other obscure journals—in Japan, Czechoslovakia, Switzerland, Britain. At least once he had stolen from a departmental mailbox the manuscript of a paper sent by a journal for refereeing, touched it up slightly, and sent it to a Japanese journal, where it appeared before the original got published. Plagiarism, as we'll see in other, more complex cases, sits at one end of the spectrum of types of fraud: it is usually simpler, less creative, and more solitary. Alsabti had no senior figure backing him; indeed suspicions and exposures had harried him from institution to institution. But only in the spring and summer of 1980 did they become public. Then his claim to a medical degree turned up false and his PhD unaccredited. Alsabti disappeared.

The case of Soman and Felig: A tangled mess came to fruition that summer of 1980, involving three scientists and a number of administrators; the editors of two journals; three institutions, the medical schools of Yale and Columbia universities, and the National Institutes of Health—and all three of the forms of fraud in the canonical definition, fabrication, falsification, and plagiarism. The case suggests a lot about the refereeing of manuscripts, and something about whistle-blowers and administrative response.

It was set off in the fall of 1978 by a dramatic set of coincidences that grew to involve a huge cast. Philip Felig, then forty-four, was vice chairman of the Department of Medicine at Yale, and headed a laboratory in a building two blocks away from his office. Working in that lab was Vijay Soman, thirty-seven, from Poona, India, a rising assistant professor, productive and well thought of. The two men had collaborated on a number of papers.

Helena Wachslicht-Rodbard, then twenty-nine, worked in Bethesda at the National Institute of Arthritis and Metabolic Diseases, in the laboratory of Jesse Roth, chief of the diabetes branch. She was lead author, with Roth and another, on a paper about insulin receptors of patients with anorexia nervosa, the eating disorder whose sufferers become pathologically thin. On 9 November 1978, she submitted the paper to *The New England Journal of Medicine,* whose editor was Arnold Relman. Routinely, the journal sent the paper to two scientists for anonymous reviews.

The world of a scientific specialty is often small. One referee was Felig, who passed the manuscript and the chore to Soman. Two years earlier, Soman had got approval to do an essentially identical study, but had not written a paper. The reviews (with the copies of the paper) came back to the *New England Journal,* one in favor but Felig's against, so the editors sent the paper to a third referee. Felig did not tell the journal that Soman had written the review and had long planned a similar investigation. He did not know that Soman had taken the opportunity to make a copy of Wachslicht-Rodbard's paper. At the end of January, after almost three months' delay, Relman at the *New England Journal* wrote to Wachslicht-Rodbard to say that of three reviews one had been negative and that the paper could not be accepted unless revised. Routinely, he did not say where the negative review had come from.

Meanwhile, towards the end of December, some six weeks after Wachslicht-Rodbard et al. reached him, Soman had submitted a paper on insulin and anorexia, with Felig as co-author, to *The American Journal of Medicine.* Felig was an associate editor there. The editors of the journal routinely sent the Soman and Felig manuscript out for review. As it happened, a copy went to Jesse Roth, who passed it on to Wachslicht-Rodbard. She saw instantly that this paper was the near-twin of her own—indeed, with some passages in her original language. Plautus or Shakespeare could not have devised a more symmetrical comedy.

The development and dénouement were inevitable, yet protracted and tangled. The reactions of the protagonists were predictable, their fates grim. (In October of 1980, while the case was still hot, William J. Broad reconstructed the chronology in full detail in a pair of long articles in *Science.*) In the upshot, Soman was found to have fabricated results; audited, he could not produce essential original data for this and many other papers, and was forced to retract, resign, and return to India. Helena Wachslicht-Rodbard left the NIH and research. Felig, meanwhile, in the spring of 1980 had started a new job as chief of the Department of Medicine—at medical schools, the premier department and chairmanship—at the College of Physicians and Surgeons at Columbia University. At the end of July, the affair became known at Columbia. Paul Marks, vice president for health sciences, formed a faculty committee at once. Though nobody accused Felig himself of fraud, the committee drew up seven pages of charges. Their essence was that at every stage Felig had acted inadequately and incorrectly—indeed, from the moment he did not decline refereeing the Wachslicht-Rodbard paper—and, perhaps worst, had failed to keep everybody at Columbia adequately informed. On August 1, the committee asked Felig to resign. On 8 August 1980, the scandal was on the front page of *The New York Times.*

None of these cases was trivial. The crescendo of publicity prompted the congressional hearings.

Hardly eight months later, on 31 March 1981, testifying before Gore's investigative subcommittee, Philip Handler, then president of the National Academy of Sciences, was obdurate to the point of arrogance. "The matter of falsification of data, I contend, need not be a matter of general societal concern. It is, rather, a relatively small matter which is generated in and is normally effectively managed by that smaller segment of the larger society which is the scientific community." He pointed to the "highly effective democratic, self-correcting mode" of science, particularly the peer-review system. He went on, "One can only judge the rare acts that have come to light as psychopathic behavior originating in minds that made very bad judgments—ethics aside—minds which in at least this one regard may be considered deranged." Handler testified seven weeks before John Darsee was first found out.

*

John Darsee was a prodigy, and at a top laboratory in a top institution; his exposure drew top media attention. The case was also a demonstration of the arrogance of power. And it provoked a remarkable analysis of the complicity of co-authors. Until the spring of 1981, Darsee had been admired and envied in his field as an able and prodigiously productive young scientist: by the age of thirty-three, working first at the medical school of Emory University and then at

Harvard Medical School, he had published more than one hundred and twenty-five research articles, book chapters, shorter papers, and abstracts. His specialty had been the causes of damage to heart muscle in heart attacks, and the prospects for recovery in relation to the severity of the damage. Darsee had been hired from Emory to Harvard in July of 1979. He came at the invitation of Eugene Braunwald, aged fifty that year, a uniquely powerful cardiologist, research scientist, and administrator, his name on some six hundred papers, Hersey Professor of the Theory and Practice of Physic, one of the committee directing the Department of Medicine at Harvard Medical School, and physician in chief of Brigham and Women's Hospital, an affiliate of the school. By that spring of 1981, Braunwald was running nearly a score of research scientists and upwards of three and a third million dollars in NIH grants, at two laboratories, one at Brigham and Women's, the other a couple of blocks away at Beth Israel Hospital. Darsee was working in the cardiac research laboratory at Brigham. He had been co-author with Braunwald on a number of papers by then, and on a chapter in Braunwald's massive, definitive textbook of cardiology. He was up for promotion to assistant professor. Braunwald was pushing his protégé's career, and later defended him as the most brilliant, creative, and hard-working junior colleague he had ever known. They talked science as equals. The pattern is familiar: the interaction of the productive, clever, plausible junior colleague and his patron or mentor. Yet, of course, such interactions are normally the mode by which careers are launched (and not just in the sciences) and good science is fostered.

Darsee's unmasking was chronicled by a meticulous account by William J. Broad that appeared in *Science* the next January and later by the reports of three separate investigations and by a synopsis of events eventually distributed to faculty of Harvard Medical School.

Over a period of months, three of his day-to-day colleagues had been growing suspicious—a postdoctoral fellow named Edward Brown, another postdoc, and a technician. On 18 May 1981, the trio went to the immediate head of the laboratory, Robert Kloner, to say that Darsee was completing for publication an abstract about some experiments with dogs, in advance of a full paper to be presented at a meeting in the fall, and that they believed that he had fabricated the research for this—invented the data outright. Kloner, incredulous, alarmed—he, too, had his name as co-author on a number of Darsee's publications—asked to see the raw data for the abstract. Darsee said he would pull some together. On Thursday, May 21, at the lab, he began taking experimental readings on a dog, and as the results came out he marked them successively Day 1, Day 2, and so on, making it look as though they had been collected over a period of two weeks. He did this in full view of colleagues, who watched incredulously.

The psychology of the unmasking is outré: Had Darsee reached the point, like some other but more morbid serial criminals, where he wanted to be caught—"Catch me before I fake more"? Be that as it may, when Kloner confronted him he claimed he had manufactured data because he had thrown the original data away—this for work supposedly done recently. Darsee maintained then and for months that this was his sole act of fabrication. The next day, Friday, Kloner reported the matter to Braunwald. Soon after, Brown and the others told Kloner they thought Darsee had been cheating on papers regularly.

Harvard acted both quickly and slowly, in the very pattern of ambivalent institutional response: *Can it be true, he is so talented, something must be done—but can we smother scandal?* Darsee was relieved of his academic and clinical appointments. Plans for his promotion were killed. Braunwald and Kloner began to analyze Darsee's research work. In June, the medical school terminated his fellowship from the National Institutes of Health—yet without mention of the misconduct. Darsee was not stopped from publishing: at the end of May, for example, the journal *Circulation* accepted from him ten new abstracts of papers to be presented that fall. He was allowed to continue working on other major research projects.

Only late in October—it took five months—did Braunwald and Kloner face the fact that Darsee's fabrications had been extensive, and had continued even after May 21. Only on November 3 were any of Darsee's papers and abstracts withdrawn. Only in that month did the dean of the medical school, Daniel Tosteson, appoint a formal committee to investigate—though when he did so it was formidable, with three members from other departments of the medical school, a member from the university's faculty of arts and sciences and another from the law school, three from institutions outside Harvard, and, as chairman, one of the grandees of the profession, Richard Ross, dean of the Johns Hopkins University School of Medicine. Only in mid-November did Braunwald notify his opposite numbers at Emory University of the problems and advise them to audit Darsee's work there. In December, the National Heart, Lung, and Blood Institute began an investigation; still later, Emory appointed an investigating committee of its own.

Braunwald's handling of the problem was conspicuously slow. Five months passed from the day he first learned of Darsee's fabrication of data until Kloner told the heart institute, at a meeting in Bethesda, what had been found. During that time, Darsee continued to work in Braunwald's lab. Broad made clear, at every step in his detailed account in *Science,* that the delay was questionable. The contrast was extreme with the case of the painted mice, where Robert Good had suspended Summerlin the day he confessed. Broad quoted

Tosteson as saying the delay was due to Braunwald's wish to "manage" the situation. Though Broad did not need to press the point, some would take such language as a euphemism for cover-up. Indeed, one who was chief of another department at the Brigham, and therefore sat with the others of that rank on the executive committee and was in a position to see the scandal developing in real time, told me in tones of persistent outrage, "At the outset, Braunwald tried to sweep it under the rug." Requesting that I not use his name, he wrote me this in a long e-memorandum in February of 2000 and repeated it in a telephone conversation in March of 2004. "Perhaps Braunwald's biggest lapse from ethical standards was his failure to inform Darsee's collaborators"—including at least four from other departments—"that they were at risk," he said. "Some of them continued to work with Darsee during the summer," writing papers together.

The investigation started by the heart institute at the NIH in December of 1981 produced a report fourteen months later that criticized Harvard scathingly, and rebuked Braunwald and Tosteson in particular, for mishandling the case. Braunwald's lab was high in pressure to publish, low in supervision by senior scientists. The leisurely audit Braunwald and Kloner had carried out, the report said, "was insufficiently rigorous and not definitive." Instead, an independent panel of investigators should have been appointed promptly. The NIH should have been notified at once; Emory University should have been informed earlier. The NIH barred Darsee from funding for ten years and demanded that Harvard give back $122,371 it had received for one of the studies Darsee's work had rendered worthless. In an unprecedented move, James Wyngaarden, the director of the NIH, put Braunwald's laboratory on one year's probation to make sure that recordkeeping and supervision were adequate, before further federal money would be authorized.

Eventually, Darsee was shown to have fabricated data for scores of publications at both institutions. One hundred and nine of the papers had co-authors. The total was forty-seven co-authors, all of whom had worked at either Emory or Harvard. No one has ever pointed to any of these men as knowingly party to Darsee's frauds. The question remained: Why had none of the co-authors caught on?

That was a question that provoked the curiosity and intervention of two of the strangest characters in the history of recent scientific fraud: Walter Stewart and Ned Feder. They were research scientists at what was the catch-all National Institute of Arthritis, Diabetes, and Digestive and Kidney Diseases, one of the National Institutes of Health, in Bethesda, Maryland. (It has since been split in two, one for allergy and infectious diseases, the logic being that both involve the immune response, and the other keeping diabetes and digestive and kidney diseases.) They were investigating misconduct out of private curiosity that grew to be a passion, with no official sanction.

One's immediate impression of Walter Stewart was the cheery, swinging vigor of his voice and greeting, and then the velocity of his speech, ideas tumbling over each other in every direction when insisting on a point about science or fraud. His hair was dark and thick, skin pale, forehead low and jaw strong, his mouth large and lippy. He had graduated from Harvard *summa cum laude,* had started graduate work at the Rockefeller University but moved to the NIH; he never earned the doctorate. In his fraud-busting days he was likely to be found sitting around in the laboratory, air-conditioned as it was, in shorts, skimpy shirt, and sandals, playing with a telegraph key, practicing morse code—which useful accomplishment he was insisting his children acquire while he was himself teaching them to read. Seen, if rarely, in jacket and tie he had a touch of the dressed-up neanderthal about him. His high intelligence was unquestioned even by scientists who detested him and his works. His eccentricity, and a big dose of stubborn fanaticism, were recognized even by his friends and allies.

Ned Feder was eighteen years older, leaner, taller, quieter, a man who blended into the crowd. Born in Minneapolis, he took his bachelor's degree in organic chemistry at Harvard, then a medical degree at Harvard Medical School, wound up there as junior faculty, did not make tenure, and moved to the NIH in 1967. Stewart had been a student of his at Harvard, "good and unusual," Feder judged. Though more soft-spoken, capable of irony, Feder had a view of scientific probity as stiff-necked—some say fanatical—as Stewart's. The interaction between the two men was close and unusual, somewhat like father and son, with the older man's steady determination steeling the younger's impulsive genius logistically, administratively, often even as a personal assistant.

Daytimes, Feder and Stewart ran a two-man laboratory investigating the genetics that determine the shapes of nerve cells in snails. They had discussed the Darsee case at length and in increasing outrage while bending over their snails. The hundred and nine publications stood in the literature, of course, for anyone who cared to dig them out. A great deal about the papers and colleagues had come to light during those three separate investigations. The snails were going nowhere fast. In the spring of 1983, with Feder's encouragement and his help in finding and copying the publications, Stewart began to scrutinize them to see what clues each might reveal on its face that co-authors, even though unaware of fraud, ought to have remarked.

They found anomalies. Many were borderline, like disagreements between a numerical value in the text and in a table covering the same material. Even those, of course, could have been caught by any collaborator paying attention. Some were serious. Stewart and

Feder wrote up their findings. They reported errors "so glaring as to offend common sense." The most flagrant example was in a paper from Emory where Darsee and co-authors had published "an arresting pedigree depicting a family with a high incidence of an unusual form of heart disease."

> Inspection of the pedigree . . . reveals that a 17-year-old male had 4 children, ages 8, 7, 5 and 4. (It is unlikely that the father's age is a misprint: it appears in the figure and in two places in the text.) Thus the father was 8 or 9 (probably 8) when he impregnated the mother of his first child and 9 or 10 when he impregnated the mother of his second. This bizarre feature of the pedigree, which perhaps could have raised questions about the validity of the entire paper, was apparently not noticed by co-authors or referees, nor were the following unlikely groupings of ages: his sister, brother and first cousin had their first children at ages 16, 15 and 15, and three women in the preceding generation had their last children at ages 41, 45 and 52.

They went on to point out a dozen more discrepancies in this paper alone: data that differed between the summary and the text and a table, so markedly that "Simple inspection suggests that these two sets of numbers cannot simultaneously be valid"; differences in data between the paper and abstracts of the same work published earlier; three family members that the published abstracts had said were living whom the paper showed to have died before the study began.

On a second paper that Stewart and Feder scrutinized from Darsee's Emory days, he had had five co-authors. It reported the results of a surgical procedure carried out on five patients. "Almost all the data in this paper are contained in three figures which cannot be reconciled with each other." Furthermore, results shown in those figures were statistically implausible, nor could they be reconciled with data scattered through the text. The introductory section of the paper described four of the patients as "uncomplicated," yet the text said that one of these suffered severe complications. Two previous abstracts had described those same patients, offering data that made each of the three reports significantly at variance with both of the others.

In all, Stewart and Feder reported, among the eighteen research papers they inspected "there were as many as 39 errors and discrepancies in a single research paper, with an average of about 12 per paper. Of the 22 scientists who were co-authors of a research paper, 19 were co-authors of at least one research paper containing 10 or more errors or discrepancies."

Then the two turned to the reports of the several investigating committees. There they found a remarkable catalogue of sloppiness and worse. Much of the time, co-authors had not retained the experimental or clinical data underlying the papers. When Stewart and Feder looked into the actual involvement of co-authors with the research or the writing-up of the papers bearing their names, they found that fully a third of the time co-authors had so little to do with the outcome that they had to be classed as honorary authors. The term is pejorative. Furthermore, "The 13 papers with at least one honorary author had more errors and discrepancies than the 5 without an honorary author."

The papers from Harvard mostly dealt with experiments on dogs. Such experiments of course included animals used as controls. Although the papers did not say so, in many instances the same dogs were counted as controls for different papers, even when the experiments in which they appeared had been performed months apart.

Such historical controls though never ideal may sometimes be justified, Stewart and Feder acknowledged, not least because they economize the use of animals. What they decried was that readers were nowhere informed of the practice and so could not judge the actual experimental procedures. But more: one paper stated that all of the dogs subjected to an experimental operation were killed by overdoses of barbiturates and that "the hearts were excised rapidly and placed on ice." Yet some of those same animals showed up in reports of later experiments. "We have assumed that the two co-authors in our sample were unaware of this anomaly at the time of publication," Stewart and Feder wrote. The lifted eyebrow, the tone of bland irony—the cumulative effect is savage.

Not only controls were cannibalized. Several times, Darsee had swelled his list by publishing the same data in separate research papers, without cross-citation. Frequently he published the same abstract in several different journals. "Of the 88 abstracts, only 47 were published only once; all the rest involve duplicate or triplicate publication." Sometimes the abstracts were modified, sometimes hardly more was changed than the title. For example, "Persistent myocardial abnormalities following brief periods of temporary coronary occlusion not associated with necrosis" also showed up as "Prolonged metabolic, functional, and ultrastructural abnormalities following transient myocardial ischemia without infarction." The authors were the same, the text and numerical data all but identical.

Drily, Stewart and Feder offered alternative conclusions:

> The evidence presented here indicates a surprisingly high frequency of lapses from accepted standards. One possible explanation is the association of the co-authors of the publications in our sample with Darsee: he may have induced them to behave in

> ways not typical of them. A second possibility is that the co-authors are not representative of their field. A third possibility is that lapses from accepted standards may be unusually common in the specialty in which these scientists worked, in clinical research generally or in research carried out in medical schools or by physicians. (The co-authors of the publications in our sample were mainly physicians, and the research was carried out exclusively in medical schools.) A fourth and disturbing possibility is that lapses from accepted standards may be more common among biomedical scientists than is currently appreciated.

A bizarre and totally unpredictable consequence of Stewart and Feder's intervention was the concatenation of events that brought the Baltimore affair to light.

*

The case of Mark Spector fits the pattern of the mentor seduced. Spector, another prodigy, was thought by all who knew him to be brilliant, charming, and multiply talented. He painted, wrote poetry, and was a virtuoso at the bench. Efraim Racker, who become his mentor and close friend, later wrote, "A young scientist who watched him perform experiments at the National Institutes of Health commented that it was like watching Beethoven play the piano." Racker went on, "Spector's lectures were spellbinders, and as a graduate student he received lecture invitations from many places, including NIH. He was also an excellent teacher." In 1980, Spector came to Racker's laboratory, at Cornell University, as a graduate student. Racker, then in his late sixties, was a biochemist. He had been born and trained in Vienna and had escaped to the United States before the second world war. His specialty was the biochemical interactions of an enzyme, sodium-potassium ATPase, that sits in the membrane of cells and that controls essential metabolic functions. Since the 1920s, the classic era of biochemistry, scientists had known that cancer cells function inefficiently: Racker's enzyme was centrally involved.

Within a year of arriving, Spector had purified the enzyme. Over the next months he developed evidence that in cancer cells its efficiency was depressed by the action of a sequence of three other enzymes, all new to science. Working with immense energy, often staying in the lab late into the night, he identified and isolated these. His methods were standard. He put radioactive phosphorus into his biochemical system, he said, where it was incorporated as the system built the protein. The product was a solution of many sorts of unknown molecules, each in tiny quantities. He separated his enzyme out by electrophoresis: along one edge of a slab of gelatin he spread a line of the solution, and then applied a weak electrical field, which drove molecules towards the other edge, some pushed farther along than others depending on minute differences in electrical charge between the various sorts. The resulting gel, so called, he then placed against a sheet of x-ray film; spots on the film identified radioactive molecules. The gels were elegant evidence; the x-ray films made convincing illustrations for papers, slides for talks. Pressing on, he showed that his enzymes were related to the genetics of cancer cells.

The possibilities electrified the international biochemical community. This was main-line stuff. Over the years, the classic research on which Racker had built his own career had generated three Nobel prizes. By the spring of 1981, Spector and Racker had written papers together, and although not many had yet made it through the publishing process they were distributing preprints and giving talks all over, to packed auditoriums. Spector began to get offers of faculty jobs.

Some biochemists were skeptical. Spector's enzyme system did not fit well with certain other data. He had been amazingly productive, in two years accomplishing work that paralleled classic discoveries which had taken twenty. On several occasions, though, others were unable to duplicate results; yet when Spector supplied the materials and took part in the work, the experiments were successful. Anomalies accumulated. Then Volker Vogt, a tumor virologist also in the department at Cornell, was unable to reproduce an experiment and demanded to see Spector's gels. Examining these, by happenstance he placed a glass plate on top of one—and noticed that a monitor was still detecting counts. But the radioactivity of phosphorus would have been too weak to penetrate the glass. Spector's supposed enzyme preparations contained not radioactive phosphorus but radioactive iodine. Confronted with the evidence, Racker had to act.

Telling the story in an article in *Nature* seven years later, Racker called Spector an outstanding example of the true "professional" fraud. It turned out that he had regularly spiked his experiments in various ways, planting the phenomena that would then emerge as evidence for his claims. Spector was fired. Racker withdrew manuscripts, retracted papers. Investigation now found that Spector had never received an undergraduate degree—and that he had once been convicted of forging checks. Racker later wistfully told a longtime friend that he had treated Spector like a son.

Next, another mentor seduced—and a whistle-blower punished. Sometime in 1979, Robert Sprague met Stephen Breuning. Sprague was forty-eight, a psychologist at the University of Illinois who for fifteen years had been studying the effects of neuroleptics, which are a class of tranquilizers that were widely used with severely mentally retarded patients who harm themselves, for example by repeatedly banging their heads against a wall. Neuroleptics calm them. However, with long use they have a serious side effect, tardive dyskinesia, involuntary spasmodic twitching and jerking, sometimes relatively harmless though distressing, but in rare cases fatal. Breuning was in his early thirties, two years out from his PhD in psychology from the Illinois Institute of Technology, in Chicago. He struck Sprague as "a very capable, articulate young man," intelligent and productive. At that time, Breuning was working at the Coldwater Regional Center, in Michigan, a residential facility for mentally retarded people, where a number of other young psychologists were doing research. Sprague's and Breuning's interests overlapped; Sprague needed colleagues and patients, and in the summer of 1979 moved his research grant to the Coldwater center. Funded under that grant, Breuning began writing paper after paper, sending them to his new friend and mentor. Sprague found them "superb studies using excellent methodology with a large number of subjects," he said at a colloquium in April of 1992; they "contained very important results."

Breuning's papers were so good and so many that after two years of the collaboration Sprague became suspicious. "I knew that I could not possibly produce the kind of studies he was reporting," and "The more I thought about this, the more apparent it became that if I could not possibly produce what he was reporting, then possibly he could not either." He began to check.

At the turn of the year 1981, Breuning moved to the Western Psychiatric Institute, at the University of Pittsburgh, well regarded for research. He obtained his own grant from the National Institute of Mental Health, administered through the university. Collaboration continued by telephone and mail; they exchanged visits every couple of months. Sprague began to think that some of Breuning's claims were patently impossible.

Towards the end of 1981, Breuning gave Sprague a copy of his first annual report. There he said he had conducted seven studies in the year. Sprague knew that these could only have been done consecutively, not concurrently. He knew that Breuning never ran his lab on Saturdays, Sundays, or holidays. When he added up the number of days reported for these studies, they totalled 273. But the year had only 263 working days. "In the real world," Sprague said, "subjects become ill, equipment breaks down, graduate assistants sometimes fail to operate the equipment correctly, etc. Thus, even in the best of circumstances, a researcher can only reach eighty to eighty-five per cent efficiency considering the complexity of these studies."

Breuning's principal claim was that the neuroleptics were often doing patients more harm than good. His research supported this; one study even showed that when patients were taken off the drugs their I.Q.s, startlingly, doubled. He won great praise for alerting psychiatrists to these dangers. He was changing clinical practice. Sprague's apprehension grew. A technical problem was to quantify the types and severity of spasmodic movements patients were making. The nurses observing them needed a rating scale to standardize their reports. But use of the scale showed, as one might expect, that nurses could not always agree on their observations: when Sprague worked with a different colleague, in Minnesota, the best agreement between judges they could achieve was about seventy-eight per cent. Then on a visit to Pittsburgh in September of 1983, Breuning's girl friend asked Sprague what was wrong, "since they were obtaining 100% reliability from the nurses on their wards. When she told me this, I nearly fell out of my chair because I knew Breuning was lying. No matter who you are or how good a researcher you are, it is impossible to obtain 100% agreement between nurses making these very difficult evaluations."

Sprague spent the next three months examining Breuning's work. In the late summer of 1983, he was planning a symposium on tardive dyskinesia. He asked participants to provide abstracts. In November he got Breuning's. It referred to a study Breuning and co-authors had published about two years' work with fifty-seven patients at Coldwater, which had ended late in 1980. The new abstract was about a follow-up of forty-five patients there, for two years that had to have begun at the time Breuning moved to Pittsburgh. He claimed to have examined each subject at six-month intervals—a total of one hundred and eighty evaluations. Sprague telephoned Neal Davidson, a psychologist who had been at Coldwater the entire time Breuning had worked there and for the years after his departure for Pittsburgh. "Davidson told me that there was no way that Breuning could have conducted the study since he left no assistants and never returned to Coldwater." Sprague called Breuning and demanded that the raw data to support the one hundred and eighty evaluations be sent him within forty-eight hours. "He was quite startled by this request. For the first time since I had known him, he became agitated, upset, and inarticulate." Three days later, he received "24 pieces of paper which represented 24 evaluations. Even some of those evaluations were suspect since none contained the initials of the person who had done the evaluation, the dates were not compatible with the known chronology," and other aspects were dubious. "The main point was that any one who claims to have conducted 180 evaluations but only has data for 24 is lying." On 20 December 1983, Sprague reported his friend's conduct to his monitor at the National Institute of Mental Health—a letter of eight single-spaced pages with forty-four pages of appendices.

The sequel was drawn out and difficult. Pressed by the National Institute of Mental Health, in February of 1984 the University of

Pittsburgh set up a faculty committee. Breuning confessed the Coldwater fabrication to them, and resigned in March. However, although Sprague had documented Breuning's continuing frauds at Pittsburgh, the committee did not look at any of his work there. Indeed, in July of 1984 the dean of the university's school of medicine wrote to the NIH that "I have no grounds to take action" against Breuning.

Then the system rounded on Sprague. The acting director of the National Institute of Mental Health appointed an investigator—whose first move was to spend two weeks investigating Sprague. For more than two years, no further action was taken against the University of Pittsburgh. When Sprague then went public, generating publicity including an article in *Science,* the first result was that after seventeen years of federal funding his grants were not renewed. The case wound on. Three times in 1988, Sprague was called to testify before congressional committees. On April 12, he told John Dingell's Subcommittee on Oversight and Investigations about the case, including the University of Pittsburgh's cover-up. (That was a long day of testimony otherwise given over to the Baltimore–Imanishi-Kari affair.) Sprague promptly received a letter most dire from George Huber, a vice president and counsel of the university's medical division, threatening a libel suit. Sprague turned the letter over to the subcommittee. Dingell, who has always acted promptly and vigorously to protect whistle-blowers, wrote to Wesley Posvar, president of the university. This in turn produced a letter from Posvar to Sprague, apologizing.

The case of Robert Slutsky, in itself, was comparatively uncomplicated and was handled expeditiously and effectively. The syndrome is familiar, yes; its exposure engendered an analytical account of depth and originality, and therein lies its importance. Slutsky was a young physician, a radiology resident and an associate clinical professor (this second appointment was without salary) in the Department of Radiology at the University of California, San Diego. He seemed able and certainly productive, having published one hundred and thirty-seven papers and reviews in thirty-four different journals, major and minor, in seven years. Early in 1985, he was up for promotion to associate professor of radiology in the School of Medicine. Routinely, an ad hoc committee was set up to evaluate his work. A member of the committee noticed that Slutsky had duplicated data between two of his published articles. A second ad hoc committee was appointed. Its members scrutinized Slutsky's published and recently submitted papers and his logs and notebooks, and quizzed laboratory personnel and postdocs. The problem broadened. The evidence showed that three recent manuscripts described experiments that had never been done. They asked that the two published papers that duplicated data be retracted. They called for a still fuller investigation. Within a month of the initial charges, Slutsky resigned. The probe continued; while it was under way, his lawyer asked that fifteen published papers, in eight journals, be retracted, saying that "information had recently come to Dr. Slutsky's attention which cast some doubt on the results."

In November of 1987, *The New England Journal of Medicine* published an analysis of the case. Its authors were Robert L. Engler, James W. Covell, Paul J. Friedman, Philip S. Kitcher, and Richard M. Peters. All had been members of one or another of the committees that investigated the case, save Kitcher, who is a philosopher and historian of science then also at the University of California, San Diego (now at Columbia University), whom the others asked to join their deliberations. "[A] committee of 10 faculty members investigated Slutsky's entire bibliography," they wrote; "77 were classified as valid, 48 were judged questionable, and 12 were deemed fraudulent."

Frauds included "numerous experiments that were never performed," measurements never made, statistical analyses described but never carried out. The investigating committee wrote to some of Slutsky's co-authors—these totalled ninety-three—and interviewed senior faculty in the department to determine "how this young investigator's fabrications had escaped detection for so long."

The authors of the analysis in the *New England Journal* recognized what they called "excessive publication." The man's role models themselves had had a great many papers published. "Slutsky's incredible productivity (one paper every 10 days while he was simultaneously a radiology resident and an associate clinical professor of radiology) failed to raise alarm." They explained the failure:

> First, his papers were published in a variety of journals in three separate subspecialty areas, so that no one group reviewed all his work. Second, Slutsky created an appearance of high collaborative productivity by putting the names of colleagues and students on papers to which they had made little or no contribution.

Instances of that were barefaced, even bizarre. Some co-authors were not even aware of a paper until the investigating committee questioned them. Slutsky forged signatures on copyright forms. More:

> One new research trainee was presented with a publication in which she was listed as a coauthor on her arrival at the laboratory. Some colleagues and trainees were either too flattered or too embarrassed to have their names removed from papers to which they had not contributed. . . . On the other hand, some faculty members expected their names to be used even though they had provided only the facilities for a project, without mak-

> ing a substantive contribution or having a knowledge of its validity. Thus, gift coauthorship was effectively a disguise for an impossible amount of work by one person.

Gift authorship leads to failure to detect error and, they pointed out, in itself "is also a culpable act of deliberate misrepresentation."

Slutsky got away with fraud for as long as he did, among other reasons, because a number of his immediate co-workers, though they became suspicious about the validity of individual manuscripts, or his productivity, or his occasional carelessness in experiments, none the less "failed to communicate their concerns to the responsible department official (the chairperson), the responsible School of Medicine official (the associate dean), or other supervisors."

One could easily quadruple the number of such tales of turpitude that have become public since Summerlin. But the Slutsky case is remarkable for the quality of that report in the *New England Journal.* Six pages plus notes, it is dense with facts and recommendations, obvious and subtle by turns, practical, exciting to read. (One detects the hand of Philip Kitcher, the philosopher and historian, for he will have brought to his colleagues a different kind and level of reasoning.)

*

The lessons prove hard to learn. Six years after Philip Handler's obtuse testimony before Gore's investigative subcommittee, Daniel E. Koshland, Jr, editor of *Science,* wrote a lead editorial for the issue of 9 January 1987 on "Fraud in Science." Koshland had been appointed that journal's editor two years earlier. He remained a practicing biochemist at the University of California, Berkeley, and had formerly been chairman of the department there. Before *Science* he had been editor of *Proceedings of the National Academy of Sciences.* In short, he was a scientist of grand standing. Although the Baltimore affair had not come to public attention by January of 1987, the Darsee case was now freshly publicized, and as we have seen others were coming to light. Alarm was mounting among scientists, among those in Congress who concerned themselves with science and its funding, and among journalists. In his editorial, Koshland said things that were sensible though obvious, but others that were, at best, wishfully optimistic. "Even good intentions are not enough. Sloppy experimentation and poor scholarship are condemned. Outright fraud is intolerable." But then, "the cumulative nature of science means inevitable exposure, usually in a rather short time." He concluded that having acknowledged some areas of concern, none the less "we must recognize that 99.9999 percent of reports are accurate and truthful, often in rapidly advancing frontiers where data are hard to collect." One can only suppose that Koshland thought he had to write to protect the scientific community. The world of science in Babbage's day was small, open, and close-knit. Perhaps *his* confidence had been justified. Koshland's assertion was recognized at once as egregiously foolish, for it can have no basis in fact, either way.

*

Fraud is hardly peculiar to American science. Cases have been reported in Australia, Britain, France, Germany, India, Poland, Sweden, elsewhere. To be sure, comparisons across national boundaries are tricky—from blowing of whistle through investigation or the lack of it to adjudication and remedy. Scientists and laboratories outside the United States are fewer, funding generally less. More subtly, in Europe research projects are typically conceived and funded for the longer term. Thus, the questions differ and so do the pressures. Organization of science, the various sciences, and individual laboratories at every level are more hierarchical. Research laboratories are often separate from universities; they lack the coffles of graduate students—slaves, but trainees—that in the United States scientists take for granted. In Germany, as just one of multitudinous differences, many laboratories hold to the tradition that the professor automatically has his name as co-author on every paper—though the practice has been falling into disrepute. Cases of fraud have received far less publicity; institutional responses tend to be far more controlling. The French, for example, are massively skillful at cover-up, with the consequence that in the rare instance when the lid is pried off the stench of corruption shakes governments. Yet viewed with caution, some cases present intriguing common features.

One of the most pungent bubbled to the surface in April of 1989, when *Nature* published a report by John Talent, an Australian paleontologist expert in the geology of the Himalayas, particularly the stratigraphy and fossils. His paper, "The case of the peripatetic fossils," laid out the story of twenty-five years of publications, more than three hundred, that had reported remarkable fossil discoveries, "an avalanche of palaeontological data," across a swathe of the high Himalayan foothills from Kashmir through northernmost India and Nepal to Bhutan in the east. "India is famed for its long tradition of excellence in geology and palaeontology enshrined in more than a century's publications of its geological survey," Talent wrote. The new findings offered "biogeographically astonishing additions to the faunal lists" of species that had been present in ancient ages. In one series of published papers, specimens were offered from northeast India and Nepal of a well-known kind of conodont—small, conical, glistening, toothlike remnants of an otherwise decayed creature—previously found uniquely in a quarry in upstate New York. Near to those, ammonoids—fossil creatures related to the nautilus—were found that were identical to specimens well known from Morocco. The discoveries also upset many established stratigraphic relationships, throwing Himalayan geology as well as the paleontology into uproar. All of this was the work of one man, Professor Viswat Jit Gupta of the Punjab University of Chandighar. For that quarter-century and those three hundred papers Gupta had been salting that vast crescent of territory with samples apparently taken from teaching collections. The difficulty, labor, and time that have been required to clear up the mess are incalculable. A residue of doubt will long shadow later work.

British scientists and journal editors had viewed the long series of American cases of fraud with some smugness. Complacency was shaken by the affair of Malcolm Pearce. In August of 1994, the *British Journal of Obstetrics and Gynaecology* ran a report by Pearce, a consultant—British for senior specialist—obstetrician at St George's Hospital, London, in which he said that he had successfully treated a woman with an ectopic pregnancy by a radical new method that had enabled her to give birth to a healthy baby girl. Pearce was an editor of the journal. The report drew immediate worldwide attention. Ectopic pregnancies are those where the embryo implants outside the uterus, typically in a fallopian tube. They always fail, and are dangerous. The first symptom is usually intense abdominal pain. The woman may bleed to death. In the United States, on the order of a hundred thousand women are hospitalized with ectopic pregnancies each year. Pearce claimed that he had treated Patient X, described as an African aged twenty-nine with a five-week-old ectopic embryo, by removing the embryo surgically from the fallopian tube and squirting it by way of the cervix into the uterus, where it implanted successfully. "The patient was discharged home on the fourth post-operative day." The report listed as co-author Professor Geoffrey Chamberlain—who was head of Pearce's department at St George's as well as president of the Royal College of Obstetricians and Gynaecologists and editor in chief of the journal. The third co-author was Isaac Manyonda, a junior physician.

In the same issue of the journal, Pearce published a paper that claimed he had carried out a three-year, double-blind, randomized trial on a group of one hundred and ninety-one women who had histories of miscarriages and multiple ovarian cysts. They had received either a hormone called human chorionic gonadotropin or a placebo. The paper reported that the women getting the hormone were less likely to miscarry. Another physician subordinate to Pearce, Rosoel Hamid, was co-author of this paper.

The condition for which Pearce said those women were treated is in fact rare. Stephen Lock, then editor of the *British Medical Journal,* commented later on Chamberlain's evident conflict: "In this case a more disinterested editor might have questioned the fact that over three years Pearce purported to have collected 191 women with a syndrome so uncommon that a major referral centre was seeing only one or two new cases a month. Moreover, all of them had had a battery of complex tests, including karyotyping"—examination of the chromosomes—"of both partners."

Only after Pearce's article was published did that extraordinary number awake suspicions among others at St George's Hospital. No other doctors there had heard of the research—nor of the success with ectopic Patient X, which shocked them because of its importance if true. When Chamberlain ordered an inquiry, Pearce was unable to produce case notes, consent forms, or any other records of the hormone trial. Furthermore, computer records of Patient X had been falsified: she had miscarried. Pearce then told Chamberlain that the real mother had been a Patient Y, who feared that the fact she had had a previous abortion would be discovered. But Patient Y had been manufactured, too—and evidently in haste, for the computer revealed that she had been born in 1910 and was dead at the time of the pioneering operation.

In the United Kingdom, qualifications and ethical standards for the medical profession are set and enforced by the General Medical Council, which keeps the medical register, the roster of physicians licensed to practice, and which has a disciplinary committee. Within nine months, early in June of 1995, the council found Pearce guilty of "serious professional misconduct" and struck him off the medical register. He was dismissed from the hospital and as an editor of the journal. Chamberlain resigned as president of the Royal College of Obstetricians and Gynaecologists and as editor in chief of the journal, but stayed on as head of the department at St George's. He and the two junior physicians were reprimanded for allowing themselves to be guest authors without scrutinizing the papers. Chamberlain told Owen Dyer, a British free-lance medical journalist who immediately wrote a news report of the case for the *British Medical Journal,* "I rubber stamped this paper out of politeness and because he asked me to as head of the department." In November of that year, the *British Journal of Obstetrics and Gynaecology* retracted four Pearce articles. Of two, including the miscarriage study, "No evidence could be found" that they had been carried out. The other two were "unsubstantiated and could not be relied upon."

Chamberlain, quintessentially English—listen for the accent—told Dyer, "Obviously Malcolm has been extremely silly on this occasion, but in the past he has done a lot of good."

In a measured editorial in the *British Medical Journal,* Stephen Lock drew the grim lesson of the affair. "Belatedly, Britain should abandon its lax approach to scientific fraud," the editorial said, and he went on, in part:

> Last week Malcolm Pearce, a British gynaecologist, was removed from the medical register for fraud: he had published two papers in the British Journal of Obstetrics and Gynaecology describing work that had never taken place. Less than nine months had elapsed between the whistle being blown on Pearce and his removal from the register. Outside observers might therefore conclude that, like other countries, Britain has established methods of preventing, detecting, and managing misconduct in research. They would be wrong. That the Pearce affair was handled well was unusual: the principal of Pearce's medical school knew what to do and was determined to do it—speedily and while protecting the rights of both the accused and the whistleblower. In most other medical institutions in Britain nothing would have happened; the affair would have been brushed under the carpet, and the whistleblower would probably have been hounded out of his or her job.
>
> Despite a report from the Royal College of Physicians, Britain has learnt little about handling fraud since the Darsee affair in the United States first brought the subject into prominence in 1983.

Indeed, the Pearce case was no isolated instance. In the mid-1990s, more came to light. In February of 1994, a television documentary accused Peter Nixon, a consultant cardiologist in London, of fraud in published papers and of serious mismanagement of patients. Nixon sued for libel. In court in May of 1997, cross-examination drove him to admit having made results up in a number of papers and having carried out diagnostic tests on patients that could be dangerous, even fatal. He withdrew the entire libel action. The court ordered him to pay the television channel three-quarters of a million pounds (then over one million, one hundred thousand dollars) in costs.

John Anderton, a senior kidney specialist in Edinburgh, conducted a clinical study on the efficacy of a drug, sponsored by its makers, Pfizer. The study ran for fifteen months, and cost forty-two thousand pounds, or seventy thousand dollars (though none of this profited Anderton personally). A clinical-trials reviewer at the drug company spotted problems in Anderton's report. The matter was pursued by a private inquiry firm, Medicolegal Investigation, which had been set up by Frank Wells, formerly the medical director of the Association of the British Pharmaceutical Industry, together with a retired detective chief inspector. They found that Anderton had faked echocardiographic and magnetic-resonance imaging data for patients and had got his assistant to invent data and to attest to forged patient-consent forms. During the investigation, he claimed that falsified data had been supplied by one Dr Shaffick—who apparently did not exist. In July of 1997, Anderton was struck off the medical register.

By that time, Wells's inquiry firm had referred seventeen other cases to the General Medical Council, all involving serious misconduct, and were working another twelve.

In February of 1998, the case of Mark Williams came to a similar conclusion. He was a senior lecturer in public-health medicine at the University of Bristol. In 1994, statistics in an abstract of his for a paper to be delivered at a conference in Sweden came under suspicion, and it was noticed that he had claimed falsely that he held an MD and a PhD from the University of Cambridge. He explained the false credentials as typist's errors and admitted that the statistics were ambiguous; the departmental professor, Stephen Frankel, let the matter pass. In September of 1995, data Williams had prepared for a paper to be submitted to the *BMJ* roused Frankel's suspicions. Frankel told Clare Dyer, legal correspondent for the *BMJ* (and mother of Owen Dyer), "We reanalyzed it ourselves. That led to a whole set of consequences. We confronted him, and he admitted what he had done. Once I had suspicions about what might have happened, we looked extremely carefully at everything he had done." One discovery was that Williams had faked his qualifications in 1991 for his first appointment at the university and again two years later for a promotion. He resigned, taking a job as a locum (*locum tenens,* or substitute) physician in another city—which of course he lost when struck off.

From Germany next, something of a rarity in the sciences (in contrast to the great world of commerce), a case of collusion in fraud—and on a heroic scale. Friedhelm Herrmann and Marion Brach, German molecular biologists, shared a notably successful scientific career. It began at Harvard, where they worked and lived together, he in his mid-thirties, she eleven years younger. They returned to Germany, to the University of Freiburg, and then in 1992 moved to Berlin. Herrmann became prominent in attempts to develop gene therapy, and headed a research group at the Max Delbrück Center for Molecular Medicine, which had been established in eastern Berlin after reunification. Brach led one of his four research groups there; her specialty was cancer therapy, in particular the attack on cancers with cellular substances called cytokines, one of them tumor-necrosis factor, which seemed to have enormous therapeutic potential. Early in 1996, they moved again, to the University of Ulm in western Germany. Soon after, they broke up, and Brach accepted a professorship at the University of Lübeck, near Hamburg, and the direction of a new Institute of Molecular Medicine there.

Early in 1997, Eberhardt Hildt, a young scientist who had worked at the Delbrück center but was there no longer, went public with the assertion that data in certain papers had been fabricated. He and other junior colleagues had known of the fakery but had been afraid to speak. Herrmann and Brach had long maintained a united front, threatening students with ruin, protecting each other. After their breakup, though, they began to comment on each other. Investigations started in March, at the universities of Ulm and Lübeck and at the Delbrück center. Brach soon admitted that while in Berlin she had falsified materials in four research papers. One appeared in *The Journal of Experimental Medicine* in 1995; this is a journal of prestige and authority. In the paper, was a figure purported to be an autoradiogram showing responses in cancer cells to tumor-necrosis factor. Brach had created the figure by computer, manipulating portions of three radiograms from other work. Herrmann, though co-author of all four papers, insisted he had no knowledge of fraud. That spring, the director of a Max Planck institute near Munich, in whom Hildt had confided, spoke to several journalists. In mid-May, a local daily paper and the German news weekly *Focus,* then just three years old but running hard, broke the scandal. *Nature's* chief European correspondent, Alison Abbott (a journalist of avid energy and good sense, based in Munich) launched the story internationally.

The scandal grew with the inexorability of a lava flow. Within weeks, the chairmen of the local investigations formed a joint committee in Bonn and heard more evidence from Brach. Herrmann still protested his innocence and refused to appear at the Bonn gathering. How to punish the two, if proven guilty, was a problem: in Germany, university professors are civil servants, employed by local *Länder,* or states. The Max-Planck-Gesellschaft, or society, which has some eighty research institutes in various sciences and humanities all over

Germany, most of them independent of universities, had set up a committee the previous year to recommend rules for dealing with fraud. Its chairman, Albin Eser, a specialist in international criminal law at a Max Planck institute in Freiburg, said that it was proving difficult to draft procedures compatible with German law. The society had not funded Herrmann and Brach. Their support had come from a German cancer foundation and from the Deutsche Forschungsgemeinschaft, or research society, which is the principal conduit for government grants to universities, and these agencies also began inquiries.

By summer's end, Brach had resigned her professorship and Herrmann had been suspended from his. More than thirty papers were said to have used fraudulent data. Their *modus operandi* had usually been to generate plausible data by computer, then place papers in mid-ranked journals; that paper in *The Journal of Experimental Medicine* had been an ambitious exception. The University of Freiburg, where Herrmann and Brach had worked before Berlin, had now begun its own investigation, headed by Eser of the Max-Planck-Gesellschaft: twelve papers from that period were also under suspicion. The chairman of the clinical department there was Roland Mertelsmann, and he was co-author on twenty-five of the tainted papers. Eser's committee did not accuse him of collusion, but said that he bore a share of the responsibility for fraud in his department. The first week in September, Herrmann announced that, his career in ruins, he was going to sue the dean of medicine at Ulm and all the local investigating groups for ten million Deutsche marks, five million, six hundred thousand dollars. His lawyer said all responsibility for fakery lay with Brach. By the end of November, the count of fraudulent papers had grown to at least forty-seven. Brach insisted that Herrmann had known all about it and had pressed her to cheat.

In April of 1998, she wrote to *Nature* to protest being made scapegoat, saying, in part:

> I have very readily admitted to having been involved in falsification of scientific papers, an achievement of which I am not proud. I resent, however, the fact that, because I alone have admitted my mistakes at an early stage of the investigations, official bodies have found it expedient to imply that I was the major or conceivably the only culprit and that I was responsible for false data appearing in numerous papers, on some of which I was not even a co-author.

In September, Herrmann resigned his professorship and set up in private medical practice in Munich. A new task force started to scrutinize some three hundred and forty-seven journal articles, which had appeared in several dozen journals, and eighty-odd book chapters, written by various combinations of Herrmann, Brach, and colleagues. The task force published its report in June of 2000. They believed that ninety-four papers contained data definitely or very probably manipulated. Fifty-three of these were by Herrmann and Brach; fifty-nine included Mertelsmann as a co-author. Mertelsmann had put his name on all papers emerging from the laboratory; only with this scandal did the century-old German practice fall into disrepute. The task force also looked at five papers that bore Mertelsmann's name but not Herrmann's. At least one of these, published in the journal *Blood,* bore indications that "data had been improperly handled." Two other scientists, Lothar Kanz and Wolfram Brugger at the University of Tübingen, were co-authors of this one, so the Deutsche Forschungsgemeinschaft launched an investigation of *them.* The only one to come out altogether clean has been Eberhardt

Hildt, the original whistle-blower, who headed a research team at the University of Munich and more recently moved to the Robert Koch Institute, in Berlin.

Two years after the task-force report, *Nature* surveyed twenty-nine journals that had published papers in question. Twenty replied. Fourteen had not retracted any, and indeed "seven said they were unaware that the investigation had taken place, and four more did not know it had been completed."

As in the United States and Britain, when a major case of fraud becomes public, suddenly other cases began to come to light. Biggest of these was at the Max Planck Institute for Plant Breeding, in Cologne. Early in March of 1998, a technician there, Inge Czaja, resigned, as did the head of her laboratory, Richard Walden. She admitted fabricating data in a paper on an enzyme involved in plant-cell growth. Others in her laboratory had had trouble reproducing her experiments. By the end of May, tests had shown that experiments in six or more papers could not be replicated; thirty papers were under suspicion, and investigators faced the prospect that results had been faked over a period of six years or more. In contrast to the usual practice of Herrmann and Brach, Walden had made big claims and placed them in top journals.

In the spring of 1999, in San Francisco, a lawsuit came to trial brought by the University of California against Genentech, Inc., for patent infringement in production by genetic techniques of human growth hormone. In a paper in *Nature* in 1979, scientists at Genentech had announced the first production of human growth hormone free of contaminants, and that paper figured in the lawsuit. In court on April 20 and 21, Peter Seeburg, by then director of a research division at the Max Planck Institute for Medical Research in Heidelberg, testified that as a co-author he had misrepresented data. In May, in response to reports of his testimony, Seeburg wrote to *Science* and *Nature* admitting the misrepresentation but claiming it was a minor technical detail. The Max-Planck-Gesellschaft named a commission to investigate. The outcome was a slap on the wrist: Seeburg had admitted misconduct, the commission said, but that was twenty years ago. He was not disciplined.

Then in a comic footnote, in October of 1999 the Humboldt University (in olden days the University of Berlin) launched a formal investigation of a charge of fabrication in a paper recently published by Holger Kiesewetter, who claimed that taking high-dose pills of garlic for four years can slow the deposit of atherosclerotic plaque or even cause it to regress. Many Germans believe fervently that garlic has remarkable therapeutic properties. A commission sniffed around, but its report has not been aired.

*

It's time to step back to analyze in more detail the characteristics of the recent cases. Caution is called for. These are histories of idiosyncratic behavior, worked up from reports that vary greatly in detail and point of view. None the less, some simple traits are obviously common to many of them. Compare Summerlin in 1974, Soman coming to light in 1980, a year later Darsee and then Spector, Breuning exposed in 1983. Often, a strong, productive, demanding mentor with an empire and little time for close supervision, or alternatively (as with Spector and Breuning) a mentor dazzled by the fit of his junior's findings to his own hopes. A younger scientist of brilliance, charm, and plausibility with a record of publications and exciting research elsewhere, taken on as a protégé and perhaps becoming a close friend: Racker's treating Spector "like a son" was hardly unique but rather a recurring extreme of the interaction of mentor and prodigy. Gift authorship, which is more than just a sign of trouble, for it leads to failure to detect error and is itself misrepresentation. Discrepancies in data not caught, even when glaring. These traits emerge as generic, diagnostic.

The narratives of exposure display a related set of traits. Often a colleague (in Soman's case a competitor) saw a problem, grew suspicious. Occasionally, attempts were made to replicate claimed results, and failed. Statistical treatments were seen to be naïve. The perpetrator grew overconfident (how could Cyril Burt have been statistically naïve?) or even reckless, sometimes as though wanting to be found out. Duplicate publication was noticed. Data was discovered to be missing or poorly maintained. Inconsistencies within and between papers and manuscripts, often blatant yet long unnoticed, were at last caught. Typically the realization dawned that the problems extended back across years. Yet considering how egregious these frauds turned out to be, what's especially alarming is how their exposure turned on happenstance, on casual accidents. No systematic safeguards appear to be in place. Peer review, for a leading example, so praised by the defenders of the scientific enterprise, never detected the problems.

One warning sign, evidently, is productivity too good to be true. Alsabti, in his mid-twenties: sixty publications. Darsee, at thirty-three: more than one hundred and twenty-five publications, while even in the week he was first exposed the journal *Circulation* accepted ten new abstracts of papers to be presented in the coming months. Spector, late twenties: described by some as producing the work of a consortium of graduate students—a consortium of professors. Slutsky: one hundred and thirty-seven papers in seven years. Too-good-to-be-true takes other forms, as well. Summerlin: surprising, even revolutionary results, not confirmed by others' work.

Must we hold success, apparent high achievement, suspect in itself? Surely it invites astonishment: *Marvellous! How does he do it?* In these and similar cases, as we look back after their exposure, the remarkable productivity is but the most conspicuous feature of the complex social settings we have examined. The increasing competitiveness of present-day scientific communities is routinely decried, routinely cited as explanation for misconduct. But competitiveness is of at least two types. One is structural, which is to say, built into the funding, the job markets, career ladders, problem choices, publication strategies and journal procedures, systems of rewards and honors. The other is personal, that is, the pressure from the attitude and behavior of the laboratory director, and the atmosphere among the technicians, graduate students, and postdocs.

So consider the productivity, or just sheer busyness, of the boss, the mentor, the example to be emulated and by emulating propitiated. Consider the elementary matter of the size of his enterprise. Robert Good, by the time of his death, had been author or co-author of more than two thousand publications and had written or edited some fifty books. Eugene Braunwald, Darsee's sponsor at Harvard, had his name on upwards of six hundred publications. He ran two labs, at two different research hospitals. None of these men has himself publicly been implicated in fraud. Yet they had earned reputations as hard-driven, rough, tough and demanding superiors—but we hardly need characterize their personality types when the numbers, crude measure though they be, speak so clearly to the nature of their ambitions, their standards of achievement, their preoccupations, and their capacity or willingness to supervise close-in. Robert Sprague was exceptional, not measuring his own productivity by such numbers. But of course he was exceptional in a far more important respect: alone among the mentors in the cases we have sketched—indeed, all but uniquely among scientists senior to individuals charged with misconduct—Sprague blew the whistle. Almost universally, higher-ups react to charges in ways diametrically the opposite.

Error can of course be honest, or it can result from sloppiness, or it can be deliberate. The distinctions, self-evident in principle, in practice present difficulties. Honest error is of course an integral, necessary part of science—and, of course, is typically claimed in defense against a charge of fraud. Scientists concerned with these problems protest that attempts to define and expose misconduct will stifle science by penalizing honest error. Howard Schachman, for example, a molecular biologist now retired from the University of California, Berkeley, an articulate defender of the freedom of the sciences, has been deeply concerned to make clear that "'misconduct in science' does not include factors intrinsic to the process of science, such as error, conflicts in data, or differences in interpretation or judgments of data or experimental design."

Yet such protests often seem disproportionate to any real danger, and to be advanced in part to whip up resistance to measures that might threaten the self-government of the sciences. The serious problem lies elsewhere. Genuine fakery, fraud itself, in common with other felonies carries with it the element of intent. It must be shown to be deliberate: in science that may be not only hard to prove but irrelevant to the protection of the process and the record.

Between honest and dishonest error lies sloppiness—and this is profoundly ambiguous. Max Delbrück famously advanced the notion that when you don't know what you are looking for, sometimes useful results emerge from what, with cocked eyebrow, he called "the principle of limited sloppiness." Sloppiness is offered in defense, as it was in the case of Imanishi-Kari. Yet that seems desperate, for surely gross sloppiness (as the authors of the Slutsky analysis said) is not a defense but a serious charge. As a charge, it neatly sidesteps that problem of proving intent. A definition of misconduct that held gross sloppiness culpable would be strong.

Efraim Racker, looking back at the Spector case and certain others, offered a further distinction among perpetrators of deliberate fraud.

> What did these scientists have in common? They had an outstanding intellect, they were all informed in their research field and recognized the missing links that if solved would represent a breakthrough. They were skilled experimenters and joined laboratories that were in the forefront of a particular research field and that were headed by an investigator with a reputation of integrity. . . . These are trademarks of the 'professionals' and are not universal. 'Amateurs', who for example publish data on patients that do not exist, are usually more readily discovered by colleagues and students.

The authors of the Slutsky analysis, attempting to put fraud in its immediate psychological and social context, located problems in a conflict between two kinds of motives that drive scientists. "Researchers are typically moved partly by intellectual motives—notably, curiosity and the desire to acquire and disseminate new information. But personal motives—the desire for fame and fortune in their various degrees or for the advancement of some cause—cannot be ignored." When the two are in harmony, the congruence "benefits both society and the investigator. Negligence and fraud arise when this happy congruence breaks down."

From this it follows, they said, that the two most common explanations for fraud are equally inadequate. "It is tempting to explain instances of culpable misrepresentation by the 'rotten-apple' theory. The scientist who is at fault has subordinated intellectual ends to personal motives as a result of character flaws." Indeed, this is the favored diagnosis of scientists high in the establishment. One recalls Philip Handler insisting that fraud is rare and must be judged "as psychopathic behavior originating in minds that . . . in at least this one regard may be considered deranged."

But the complementary explanation won't do, either.

> On the other hand, the "rotten-system" theory suggests that the organization of scientific institutions is responsible. In our judgment, each of these theories is simplistic: broad social factors may generate situations in which personal motives conflict with intellectual motives, and the individual characteristics of scientists determine the relative priority they give to personal ends and the means they use to achieve them.

Reading this, one recalls Robert Merton's hope that the social system of science keeps its practitioners honest. In the flickering light of the multiple instances and kinds of fraudulent behavior that have been found out in the last quarter-century, that hope seems poignantly naïve. The Slutsky analysts turned it on its head. The lesson must be that the laboratory, its hierarchy, and the larger institutional settings—factors that can be analyzed, anticipated, perhaps modified—may make such conflicts between an individual's intellectual and personal motives more likely, or less likely.

The victims, when fraud is in clinical research, are in the first place potentially the patients themselves whose course of treatment may be affected by publication of misleading data. Among the cases recounted here, the painful example is provided by Breuning's fabrications of evidence that neuroleptics used to treat mentally deficient, self-damaging patients do more harm than good. What if a fraud of that kind were committed about, say, a common cancer, and not discovered? Others harmed are those who were colleagues of the perpetrator. In every case we have looked at, co-authors at all levels have been injured. The analysts of the Slutsky affair again:

> Young investigators who were trained in the laboratory at the time of the fraud have had to remove collaborative publications from their bibliographies and to live with the onus of having worked with a perpetrator of fraud. Among the most vulnerable were younger faculty members who were independent investigators at the time the fraudulent work was performed but whose careers were not firmly established. Senior faculty may have their future opportunities for positions or funding compromised.
>
> Thus, research fraud affects the careers of coauthors with varying levels of responsibility or culpability.

All this applies, of course, to virtually every recent case we have looked at. In sum, fraud and its consequences cannot be understood except by addressing the social dynamics of laboratory life, and then their setting in the larger and evolving communities of the enterprise of science.

Questions then arise as to how fraud in the sciences is to be prevented, anticipated, discovered, dealt with. To begin, how widespread is it? Can we measure the incidence? Is it even possible to define scientific fraud in a way that makes effective response possible?

The problem is not going away. On 22 May 2002, *The New York Times* reported Bell Labs Forms Panel to Study Claims of Research Misconduct. Bell Laboratories has been for decades one of the most productive, pristine private research institutions in the world. (It was a division of AT&T, and is now a part of Lucent Technologies.) Bell Labs has brought forth six Nobel prizes in physics for eleven scientists—and had looked as though it was gestating another, in the work of Jan Hendrik Schön in nanoscale electronics, down in the region of a few molecules, work that astonished physicists worldwide. The *Times* reporter, Kenneth Chang, wrote that in question was "the validity of some recent impressive experiments, including a claim last fall that it created a transistor in which the electronic switch consists of a single molecule."

Chang offered more details the next day. The case has the classic signs of the prodigy.

> What had been hailed a few months ago as a breakthrough in molecule-size electronics is now in doubt, and a rising star at Bell Laboratories is under suspicion of improperly manipulating data in research papers published in prestigious scientific journals.
>
> The accusations, by scientists not connected with the research, came to light this week, when Bell Labs appointed an independent panel to look into them. Yesterday, the scientists said that their concerns focussed on graphs that were nearly identical even though they appeared in different scientific papers and represented data from different devices. In some graphs, even the tiny squiggles that should arise from purely random fluctuations matched exactly.

The panel was made up of outside scientists, headed by Malcolm R. Beasley, a professor of applied physics at Stanford University. Bell Laboratories had forwarded five reports to the panel for scrutiny.

> The lead author of all five is Dr. J. Hendrik Schön, 31, a Bell Labs physicist in Murray Hill, N.J., who has produced an

> extraordinary body of work in the last two and a half years, including seven articles each in *Science* and *Nature.*

That week, *Nature* and *Science* rushed in with detailed accounts. Fraud was first suspected late in April when a physicist at Princeton University, Lydia Sohn, compared graphs in two papers published late the previous year, one in each of those two journals, by Schön and two colleagues. Sohn's own work is also in nanoscale electronics. The two graphs, representing different experiments, were identical. Then Paul McEuen, a colleague of Sohn's, at Cornell, searched through earlier papers of Schön's and found a graph so closely similar to the others that even the little wiggles called noise, supposedly caused by random background fluctuations, were the same. Sohn and McEuen promptly took the problem to Schön, to his superiors at Bell Laboratories, and to editors at the journals. Schön explained that by mistake he had included the wrong graphs in some of the papers; he stood by his work. As is typical of such cases, when Sohn inspected still more of Schön's publications she found further duplications in data supposedly from diverse experiments. The number of suspect papers grew from five to seven to eleven. Then three more were implicated, the ones that had been thought sure, if verified, to win Schön a Nobel.

And so, so predictably, forth. By May 31, the number of papers with suspect graphs had grown to thirteen. Chang reported that "the investigation casts a pall over all of Dr. Schön's research. He has been an author on more than 70 scientific papers in the last two and a half years—a remarkably prodigious output." In 2001, Schön had averaged one scientific paper every eight days.

The one unusual aspect of the case was the alacrity and openness with which Bell Laboratories and the journals responded to the charges. From Lydia Sohn's first suspicions to the first press accounts took one month. The Beasley committee reported on 25 September 2002. Seventeen of Schön's papers were based on data that had been manipulated, even fabricated. He had had various co-authors, a total of twenty: these the committee entirely exonerated. That same day, Bell Laboratories fired Schön.

Altogether typical, however, was the institutional summing-up. "This is an individual case performed by an individual," the president of Bell Laboratories told Chang. "In this case, we had an individual who didn't live up to the scientific requirement for integrity." Yet the discoveries claimed were remarkable, surely demanding scrutiny, and the discrepancies in some of the papers were evident on careful inspection. The resemblance, the near identity, of the Schön affair to that of John Darsee is inescapable.

Our litany and simple analysis of representative frauds in science is neatly bracketed by another case from journalism. On 6 January 2004, Jack Kelley, aged forty-three, a reporter for *USA Today* for twenty-one years since its founding, the last ten of those as a world-ranging and highly regarded foreign correspondent, resigned. Handsome, open-faced, confident, Kelley had filed stories that had often been sensational: a personal encounter with a suicide bomber in Jerusalem, a pursuit of Osama bin Laden, an interview with a leader of resistance in Afghanistan obtained in a tiny village on the Pakistan-Afghan border. The paper had nominated him for a Pulitzer prize five times, and in 2002 he had been a finalist. Suspicions had been growing, though; colleagues had repeatedly complained to editors about discrepancies. Then after the Jayson Blair scandal broke at the *Times*, an anonymous complaint prompted an investigation. On March 19, the paper's front page proclaimed Ex-USA TODAY Reporter

FAKED MAJOR STORIES. With that article and two full pages inside, *USA Today* announced that Kelley had regularly been fabricating articles, including many for which he was most celebrated, and had plagiarized "dozens" of times from *The Washington Post* and other newspapers in a list that grew to at least ten. The Kelley scandal developed as an almost identical reprise of Blair. The paper's coverage of itself, given its resources, was honorably comparable to what the *Times* had done. Sure enough, for example, scrutiny of expense accounts showed that Kelley could not have been at places certain stories had described—notably, that village on the Pakistan border. On April 20, Karen Jurgensen, the paper's top editor since 1999, resigned.

Jane Austen remarks in *Pride and Prejudice* that there are "no limits to the impudence of an impudent man." In March of 2004, Jayson Blair published a memoir, *Burning Down My Father's House: My Life at The New York Times.*

4

HARD TO MEASURE, HARD TO DEFINE

THE INCIDENCE OF SCIENTIFIC FRAUD AND THE STRUGGLE OVER ITS DEFINITION

Anecdotes abound. Cases accumulate. Most of those who have dealt with fraud extensively are convinced that it is considerably more widespread than much of the scientific community is willing to recognize. We have not yet found a way of getting at the true incidence of fraud in science. Like other secret acts that may be regarded as shameful—like some sexual behavior, for example—fraud is difficult to survey: the very absence of public data allows scientists in power, defenders of the system, to dismiss reported cases as unfortunate aberrations. Among senior scientists the instinct has always been strong to avoid alarming the community and frightening Congress, and so to downplay the seriousness and frequency of misconduct.

To define scientific misconduct, fraud, has turned out to be difficult and bitterly contentious. The deep reason is that any fully articulated definition carries with it an entire set of views of the sociology of the sciences. Deeper still, such a definition necessarily threatens that shibboleth, the independence of the sciences. Fraud is a felony: any but the most cursory definition of scientific fraud brings to the fore the antithetical relationship between the needs of the scientific community and the American (or British) legal system, adversarial as we all know it to be, centered on the guilt or innocence of individuals. Any serious definition of scientific fraud must mediate between the ways of science and the procedures and presumptions of the law—and these may well be incommensurate. So such a definition calls up the specter of governmental interference, supervision, regulation. Make no mistake, the contention has been fierce.

Ignorance walks with incredulity throughout the controversies. With the possible exception of brushes with plagiarism, few scientists have encountered many live cases. Department heads, deans, other administrators, though they may have had to deal with charges of misconduct, are almost always limited in experience to one institution. As the federal government pays for most research in the hard sciences and much in the soft, mechanisms for intervening in more serious cases have been established, yet those responsible for dealing with them in detail are the relatively small staff of the Office of Research Integrity, in the Public Health Service, Department of Health and Human Services, and a still smaller staff that is part of the Office of Inspector General of the National Science Foundation.

Yet certain people have made themselves experts, one way or another. Some science journalists have made fraud a specialty, notably John Crewdson, of the *Chicago Tribune,* and, as occasion offers, Nicholas Wade, of *The New York Times.* Among journal editors, Stephen Lock, formerly of the *British Medical Journal,* was the earliest to acknowledge that the plague cannot be dismissed as merely an American problem, and to think out the consequences for British science. His successor, Richard Smith, the present editor, has been a vigorous and persistent advocate of reform in the British handling of fraud and particularly of conflict of interest in research.

Among those who have gone deeply into the analysis and handling of fraud in science two are prominent for their intellectual vigor, rigor, and clarity. C. K. (Tina) Gunsalus was for some years an associate provost at the University of Illinois in Urbana-Champaign; her father was a noted chemist and her several brothers and sisters scientists, but she is a lawyer by training and indeed is now in the law school there. She was the person at that university to whom came allegations of academic misconduct of any kind, in the humanities, in the professional schools, in the sciences soft and hard. She is nationally known as the leading expert on the problems and reform of institutional response to misconduct charges.

But today and for two decades, the most relentless, wide-ranging, original, and effective figure in getting at the root causes of fraud and other misconduct, particularly in the biomedical sciences, is Drummond Rennie, from his vantage point as deputy editor (west) of *JAMA, The Journal of the American Medical Association.* Rennie is a tall, bulky, shambling man, bespectacled, with white hair and white beard and lugubrious rhythms of speech. In his younger days he was a noted alpinist—and physicians who are skillful climbers are sought after—with Everest in his sights until a rescue, bringing a fellow climber down a mountain, left him with a lame hip. Of all his research, he values most highly his extensive studies of the physiology of hypoxia, oxygen deficiency, in the high mountains, the Andes, the Alps, the Yukon and Alaska, the Himalayas. He was born in the north of England, got his medical degree at Cambridge University, was briefly on the staff of Rush-Presbyterian-St Lukes Medical Center and its associated medical school, in Chicago, before becoming deputy editor of *The New England Journal of Medicine* in 1977. He was also an associate professor of medicine at Harvard Medical School, working at Brigham and Women's Hospital. He got his first direct exposures to fraud in 1979, when the Felig and Soman case hit the *New England Journal*; in 1980 when John Long, at Massachusetts General Hospital, was caught; and then when the Darsee affair broke, at Brigham, in 1981. Rennie moved to *JAMA* in the fall of 1983: George Lundberg, its editor, was beginning a long tough process of building a debilitated and inconsequential journal into the powerful and engaged publication it is today. Rennie is based at the University of California, San Francisco, a top medical school. There, among much else, he is co-director of the local Cochrane Center, part of an international network of individuals and groups who prepare scrupulous evaluations and statistical analyses of the biomedical literature on particular diseases, with the aim of letting practitioners know what the best evidence has established as preferred treatment. Rennie apparently cannot be easy of mind unless working at several jobs at once, travelling incessantly, writing and speaking on clinical trials, research integrity, authorship, electronic publication, team science—and editorial peer review, about which, beginning in 1989 in Chicago, he has organized a series of international research conferences, the fifth to be held in Chicago in September of 2005. Gunsalus and Rennie are old comrades-in-arms. His transatlantic liaison with Stephen Lock and Richard Smith has been strong. There's a network here, with Rennie at the hub.

*

We can be sure, at least, that incidence varies greatly along the spectrum of the sciences. At one extreme, fraud in mathematics ought to be impossible: yet even in mathematics a whiff of sulfur drifts past from time to time. In some areas of mathematics over the past quarter-century, proofs have become so hugely difficult that they can only be carried out by high-speed computers iterating thousands of steps. Such proofs are not well liked. They are ugly, and many mathematicians pride themselves on bringing an aesthetic sense to their work. Worse, even when all the steps are published they can be hard to verify. From time to time, such a proof will be withdrawn when the author or others note an error or a weak length of the chain. A while ago, a young mathematician—never mind name or nationality—fell under suspicion: he had published an immense computer-derived proof of a contentious conjecture, but after a few weeks had withdrawn it and has not republished. The speculation circulated that he had known all along that the proof was faulty, but wanted at least to get credit for working on this difficult problem. Scientific ambition takes odd forms.

At the other end of the spectrum, in clinical research, as our litany of cases suggests, instances of fraud have turned up with scary frequency. In the fall of 1988, Patricia Woolf, a sociologist associated with Princeton University, published an analysis of latter-day scientific misconduct. Among much else, she listed twenty-six cases that had become public from 1980 through 1987. While warning of the obvious impossibility of generalizing from such a small sample, a sample that surely was skewed, she wrote of the twenty-six that "the outstanding feature that emerges is that all but four are medical or related to medical purposes." Furthermore, "Twenty-two individuals had M.D. degrees, and/or were connected with medical schools, hospitals, or medical research institutions." Most of the "allegedly faulty research" took place at top medical schools or hospitals; many of the perpetrators had their MDs from similarly well regarded places.

The problems are hardly unique to the sciences in the United States. Science is worldwide, after all. Britain offers the most open research environment overseas. Experience there is instructive. In the nineteen-eighties and nineties British scientists, journal editors, and civil servants began to realize that they could not smugly take for granted their ethical superiority over their cousins across the pond. In Washington in October of 1988, at the National Academy of Sciences, the Council of Biology Editors—an international organization—convened a conference on ethics in scientific research. Stephen Lock addressed the assembled editors. He said, among much else, "In Britain any study of misconduct is more difficult than in the United States. In particular, we have no Freedom of Information Act and very stringent laws of libel; the smallness of the country means that any accusation is likely to be heard in every medical school, with serious implications." Thinking about biomedical misconduct, he said, he had been able to remember only five recent documented cases. (One was the champion plagiarist Alsabti.) And yet:

> Perhaps there were reasons why the pattern was different in Britain from that in the United States. In casual conversation British doctors would cite the hectic research pace in the States, with its emphasis on positive results, hinting that the more civilised, relaxed demands in Britain make malpractice less likely. Even so, I was struck when I talked, say, at postgraduate centres about misconduct in medical research how doctors would sidle up to me after the lecture and tell me about an unpublicised instance, sometimes but not always in some detail. Physicians concerned with or in drug companies seemed to be particularly cynical.

Ten years later, in the issue of 6 June 1998, the *BMJ* ran a special section on fraud in research. In an editorial introducing the section, Richard Smith wrote, in part:

> The British medical research community is busy assembling its response to research misconduct. The question is no longer, "Do we have a problem?" but rather, "How can we best respond?" The BMJ has thus commissioned five answers to the question (p 1726), two from people outside Britain with extensive experience of research misconduct. . . .
>
> The answers are published in a week when we have to retract yet another article because of probable fraud (p 1700). One of the authors of the retracted paper was recently struck off by the General Medical Council for research misconduct. He had also lied about his qualifications. Cameron Bowie, his coauthor, then started from the inevitable assumption that all of the rest of his work was fraudulent until proved otherwise and found that he could not satisfy himself that his coauthor had completed the work he said he had. Bowie describes his miserable experience in a personal view and has retracted a paper that has gained wide attention and been influential in developing policy (p 1755).

The lead and longest article in the section was by Drummond Rennie. His outrage about misconduct in research is abiding and personal. He it was who, as deputy editor, had to deal initially with the letter to the *New England Journal* in February of 1979 from Helena Wachslicht-Rodbard that charged that the paper by Vijay Soman and Philip Felig was plagiarized from hers. Then in 1980 came the scandal of John Long. Even as it broke, an issue of the *New England Journal* was closed and going to press with a paper in it by three authors, one of them Long. Rennie spent a hectic weekend yanking that article and finding another that he could trim to fit the gaping hole. He told the other two authors to rework their paper entirely, with nothing of Long in it. A year later, that revised paper was duly published—in an issue of the *New England Journal* in which the lead research article was by John Darsee and an editorial was by Robert Gallo. Rennie treasures that issue. Anyway, in the *BMJ* in 1998 Rennie told the British, in his characteristic rolling periods:

> An allegation of scientific fraud can ruin the careers of both the accused and the accuser, divide faculties, bring a research institution's functions to a halt, provide a field day for the media, and, when the scientific establishment is unprepared, result in a loss of confidence in the entire research enterprise. Yet, despite repeated demonstrations that this is the case, scientists are still reluctant to face up to such an unpleasant problem. Three years ago, at a meeting on research misconduct held by the BMJ, I warned that many extremely embarrassing incidents at a variety of institutions would be required before anyone took any action in the United Kingdom. This seems to have been borne out. At a meeting organised by the Committee on Publication Ethics (COPE) in London, I was depressed that so few seemed to have paid the slightest attention to the rich, well documented and instructive experience of the United States, where an energetic attempt to face up to the problem has been made. Such parochialism may doom the UK to repeat the many mistakes already made by others.
>
> To the American observer, the news from the UK about incidents of misconduct in research is, as baseball's greatest philosopher, Yogi Berra, remarked, "déjà vu all over again."

Rennie went on:

> The idea that the situation in the United States is uniquely bad rings hollow in the face of growing numbers of cases in Britain,

> Germany, and elsewhere. It has not been shown that scientists in Britain differ importantly from those in the US. Institutional denial in the United Kingdom is therefore no longer a sensible option.

Surveys of the incidence of fraud have of course been attempted. Results are hardly conclusive. The most suggestive have put questions to scientists not asking them to admit to any of the specified acts of scientific misconduct themselves but rather to say whether they have observed such acts by others of their colleagues. The limitations are obvious. Even though they have been guaranteed anonymity, many people out of busyness, distaste, anxiety, or guilt will be reluctant to respond. Among those who do answer from a given institution, several accounts may overlap, multiply reporting the same instances. From such attempts at surveys we can draw no hard conclusions about the incidence of acts but only, at best, about the pervasiveness of awareness of them.

Such are the psychological limitations to the real significance of estimates. Then, too, survey results will vary widely depending on the definition of misconduct. A rough base line can be set for fraud *sensu stricto,* or fabrication, falsification, and plagiarism. The Public Health Service and the National Science Foundation account for the vast bulk of federal funding of research, leaving out military research and development. In the decade to 2002, the two agencies dealt with somewhat more than two hundred cases, twenty to thirty per year. These, of course, are the most serious ones, and not resolved, or suppressed, by the research institutions. With an estimate of, say, two to three hundred thousand scientists enjoying grant money, that suggests an incidence of one in one hundred thousand. Nobody close to the problem believes that the figure is that low. Surveys of conduct that respondents say they have observed come up consistently with rates on the order of one in one hundred. What evidence exists suggests that a high proportion of instances are not reported.

Ian St James-Roberts, on the faculty of the University of London, writing in the British weekly *New Scientist* in 1976, tried a crude version of this approach. At the beginning of that September, he published a brief article on the subject "Are Researchers Trustworthy?," topped off with an eleven-item questionnaire about "intentional bias" in research—ironical euphemism—and invited readers to send in their replies. Twelve weeks later he published an analysis of the responses. (In the interval, Oliver Gillie published his exposé of Cyril Burt in *The Sunday Times.*) Two hundred and four readers replied to the survey. St James-Roberts excluded five of the responses as spoiled: "One purported to come from a laboratory rat, one from an inspector of custard pie stability," and so on. St James-Roberts had included several questions to test the overall plausibility of the distribution of responses. Thus, nearly two-thirds of the respondents were over the age of thirty, while 36 per cent were over forty. Twenty-three per cent were tenured faculty, 12 per cent were students, the rest scattered. As for the subjects in which they worked, 30 per cent were in the biological sciences, another 8 per cent in the medical sciences.

"The sample is clearly a very biased one," St James-Roberts wrote with calm candor. Be that as it may, 92 per cent of his respondents had encountered instances of intentional misstatement, directly or indirectly. Fifty-two per cent had encountered it in direct contact. Complete fabrication of data was reported rarely, in 7 per cent of the responses. Manipulation of data was most frequent, at 74 per cent of the total. Perhaps most useful was the breakdown of the various ways that misconduct was detected. A third were spotted because somebody noticed dubious data. A quarter came from difficulties of replication. In 17 per cent of the cases somebody was caught in the act. Some 14 per cent became known because someone admitted to the act—or boasted of it. Four out of five respondents believed that nothing had happened to the perpetrators.

Stephen Lock, in 1988, was impressed by "the masterly survey" carried out by Patricia Woolf. He was able to exploit his central position and inimitable connections as editor of the *British Medical Journal* to try some research of his own. He told the biology editors at that meeting in Washington that he had concluded that present-day knowledge of misconduct in Britain "resembled that in the United States less than a decade ago." He saw that the relatively small scale of biomedical research in Britain "does mean a closely knit medical society whose members can be asked in strict conference for their experience of misconduct. Hence I used this positive feature to do a non-systematic survey."

Lock wrote to one professor of medicine and one of surgery at each of twenty-nine British medical institutions, and to five other academics or other physicians who he happened to know had experience with misconduct, as well as the editors of fifteen medical journals and two scientists administering medical research—eighty people in all. He asked whether they could add any cases of fraud in Britain to those already known. He asked also whether their institution had any procedures in place for dealing with cases. He included a copy of an editorial he had written after the Slutsky report appeared in the *New England Journal*, and a post-paid reply envelope.

Seventy-nine of the eighty replied. (The one who did not was a professor of surgery.) They provided enough detail so that duplicates could be counted out. Forty-six of the respondents knew of some instance of research misconduct—for a total, excluding duplicates, of 61 distinct cases. Most of the cases, Lock said, had been "encountered firsthand, though a sizable minority were well authenticated secondhand instances—and there were a few rumours as well." The figures were 41 cases known firsthand, 26 "authenticated secondhand," and 5 he classed as rumors.

> Most reports concerned episodes in Britain (41); episodes were also reported in the United States (seven), Australia (four), and other countries (seven). In 13 cases the country was not stated. A notable feature of the cases was a surprising proportion of senior workers (18 out of 43 cases in which the grade was given . . .). Of the cases reported, more than half the results had been published, but only in six cases had subsequent retractions appeared, all in too vague terms to indicate the nature of the misconduct.

Of the twenty-nine medical institutions, only three were reported to have mechanisms for investigating misconduct. "The fate of the perpetrators was not always given," Lock said; "but in some cases they were allowed to resign, a degree was disallowed, or there was ostracism by colleagues; not infrequently a researcher of intermediate grade had moved to another department by the time of discovery," and the new department took no responsibility.

> Clearly, the informality of the survey might be criticised. There were few ways of checking many of the statements, and yet they rang true for several reasons. Firstly, I knew most of the respondents (being able to address half of them by their Christian names); secondly, five of the cases were corroborated independently by other replies; and, thirdly, given the circumstances of the inquiry, the respondents had nothing to gain or lose from their replies.

Be that as it may, Lock believed that "If anything, the results underestimate the true number of cases." He gave various reasons for supposing so. Problems of whistle-blowers, problems of overproductiveness and of gift authorship—all these were showing up in British science, and Lock was explicitly aware of the similarities to the United States. "Authorship no longer carries the responsibilities it should—the ability to justify intellectually the entire contents of an article."

Must something be done? "Firstly, fraud is a crime," he said. "Secondly, too many such felonies have gone unprevented, uninvestigated, or undetected, and this will encourage others into attempting them." And, finally and importantly, "if medicine and science no longer carry the hallmark of a profession—self correction—then official outside agencies are likely to do the correcting for them." Here again, developments in the United States were a dread warning.

What looked to be a thorough and sophisticated attempt to get at incidence, although by a method essentially the same as St James-Roberts's or Lock's, appeared late in 1993 in a paper by Judith Swazey and two colleagues. Swazey had trained at Harvard as a historian of science. In 1984, she founded the Acadia Institute, a small nonprofit center in Bar Harbor, Maine. Her colleagues in the study, Melissa Anderson and Karen Seashore Louis, were in the College of Education at the University of Minnesota. Guaranteeing anonymity, they mailed a questionnaire to two thousand doctoral candidates and two thousand faculty members, from ninety-nine of "the largest graduate departments in chemistry, civil engineering, microbiology and sociology." The questionnaire put a set of thirteen questions to faculty and graduate students; faculty got another two.

> Although our results do not measure the actual frequency of misconduct—instead, our questionnaires sought rates of *exposure* to

> perceived misconduct—they do demonstrate that such problems are more pervasive than many insiders believe. We also found significant differences among disciplines in the frequency and the types of questionable behavior observed.

The instructions said, "Have you *observed* or had other *direct evidence* of the following types of misconduct. Please indicate the number of graduate students and faculty members whose misconduct you have observed / experienced." Faculty were asked also to limit responses to the past five years and to the department with which they were currently affiliated. The questions fell into three broad categories. The first was fraud, including fabrication, falsification, and plagiarism. Second came questionable research practices like keeping poor records, withholding methods or results, or permitting honorary authorship. The third comprised a miscellaneous set of other misconduct running from sexual harassment to violating government regulations—behavior not unique to the sciences and "subject to generally applicable legal and social penalties."

Of the four thousand scientists and students, about twenty-six hundred replied—72 per cent of the faculty members, 59 per cent of the students. For any direct-mail survey, let alone one on a sensitive subject, these are remarkably high rates of return. Fully half of the faculty members and 43 per cent of the graduate students reported direct knowledge of more than one kind of misconduct. Eight per cent of the faculty, and 8 per cent of the students, said they had direct knowledge of faculty who had plagiarized. Plagiarism by students was reported at far higher rates, a third of the faculty claiming to have observed one or more instances, and a fifth of the students reporting direct knowledge of graduate-student plagiarism. Again, 6 per cent of the faculty responded that they had observed or had other direct knowledge of faculty members falsifying data, while 8 per cent of the students said the same; 13 per cent of the faculty and 15 per cent of the students had direct knowledge of graduate students who had falsified data.

Breakdowns among specialties varied. Plagiarism was reported much more highly among the engineers and the sociologists than elsewhere, with more than 40 per cent of the faculty in each of those areas having detected plagiarism by doctoral candidates; 18 per cent of the civil-engineering faculty members knew of colleagues' plagiarizing. Data falsification by faculty was reported by 12 per cent of the graduate students in microbiology. Turn about, their professors said that they had observed some 15 per cent of microbiology graduate students to have cooked research data—and chemistry students were as bad.

Of the questionable but not fraudulent research practices, 22 per cent of faculty, overall, reported instances of other faculty passing sloppy use of data, and 15 per cent said they knew of colleagues who had suppressed data that contradicted their own previous research. In another touchy area, nearly a third of faculty said they knew of cases where authorship of research papers had been assigned inappropriately.

Swazey's report has been attacked scathingly—and not by those who want to defend the purity of scientists but precisely by some who are deeply concerned to know the true extent of misconduct. One suspects that those who returned the questionnaire were more likely than others to have had misconduct to report. A debilitating deficiency is that there can be no way of knowing the extent of double counting, that is, whether single instances got reported by multiple observers. One critic at a large midwestern state university, renowned for research: "The fact is that Dr. Swazey, even at the most charitable interpretation of her study, has data only on the degree to which people report observing misconduct. She has no evidence whatever of the reliability and validity of these observations she has collected."

To be fair, Swazey and her colleagues repeatedly warned of the limitations of their approach. One caveat they raised: "Since the questionnaires did not ask the respondents to distinguish between what they believed to be instances of misbehavior and cases that had been confirmed by an official or unofficial type of investigation, their responses should be viewed as 'strongly suspected' instances." The problem of double counting they acknowledged—yet they gave it a thoughtful new interpretation:

> One cannot estimate from our data what percentage of faculty or graduate students in a given department or in the four disciplines may be engaging in a particular type of misconduct or questionable research practice. Rather, growing out of the project's focus on the ways that departments and disciplines affect the education and socialization of graduate students, our objective was to document the exposure of graduate students and faculty to what they believe is ethically wrong or problematic conduct in their departments.

Thus, the project despite its weaknesses also had a peculiar strength. Misconduct of various types was observed so frequently and pervasively that questions beyond the particular kinds of bad behavior must arise. What are the effects on observers, particularly the graduate students? What happens to their ethical standards, their sense of acceptable conduct—or of what one can get away with? In such settings, such an atmosphere, what happens to the morale of students and faculty? A new problem emerges: reverse or negative mentoring.

Other attempts have been made to get at the problem of incidence, yet surprisingly few. Nicholas Steneck, who is a historian at the University of Michigan but who spends several months a year as a consultant at the Office of Research Integrity, has compiled a list of seven studies over the quarter-century beginning with St James-Roberts's. He missed Stephen Lock's, yet included two small ones from Norway, perhaps not easy to extrapolate to the United States. By far the largest sample size was Swazey's, at four thousand, and her survey got the best response rate, too. Depending on the questions, those surveys that addressed fabrication, falsification, and plagiarism narrowly came up with incidences from three to thirty per cent, a range so wide as to cast conclusions in doubt. Taking a different approach, some analysts have carried out audits of the data on which published reports of clinical trials were based. Reviewing these, Steneck wrote in 2000 that they "have turned up 'significant problems' or 'major deviations' at and above the 10 percent level." Yet such results do not correlate with fraud as such, "since they do not take into account whether discrepancies result from deliberate actions."

More rigorous and illuminating research may be coming soon. Steneck has initiated a variety of research projects at the Office of Research Integrity, and one dealing with the incidence of misconduct was nearing completion in the summer of 2004.

*

The ruling definitions of scientific fraud and other gross misconduct are those of the National Science Foundation and of the National Institutes of Health—or, more properly, the Public Health Service, of which the institutes comprise the largest part. The two definitions are all but identical, and on casual reading seem, like Babbage's, to be simple enumerations. The language was set only in the nineteen-eighties, and was a response to the increasing number of publicized cases. Here is how they used to read:

The National Science Foundation:

> (1) Fabrication, falsification, plagiarism or other serious deviation from accepted practices in proposing, carrying out, or reporting results from activities funded by NSF; or (2) Retaliation of any kind against a person who reported or provided information about suspected or alleged misconduct and who has not acted in bad faith.

The Public Health Service:

> 'Misconduct' or 'Misconduct in Science' means fabrication, falsification, plagiarism, or other practices that seriously deviate from those that are commonly accepted within the scientific community for proposing, conducting, and reporting research. It does not include honest error or honest differences in interpretations or judgments of data.

Note, however, that the "other serious deviation" clause has recently been removed from the definitions—a contentious point to which we shall return.

Fabrication is Babbage's forging, or what biologists call "dry labbing," making the data up. Falsification includes Babbage's trimming, and means altering the data or tendentiously selecting what to report. A miniature, emblematic instance was told to me a while ago by a senior biologist, a man who after wide and international experience in immunology has been director of research for a small start-up company in northern California that is exploring traditional Chinese medicines, many in use for centuries, to see whether any contain active ingredients that may prove genuinely efficacious. At a seminar of one of the laboratory groups, he had dropped in to sit at the back of the room. A junior investigator was presenting the data from a series of experiments, in preparation for writing them up as a paper. She summarized her results as points on a graph, and offered an interpretation. The distribution of the data points looked good, much as expected—except for one result, which was far from the others and from the interpretation. She pointed this out. The lab chief, coincidentally also a woman, told her she should leave that result off the graph, out of the paper. Whereupon the research director rose in wrath. *"No."* Even as he spoke of the episode several weeks later, his voice rose and his face twisted in distress and puzzlement. "I told them, 'You can never do that! You can re-examine the experimental conditions, or repeat the experiment, to see whether something went wrong. You can do other experiments to try to explain the result. If these don't solve it, you must confront the anomaly in the discussion section of the paper. But you cannot suppress a result.'" He shook his head and said, "They must know that." He is surely right—yet anomalous data of that type are called outliers, and suppression of outliers is by no means rare and is sometimes defended as a matter of good scientific judgement, part of the art of discovery. The classic instance, of course, is Robert Millikan.

We think of plagiarism as copying, what we were warned against as schoolchildren, but it comprises all ways of representing someone else's work as one's own, and the full, formal term is theft of intellectual property. Fabrication, falsification, plagiarism—all three are fraud, but plagiarism stands in a curious relation to the other two in purpose, in methods, in its incidence, in its effects. After all, if you get away with faking the data or fudging it you pollute the scientific record with something you know to be untrue; but you would never filch someone's work you thought false. Yet what plagiarism may lack in turpitude it more than makes up in ubiquity. At the level of anecdote, anyway, rare is the senior American scientist who has not suffered it. Many scientists say they have encountered plagiarism in the process of peer review of grant applications, though as we have seen this is hard to prove. Surely more frequent is the theft of ideas, even of data and language, in the process of refereeing of journal articles. A senior scientist at Stanford University, in a conversation walking down the corridor: "It happens, but if you're productive it affects little of your work."

"Other serious deviation from accepted practices." The clause seemed tacked onto the definitions. Many scientists and lawyers found it vague. Some found it threatening. It provoked controversy. Examination of the clause will cut to the root of the controversy.

Attempts to improve the definition began early in 1989, as misconduct cases were becoming increasingly conspicuous, when the National Academy of Sciences gingerly took up the issues. The academy's loftiest organ, the Committee on Science, Engineering, and Public Policy, had taken months to come to agreement that a study was needed. Seven more months were used up finding someone who would agree to be chairman. The next and by no means minor difficulty was to find willing panelists who did not have close ties to leading scientists involved in the current controversies—notably the Baltimore affair and the case of Robert Gallo and associates at his laboratory at the National Cancer Institute, who stood accused of theft of intellectual property. Only by the spring of 1990 was the academy's panel assembled. Among its twenty-two members were eleven university scientists of various fields, two from industry, together with university administrators, ethicists, a lawyer, a historian, and the editor of a journal.

From the start, according to reports, disagreement flourished. The panel took two years to produce its study. Released in April of

1992, it was anodyne. The panel balked at the question of incidence, saying rightly enough that no one had measured it. Their definition limited scientific misconduct to fabrication, falsification, and plagiarism; they called on the National Science Foundation and the Public Health Service to eliminate the "other serious deviation" clause. Despite the safe bland character of the report, two scientists on the panel dissented, saying that among other faults it exaggerated the importance of misconduct.

One of the dissenters was Howard Schachman. He is a molecular biologist at the University of California, Berkeley (since retired), who has long, intelligently, and angrily defended science against what he considers to be the mortal danger of interference from government. He had put his argument before congressional committees as well as in the debates of the National Academy panel; he campaigned publicly, privately, pugnaciously with all who would listen. In July of 1993, in an article in *Science,* reaching a wide audience, Schachman wrote:

> Many scientists, like others in our society, are ambitious, self-serving, opportunistic, selfish, competitive, contentious, aggressive, and arrogant; but that does not mean they are crooks. It is essential to distinguish between research fraud on the one hand and irritating or careless behavioral patterns of scientists, no matter how objectionable, on the other. We must distinguish between the crooks and the jerks.
>
> . . . My concern is over vagueness of the term "misconduct in science" and how people with different orientations interpret various alleged abuses.
>
> In formulations of the term "misconduct in science" there is agreement on fabrication, falsification, and plagiarism. Scientists have emphasized that "misconduct in science" does not include factors intrinsic to the process of science, such as error, conflicts

> in data, or differences in interpretation or judgments of data or experimental design.

Schachman took particular exception to the inclusion of the phrase, from the Public Health Service, or NIH, version, "other practices that seriously deviate from those that are commonly accepted within the scientific community for proposing, conducting, and reporting research." As he read it:

> Not only is this language vague but it invites over-expansive interpretation. Also, its inclusion could discourage unorthodox, highly innovative approaches that lead to major advances in science. Brilliant, creative, pioneering research often deviates from that commonly accepted within the scientific community.

Schachman's conclusion was apocalyptic. Mindful of the Hitlerite rejection of the physics of relativity as "Jewish science" and the Stalinist consignment of mendelian genetics and geneticists to the gulag as not Marxist-Leninist science during the fifteen-year domination of Soviet biology by Trofim Lysenko, he wrote:

> History is full of examples of governmental promulgations of laws expressed in broad, open-ended terms that were elastic enough to be stretched to cover any individual action that irritated some officials. In this century alone it was a major offense in some countries to publish scientific papers that seriously deviated from accepted practice. The enforcement of such strictures virtually destroyed major areas of science in those countries. We should not expose science in this country to similar risks.

Other scientists have concurred. Among the more prominent was Bernadine Healey, who had been appointed by President George Bush *père* as director of the National Institutes of Health. She was conservative, combative, and politically ambitious, unpopular as an administrator. Healey attacked the Public Health Service version of the "other serious deviation" clause because, she said, "tinkering, bold leaps, unthinkable experimental design, and even irritating challenges to accepted dogmatic standard" could be labelled misconduct. Healey's successor, Harold Varmus, a Nobel-prize winner whom President Bill Clinton appointed in 1993, a smart, effective director liked by the scientific community as one of their own, called for dropping the clause because it is "too vague and could be used inappropriately."

The import and utility of the clause may be diametrically the reverse of the received view. What Schachman and others fail to understand is that the "other serious deviation from accepted practices" clause or something like it is essential to a definition of misconduct that can preserve the autonomy and self-governance of the sciences they think their objections are defending. Early in 1997, Karen Goldman and Montgomery Fisher published a tightly reasoned defense of the "other serious deviation" clause. The two were lawyers in the Office of Inspector General of the National Science Foundation, which handles misconduct charges serious enough to be brought to the foundation. They posed the question in its toughest form: Is the clause constitutional?

Goldman and Fisher turned the usual reading of the foundation's definition end for end: fabrication, falsification, and plagiarism, they wrote, are but three examples of misconduct, while "other serious deviation from accepted practices," which they called the OSD clause, is the overriding and most significant part, "a general standard defining misconduct."

> The OSD clause provides a legal basis for a finding of misconduct in all cases of serious breaches of scientific ethics, including cases that cannot be categorized as fabrication, falsification, or plagiarism. The clause relies on the practices of the scientific community to separate acceptable from unacceptable conduct.

This is the crucial point. The disputed clause throws the responsibility for determining standards back on the scientists themselves. Although Goldman and Fisher did not say so, this is just what those who champion the independence of the sciences should want. They went on, "Only *serious* deviations from that standard constitute misconduct." Furthermore, the critics including the academy panel had failed to address the fact that the agencies that give out money for research must "ensure that government funds are provided only to responsible individuals." For that reason, too, they wrote,

> [A] generalized, comprehensive provision such as the OSD clause is necessary because it is impossible to predict and list all the unethical actions that might warrant agency action. Conduct such as tampering with other scientists' experiments, serious exploitation of subordinates, misuse of confidential information obtained during the grant proposal review process, or misrepresentation of authorship, could, if sufficiently serious, be considered misconduct in science. Yet, such actions do not fit within the specified categories in the definition.

None the less, the law recognizes vagueness as a reason a statute or regulation may be found void. Indeed, in constitutional law a void-for-vagueness doctrine has a recognized place—for the due-process clause of the Fifth Amendment, they pointed out, "requires adequate notice of what conduct is prohibited." Thus, a law or regulation must specify "with reasonable clarity the prohibited conduct," and this prevents—in the words of a decision that Goldman and Fisher cited—"arbitrary and discriminatory enforcement." Now, lack of specificity and the likelihood of arbitrary enforcement are precisely the dangers alarming Schachman and all the other critics of the "other serious deviation" clause. Void for vagueness would be the one possible legal ground for attacking the clause.

Yet elsewhere, open-ended, general language about misconduct, relying on accepted standards within a community, has repeatedly passed constitutional scrutiny. In a case in 1974, "the Supreme Court upheld broad provisions of the Uniform Code of Military Justice that prohibit 'conduct unbecoming an officer and a gentleman' and 'all disorders and neglects to the principles of good order and discipline in the armed forces.'" This case and others, Goldman and Fisher wrote, "make it clear that broad, undefined prohibitions are not void for vagueness when they are interpreted by reference to community standards of conduct."

Many professions are subject to such standards. Lawyers are ridden especially hard. The American Bar Association has a model code of professional responsibility, with a misconduct rule that winds up by saying that a lawyer must not engage "in any other conduct that adversely reflects on his fitness to practice law." The federal rules of appellate procedure have similar language: "conduct unbecoming of a member of the bar of the court" can subject a lawyer to suspension or disbarment. A Supreme Court rule penalizes "conduct unbecoming a member of the Bar of this Court." State courts typically have similar rules.

Public-school teachers and university professors, even when tenured, can be fired on such grounds. Goldman and Fisher quote the language of several cases, including one where "Rutgers University determined that the professor had misused grant funds and exploited, defrauded, threatened, and abused foreign students carrying out research in his laboratory." Rutgers had a regulation defining adequate cause for dismissal as "failure to maintain standards of sound scholarship and competent teaching, or gross neglect of established University obligations appropriate to the appointment, or incompetence. . . ." The professor sued. On appeal, the federal Third Circuit rejected the vagueness challenge, writing in 1992 that "the academic community's shared professional standards" permitted Rutgers to apply the regulation. The Supreme Court refused to entertain a further appeal. Physicians, nurses, pharmacists, architects, engineers have been subject to similar rules, and across the country state courts have upheld them.

By far the most thoughtful and detailed inquiry into the definition of scientific misconduct was carried out in two years of meetings and consultations by a panel called the Commission on Research Integrity. In the summer of 1993, Congress passed a National Institutes of Health Revitalization Act, which among much else directed that such a commission be set up. The then Secretary of Health and Human Services, Donna Shalala, chartered the commission on 4 November 1993. Its chairman was Kenneth J. Ryan, a physician and professor in the Department of Obstetrics, Gynecology and Reproductive Biology at Harvard Medical School, and also chairman of the Ethics Committee at Brigham and Women's Hospital. Ryan was a tough-minded senior public figure. He had previously chaired the federal panel that drafted the regulations on protection of human subjects of biomedical and behavioral research, acclaimed a success. For the new commission, he recruited biomedical scientists, university administrators, others experienced in dealing with allegations of misconduct, a lawyer, and a member of that egregious new trade, an ethicist—twelve in all. Their charge was to arrive at a new definition of misconduct, to recommend procedures for institutions' responses to allegations, and to develop a regulation to protect whistle-blowers.

From 20 June 1994 to 25 October 1995, they met fifteen times, in characterless modern hotels in Washington suburbs or at Dulles Airport and with expeditions to the medical school and research complex that makes up the University of California at San Francisco, to DePaul University in Chicago, to Harvard Medical School, and to the University of Alabama at Birmingham. The meetings grew from single days to two- and sometimes three-day sessions. They were public, as the law requires of governmental commissions. They attracted few journalists, but were trailed from meeting to meeting by a fringe of followers who were known victims of misconduct, mostly those whose work had been plagiarized or who had blown the whistle and suffered retaliation. At many of the earlier meetings the commission heard from new individuals—bitter tales of careers ruined, lives blasted. In Chicago, at a break after particularly pathetic testimony from one such witness, Kenneth Ryan was seen in the men's room rinsing tears from his eyes. The panelists were earnest, hard-working, and increasingly harried, as they grew aware of the full extent of the problems and of the difficulty of educating, inciting, goading the scientific community and the senior staffs of organizations where scientists work to anticipate or respond to the problems seriously and effectively.

The definition gave trouble. They soon moved away from fabrication, falsification, plagiarism, and other serious deviation, seeking language at once more particular and more comprehensive. Over the months, too, they groped towards understanding that existing definitions of scientific misconduct have been rooted in the law, in the legal concept of fraud, and in the consequent obsession with due process, with adversarial proceedings, and with the guilt or innocence of individuals. Tina Gunsalus and Drummond Rennie were both members of the panel; they pressed implacably for a transformation in concepts and words.

The most significant moment in the evolution of the definition came at the session where Karl J. Hittelman, then the associate vice chancellor of academic affairs at the University of California, San Francisco, stated the fundamental aim of the commission and of the definition: "to protect the integrity of the scientific literature and of the scientific process." Here, in less than twenty words, was the clear assertion of the distinction between the attitudes of the law and the needs of the scientific community.

Later at that session, I offered from the floor a definition that attempted to capture the thinking of the panelists at that point in its evolution.

> Scientific misconduct is behavior that corrupts the scientific record or potentially compromises the objectivity or practices of science. All forms of scientific misconduct are inescapably and immediately the responsibility of research institutions and the scientific community. The most serious forms comprise research fraud: these are fabrication, falsification, and plagiarism, and if the research is funded by government, fraud in that research is also the concern of government. Furthermore, the testimony of whistle-blowers is crucial to exposure of fabrication, falsification, or plagiarism: therefore, effective protection of whistle-blowers is also the concern not only of the institution but of government. In the category of research fraud, fabrication and falsification can take many forms and occur at many stages. . . . Similarly,

> plagiarism has many forms. . . . In sum, plagiarism is theft or other misappropriation of intellectual property.
>
> But scientific misconduct includes other important forms of behavior that are not the concern of government yet cannot be ignored or played down. These behaviors group broadly under three headings, matters of authorship, of good practice, and of mentorship. This category of scientific misconduct includes, *inter alia,* misattribution of authorship or credit; failure to keep appropriate records; gross sloppiness in the conduct of research; unreasonable withholding of materials and results; and failure responsibly to train, supervise, and counsel research students and other junior colleagues.
>
> Finally, certain types of illegal or unethical behavior may occur in connection with research that are not scientific misconduct as such: these include, chiefly, failure to comply with rules that govern research with human subjects or with animals; sexual harassment; and fiscal misconduct. All of these—even such rare types as research sabotage—are already subject to regulations, laws, and punishments.

By this time, though, the interactions among panelists had taken on an intellectual dynamism that drove them towards a more radical reformulation. In a subtle—perhaps overly subtle—appeal to the standards of the community, their final report asserted that their "new definition introduces an ethical approach to behavior rather than serving as a vehicle for containing or expanding the basis for blame or legal action." They offered their definition as a nested set.

> Research misconduct is significant misbehavior that improperly appropriates the intellectual property or contributions of others, that intentionally impedes the progress of research, or that risks

> corrupting the scientific record or compromising the integrity of scientific practices. Such behaviors are unethical and unacceptable in proposing, conducting, or reporting research, or in reviewing the proposals or research reports of others.

To replace FF&P, they proposed three new categories, "misappropriation, interference, and misrepresentation, each a form of dishonesty or unfairness that, if sufficiently serious, violates the ethical principles on which the definition is based." The definition set these out.

> *Misappropriation:* An investigator or reviewer shall not intentionally or recklessly
>
> a. plagiarize, which shall be understood to mean the presentation of the documented words or ideas of another as his or her own, without attribution appropriate for the medium of presentation; or
>
> b. make use of any information in breach of any duty of confidentiality associated with the review of any manuscript or grant application.

The second category dealt with a kind of conduct that was turning up in a number of cases but that had not been characterized separately before this.

> *Interference:* An investigator or reviewer shall not intentionally and without authorization take or sequester or materially damage any research-related property of another, including without limitation the apparatus, reagents, biological materials, writings, data, hardware, software, or any other substance or device used or produced in the conduct of research.

Only with the third category did the commission come to the usual starting point of definitions of misconduct.

> *Misrepresentation:* An investigator or reviewer shall not with intent to deceive, or in reckless disregard for the truth,
> a. state or present a material or significant falsehood; or
> b. omit a fact so that what is stated or presented as a whole states or presents a material or significant falsehood.

In law, proof of fraud (as of any other first-degree crime) requires evidence of intent, what lawyers call a "state-of-mind" requirement—*mens rea,* in the older legal jargon. This has proved to be an almost insuperable obstacle to effective pursuit of major cases of misconduct. In the commission's definition, the first and third of the categories offered an alternative to intent, namely, reckless behavior. The report explained, "An intent to deceive is often difficult to prove." Therefore, "One commonly accepted principle, adopted by the Commission, is that an intent to deceive may be inferred from a person's acting in reckless disregard for the truth."

The commission went on to define other forms of professional misconduct, comprising obstruction of investigations—meaning cover-ups, tampering with evidence, and intimidation of whistleblowers and other witnesses—and noncompliance with regulations governing such things as research using biohazardous materials or human or animal subjects. These matters, though grave, are straightforward, even routine. The guts of the definition are in those three categories. In logic and rhetoric they are clear, concise, hierarchically inclusive, watertight. On thoughtful re-reading, the definition emerges as masterly.

Yet having gone so far, the commission failed to go far enough. As one member pointed out, despite the appeal to an ethical base the language remains lawyer-bound. This is symptomatic. The panelists recognized, for example, the terrific problem of demonstrating intent, imposed by the legal system's definition of fraud. But they failed to address—and perhaps never fully perceived—how far the needs of the scientific system, in Hittelman's formulation to protect the integrity of the scientific literature and of the scientific process, diverge from the attitudes inherent in the Anglo-American legal system, and how profoundly important the divergence is. To repeat, the two sides of that opposition may be incommensurable.

Few scientists read the definition thoughtfully. Most dismissed it, some scathingly. The president of the National Academy of Sciences, Bruce Alberts, and six other members of the academy's governing council wrote attacking the Ryan Commission report. So did Ralph Bradshaw, president of the Federation of American Societies for Experimental Biology, an umbrella organization covering fifty professional groups, which claims to represent more than two hundred and eighty-five thousand scientists. The report was referred to a new panel, established in the Department of Health and Human Services, to develop a common federal definition of research misconduct. There it was entombed.

Yet a standard federal definition of scientific misconduct is still a desideratum. In October of 1999, the Office of Science and Technology Policy—part of the Executive Office of the President, and headed by the presidential science advisor—put forward the next attempt, offering a policy for public comment. Its aim: "to protect the integrity of the research record." It had two parts, the definition itself and a rule for reaching a finding of misconduct. The policy recycled the too-familiar language, though with some improvements. Perhaps most significant of these was the implied, yet clear, separation of investigation from adjudication. The National Science Foundation, in its handling of misconduct cases, does divide the two. The Public Health Service, however, has not done so, and this failure has seriously muddled handling of several big cases—notably the Baltimore and the Gallo affairs.

The proposed definition contains sub-definitions of its own key terms.

> Research misconduct is defined as fabrication, falsification, or plagiarism in proposing, performing, or reviewing research, or in reporting research results.
>
> Research, as defined herein, includes all basic, applied, and demonstration research in all fields of science, engineering, and mathematics.
>
> - Fabrication is making up results and recording or reporting them.
> - Falsification is manipulating research materials, equipment, or processes, or changing or omitting data or results such that the research is not accurately represented in the research record.
>
> The research record is defined as the record of data or results that embody the facts resulting from scientific inquiry, and includes, for example, laboratory records, both physical and electronic, research proposals, progress reports, abstracts, theses, oral presentations, internal reports, and journal articles.

FF are back. Vanished is any sense of the needs, nature, or context of the scientific community in action. Gone, too, is any notion of "private science," the notebook as the kingdom of speculation or even confessional. And here came P:

- Plagiarism is the appropriation of another person's ideas, processes, results, or words without giving appropriate credit, including those obtained through confidential review of others' research proposals and manuscripts.
- Research misconduct does not include honest error or honest differences of opinion.

And then the rule for adjudication:

A finding of research misconduct requires that:
- There be a significant departure from accepted practices of the scientific community for maintaining the integrity of the research record;
- The misconduct be committed intentionally, or knowingly, or in reckless disregard of accepted practices; and
- The allegation be proven by a preponderance of evidence.

Although the definition itself includes nothing of that state-of-mind requirement for a charge of misconduct, intent reappears as part of what must be proven. Yet the rule allows alternatives: accepted practices could not, after all, be ignored, and although the scope of the language is narrow, applying only to FF&P, the possibility of escaping the necessity of proving intent is a distinct gain. In principle, at least, those pursuing misconduct cases will no longer have to get inside the mind of the perpetrator. And this should make much more difficult the sloppiness defense—as in Imanishi-Kari's appeal from the finding of the Office of Scientific Integrity (as it then was). Lawyers defending scientists will, of course, attempt to force the issue of intent.

That allegations be proven by a preponderance of the evidence is the standard that applies in civil cases, a lighter burden for plaintiffs than the criminal law's requirement for proof beyond a reasonable doubt. The preponderance standard is indeed applied throughout the federal government in cases where individuals or companies are barred from doing business with the government. It has in fact been the practice in adjudicating cases of scientific misconduct. It had not previously been stated explicitly as part of the policy. After public commentary and minor revisions and clarifications, the policy was made final on 6 December 1999.

On 17 November 1999, the National Academy of Sciences held a day-long conference at its headquarters in Washington, billed as a town meeting to present and discuss the new policy. Arthur Rubenstein, dean and chief executive officer of Mt Sinai School of Medicine, in New York, ran the show. The response was pretty predictable. Scientists generally loved the policy. In particular, Howard Schachman, this time representing the American Society for Biochemistry and Molecular Biology, asserted that scientists "virtually unanimously" welcomed the disappearance of the "other serious deviation" clause. Even its ghost, in the rule for adjudication, roused an objection from Julius Youngner, who had been one of the three-person panel of the Health and Human Services appeals board that had cleared Imanishi-Kari. But Anita Eisenstadt, assistant general council at the National Science Foundation, said that her agency found that language important, given the wide variation of practices across the range of scientific specialties.

Schachman and others asked that "omitting" data be dropped from the itemization of types of falsification. Equipment may fail, cultures may become contaminated—Rubenstein, from the chair, observed that "there are a hundred reasons why one might exclude data without intending to deceive." Nobody broached the matter of scientists' judgement of the validity of individual data where the question was their agreement with hypothesis, as in the Millikan case: to be sure, the issue is subtle.

The most extensive and reasoned comment was that by Tina Gunsalus. She asked for a number of improvements. The policy was intended to cover all federally funded research, yet the definition spoke only of "all fields of science, engineering and mathematics," leaving out, for example, research in education or the humanities. Also, she said, "By concentrating exclusively on the integrity of the research record, the new definition apparently excludes such acts as falsifying credentials in a research proposal submitted to a federal agency." She called also for a deadline by which all federal agencies should have put the definition and procedures in place. Further, many research projects, particularly in medicine, are carried out at multiple institutions—half a dozen cancer centers performing randomized trials of a new procedure, for example. Yet the guidelines do not say how misconduct charges in such projects are to be handled. She asked for a mechanism for designating a lead institution at the start of such collaborations. And a crucial omission was protection—detailed and effective protection—for whistle-blowers, as well as prevention of obstruction of investigations and of retaliation against those involved.

5

THE BALTIMORE AFFAIR

> *David's misconduct was— When an experiment is challenged no matter who it is challenged by, it's your responsibility to check. That is an ironclad rule of science that, when you publish something you are responsible for it. And one of the great strengths of American science, as opposed to Russian and German and Japanese science, is that even the most senior professor if challenged by the lowliest technician or graduate student, is required to treat them seriously and to consider their criticisms.*
>
> *It is one of the* most fundamental *aspects of science in America.*
>
> —*Howard Temin, 16 March 1993*

Of the scores of controversies over fraud in the sciences from the nineteen-eighties to the present day, one beyond the rest has dominated attention, and far beyond the scientific community. This was the Baltimore affair. It involved an obscure immunologist named Thereza Imanishi-Kari, an eminent virologist and molecular biologist, David Baltimore, and a postdoctoral fellow in Imanishi-Kari's laboratory, Margot O'Toole. It began in May of 1986, a few days after publication in the journal *Cell* of a report about genes that are involved in the immune systems of mice, with a title hermetic to all but the initiated: "Altered Repertoire of Endogenous Immunoglobulin Gene Expression in Transgenic Mice Containing a Rearranged

Mu Heavy Chain Gene." The authors were David Weaver, Moema H. Reis, Christopher Albanese, Frank Costantini, David Baltimore, and Thereza Imanishi-Kari. As the first author was David Weaver, the paper is cited as Weaver et al. Baltimore and Imanishi-Kari, listed last, were in effect joint senior authors. The affair centered on data reported in the paper by Imanishi-Kari—or rather, on charges raised by O'Toole, that May, that Imanishi-Kari did not have data she claimed and that in the paper she misrepresented what data she did have.

The paper was the latest in a series of collaborations between Baltimore's large science factory at the Whitehead Institute, an MIT affiliate he directed, and the much smaller laboratory Imanishi-Kari headed, with a fluctuating population of some half-dozen at any time. When the paper appeared, Imanishi-Kari had a foot in each of two canoes: she was negotiating a move to Tufts University Medical School, in Boston. O'Toole's fellowship ran out that June of 1986, and for years after that she could not get a job in science. As the controversy heated up, in the summer of 1990, Baltimore was recruited by David Rockefeller to be president of Rockefeller University, in New York—until the faculty there used his involvement in the case and the unceasing publicity it had generated to drive him from office.

The Baltimore affair achieved fragrant notoriety, and the stink has lingered. At the obvious, journalistic level, the story has vivid dramatic elements: poisonous personal enmities within a laboratory; a whistle-blower of remarkable strength of character; multiple attempts at cover-up and then just when it had seemed safely suppressed a devious, secret leak to a congressional investigating committee; explosive congressional hearings; a flawed federal investigation, a punitive decision, an appeal, a trial badly bungled, a dubious verdict—and over it all the grimly intransigent figure of David Baltimore. In our typology of fraud in recent science, the Baltimore affair epitomizes, and on a grand scale, the arrogance of power. At another level, though, and as no other single case has done, in its tortuous and protracted working out the Baltimore affair marched through every element of scientific fraud that we have been considering, save only plagiarism. At the end, it erupted into an enraged months-long public confrontation over the norms of science. Two levels, then, narrative and analysis: each illuminates the other.

In setting out the patterns of fraud in recent science, we have looked at the immediate contexts of some of the cases—Summerlin, Soman and Felig, Darsee, Spector, others—but summarily and from the middle distance. The Baltimore affair rewards closer scrutiny. At the most primitive, it was driven by the interactions among three personalities, strong, incompatible, and imbued with irreconcilably different attitudes towards the conduct of science. From the beginning, the norms of science were in play.

Here, then, is how the affair began. Mid-morning on Wednesday, 7 May 1986, in a laboratory on the ground floor of a building at the Massachusetts Institute of Technology, Margot O'Toole had reason to consult the breeding records of colonies of purebred mice maintained there to provide animals for certain sorts of experiments with their immune systems. Every element of that moment in O'Toole's life is fraught with significance—the timing; the laboratory and its scientific and social composition; the mouse colonies and breeding records; the experiments for which the mice were bred; O'Toole herself and the circumstances that had her tending the colony. She was a postdoctoral fellow, seven years out from her PhD. She had joined this laboratory on a one-year fellowship at the beginning of the previous

June. Her career until that June had hardly seemed brilliant—"I wasn't a blue-blood, pedigreed scientist," she said to me in a conversation in April of 1991. A year earlier, the new job had looked wonderfully attractive. The laboratory was small, and was directed by Thereza Imanishi-Kari, an assistant professor, who was vivacious, ambitious, brimming with scientific creativity. Imanishi-Kari was collaborating with the laboratory, much larger, headed by Baltimore—a scientist whose scientific blood ran not merely blue but deepest imperial purple. They were pursuing what promised to be an important discovery.

The two laboratories had complementary specialties, and between them had been working for several years with a purebred line of mice that had had a certain gene, a specified length of the genetic material deoxyribonucleic acid, DNA, transplanted into it from another line. Transplanting genes in mammals, in such a way that they will be inherited by the creatures' descendants, was a new and exciting technique in itself, invented only in 1980. Ten years earlier, indeed, many biologists had said such transplants were so difficult that they were a long time away. Now they spoke routinely of "transgenes" and "transgenic" creatures. The very terminology had a futuristic sheen. Imanishi-Kari, Baltimore, and their collaborators knew their transgene was present and active in this line of mice: it caused cells of the mouse immune system to produce a specific protein—an antibody molecule, otherwise called an immunoglobulin—that mice of the original line did not make. They could isolate this immunoglobulin from samples of mouse blood, and Imanishi-Kari was trained in methods for identifying particular immunoglobulins and distinguishing them from others that were closely similar. The two labs had published a couple of joint papers saying how and why they had created the mice and reporting that the transgene was expressed—present, active, its particular immunoglobulin being made.

Then Imanishi-Kari had turned up experimental evidence not merely that the transgene was expressed but, she said, that its presence somehow altered the action of certain other genes, ones that determined other immunoglobulins. And these were endogenous genes, meaning ones that mice of this lineage had already possessed before the transplant. This finding, a new gene inserted into the immune system affecting the output of an old one, was curious in itself. Its significance was possibly momentous. Some immunologists in those days, particularly in Europe, were passionately convinced that interaction among immune-system genes and with their protein products, in a vast and intricate network, must be the central means by which the system regulates itself. The network had originally been proposed in 1974 by Niels Jerne, Danish by ancestry, British by the accident of where he was born, and working in Switzerland, at the Basel Institute of Immunology, which he had built into a formidable research center. Jerne was the preëminent theorist of modern immunology. Many scientists had been on the lookout for evidence of such network regulation. None had yet found it. Imanishi-Kari said she'd got it.

That, in briefest outline, was what was going on that had made Imanishi-Kari's laboratory seem an exciting place when O'Toole joined it. "I mean, it was a big deal for me," O'Toole said. "Here I was, getting into a lab at MIT, a tip-top lab, very exciting finding, on a collaboration with the Baltimore lab—I mean, it was like I had fallen on my feet, it was great." She was hired even before the paper reporting the exciting finding had been written, to do the next experiments that the finding suggested. But she had had a difficult year.

Joining the laboratory in June of 1985, O'Toole soon realized it was an unhappy place. Imanishi-Kari proved temperamental, excitable, and secretive; towards the end of summer, she forbade O'Toole to discuss her work with others there or to share materials and methods.

None the less, at first her work had seemed to go well. That fall, with the help of Moema Reis, a longtime close junior colleague of Imanishi-Kari's, she had obtained one experimental result that apparently supported Imanishi-Kari's finding. Imanishi-Kari urged her to write up the initial success, give seminars on it, publish it. Indeed, Imanishi-Kari inserted a coy reference to O'Toole's unpublished result as the last sentence of Weaver et al. O'Toole found, though, that working alone she could not confirm that first success. Early in December of 1985, Imanishi-Kari, Baltimore, and their colleagues completed Weaver et al. and sent it off to *Cell.* She studied it doggedly. All that winter of 1985–1986, O'Toole persistently refused to publish the experiment she could not repeat, and carefully, inventively, grindingly tried to obtain the result demanded of her, the result that would have laid the groundwork for following out Imanishi-Kari's putative discovery. Imanishi-Kari railed at her—screamed at her—that she was incompetent. So, with the unforeseen consequence at the heart of this tale, in March she agreed to run out her fellowship by tending the mouse colony, relegating herself to a routine duty that had been done by Reis, who was leaving the laboratory. O'Toole's account, though clear and consistent, has been dismissed by some as the grumbling of a "discontented postdoc." Even years later, in June of 1991, the Harvard biochemist Bernard Davis characterized the controversy: "For a long time I thought it was just two women in the kitchen having a fight."

Weaver et al. appeared in the 25 April 1986 issue of *Cell.* Twelve days later, O'Toole needed for the first time to verify the pedigree of a particular mouse. Reis had kept the studbook in a pair of blue loose-leaf binders labelled "transgenic mice," stored on a convenient shelf. When O'Toole went to look for them, they were not there, but she found them in the next room. "So I pulled one of them down, and I said, 'How on earth am I going to find this mouse?' So I opened the book." When she spoke of her actions, she had a Xerox copy of the notebook on a table before her, and now opened this. "Here are the mouse numbers; I knew that these are mouse numbers, mouse typings." She flipped through slowly, pages rattling. "More typing of mice, mm-hm, mm-hm, I'm trying to find the mouse that I'm looking for—and I'm trying to get a feel for how she organized her book. OK. More mouse numbers, more mouse numbers, mouse numbers— 'Where am I going to find the mouse I'm looking for?' I was just turning the pages. Look at this: they're just mouse numbers."

O'Toole turned yet another page, and stopped, and her voice dropped. The entries here were no longer mouse numbers. "This page— Here is a result that I had got all year long, that I was told I was not to get, that I was told I ought not to be getting."

O'Toole turned from the book. "Everybody thinks that I was suspicious and started going through notebooks," she said. "What happened was that I was breeding mice and I had to look up a pedigree—and right here in the middle of these pedigrees and all this mouse testing is this page." These were records of experiments. O'Toole's first glance at the page told her that even before she had joined the laboratory, others—Reis, perhaps Imanishi-Kari—had got experimental results that prefigured her own failures and could not be reconciled with the paper submitted in December and just published.

"*Whoa!* And then *whoa* a million times." For the next thing she saw was that certain of the data, the experimental results, from this page and succeeding pages—seventeen pages—had been incorporated but in different form, a misleading form, as an essential part of the published paper. "I knew the paper by heart."

That morning in May of 1986, as O'Toole stared at those pages of scientific records, her world, or that part of it that had come to preoccupy her all but completely, all that she had done and witnessed and endured in the laboratory in the preceding eleven months, began to rearrange itself in her mind into a shape at once radically different and strangely the same.

David Baltimore's career is mostly public; a sketch adds some piquant details. By the time his collaboration with Thereza Imanishi-Kari was under way his curriculum vitae listed some three hundred publications. One, of course, was his paper in *Nature* in June of 1970 showing the existence of the enzyme we now call reverse transcriptase. (It was followed in that issue by the article by Howard Temin and Satoshi Mizutani, announcing the same finding.) Baltimore had graduated from Swarthmore College, near Philadelphia. He began doctoral work at MIT in biophysics, but after a year moved to Rockefeller University and molecular biology. He stood out for various reasons, some trivial: he wore a flowing cloak, he shocked classmates by smoking in class and addressing senior scientists by their given names. Others called him Lord Baltimore. He worked on polio virus and another, closely related. More seriously, some of the faculty found the data in his proffered dissertation of borderline quality at best, thin. Yet he was productive, and an author on a dozen papers even before he received his PhD. He served a postdoctoral fellowship in the laboratory of Renato Dulbecco, at the Salk Institute, in La Jolla, California; Dulbecco had invented ways to grow animal cells loose in culture, like bacteria, an essential technique for studying animal-cell viruses. In 1968, Baltimore returned to MIT as an associate professor. The Nobel prize in physiology or medicine, with Temin and Dulbecco, came in 1975. At MIT, he raised the millions for the Whitehead Institute, and in 1982 became its first director. The institute is separate both administratively and physically from MIT's cancer center, where Imanishi-Kari had her laboratory.

The social and scientific world in which the affair first developed was distinctly inbred. In 1973, Margot O'Toole graduated from Brandeis University, which lies west of Cambridge, with honors in biology. She entered Tufts University and went to the laboratory of Henry Wortis, at Tufts' school of medicine, in Boston. She got her PhD at Tufts in the summer of 1979, and stayed on in Wortis's laboratory to the end of the year. Then she and her husband, Peter Brodeur, moved to the Institute for Cancer Research, in Fox Chase, a suburb north of Philadelphia, where she had a postdoctoral fellowship. Just before Christmas of 1984, they moved back to Boston, where Peter started an assistant professorship at Tufts, with a laboratory down the corridor from Wortis's. By O'Toole's own admission her career until then had seemed plodding; but she was intelligent, more than competent, and implacably conscientious. The chief of one laboratory where she had worked at Fox Chase said of her, "Margot was fun to be around: volatile, assertive, opinionated. A warm person, very outgoing." Of her work, "She certainly was a stickler for controls," and, "She showed her good training: Henry Wortis did a good job." She was reputed also to be stubborn, even inflexible. O'Toole brought with her from Fox Chase the beginnings of an independent project, but obtained no funding and so no lab space. Late in March of 1985, at a party at Wortis's house, she met Thereza Imanishi-Kari for the first time. Imanishi-Kari in three years at MIT had gotten to know Wortis well. At the party, she was vivacious and brimming with enthusiasm about her important discovery. A postdoc was leaving the lab in June, and would Margot like to fill the slot? Agreed. She was hired, even before Weaver et al., reporting the exciting finding, was drafted, to do the next experiments that the finding suggested. On 1 June 1985, O'Toole reported for work.

Imanishi-Kari's background demands scrutiny. It is unusual and casts light on her approach to doing science. Beginning about 1908, some hundreds of thousands of Japanese emigrated to South America. A descendant of that outpouring, for example, is the controversial ex-president of Peru, Alberto Fujimori. The city and state of São Paulo, Brazil, received an especially large number; today, indeed, Liberdade, the Japantown of São Paulo, at nearly a million makes up the largest such community outside Japan itself. Thereza Imanishi was born in December of 1943 in the small town of Indaiatuba, eighty-three kilometres (fifty-one miles) from São Paulo, a daughter of a farming family that had crossed the Pacific a couple of generations earlier. For her the stereotypes suggested even by her name held true: she had the intelligence and determination so often attributed to the Japanese and the volatile temperament supposedly characteristic of Latins.

Biographical information she supplied to the National Institutes of Health in applications for research grants—these are documents carrying criminal penalties for false information—states that she has a BS in biological sciences from the University of São Paulo. The director of the university's Division of Academic Registration, citing volume, page, and registration number of the records, confirms that "em 19 de novembro de 1968, o diploma de Bacharel em Ciências Biológicas" was awarded by the Faculty of Philosophy, Sciences, and Letters "em nome de THEREZA IMANISHI. . . ."

She has claimed also to possess the degree of Master of Science, in developmental biology, from the University of Kyoto, 1970. She does not. In May of 1991, *The Boston Globe* reported that their Tokyo bureau had checked with the university and found that no such degree had been awarded her. At this, Imanishi-Kari's Boston lawyer at the time, Bruce Singal, produced a letter signed by two members of the Laboratory of Developmental Biology at the University of Kyoto, Mikita Kato, chairman of the Zoological Institute and professor of radiology, and Atsuyoshi Hagiwara, an assistant professor. This letter was written not in 1991 but two decades earlier, on 26 August 1970. It listed some work that Miss Thereza Imanishi had done, vaguely described—for example, "isolation and purification of gamma-globulin and its subunits"—but generally related to immunology. "The results of these works are now in manuscript for future publication," Kato and Hagiwara wrote. Her list of publications mentions none of this. Their letter concluded, "From these evidences it may be evaluated that her two year activities are equivalent in quality to complete two year Master course of our graduate school." However, in 1995 an inquiry by the Office of Research Integrity produced a letter, 18 May 1995, from Yoshio Miyabe, head officer of the Faculty of Science of the University of Kyoto. He wrote that the university had not awarded Thereza Imanishi a degree of Master of Science. Indeed, "We could not find any official evidence that she was enrolled or employed in any position or grade in Kyoto University." The obvious line of further inquiry was closed, as Kato had long retired and Hagiwara had died in 1976. Although, Miyabe wrote, Imanishi may have been privately engaged in some research activity under Hagiwara's supervision, this was "with no official position. Therefore, we can not issue an official certificate of her existence at our faculty." Miyabe concluded, "Therefore, the letter, which mentioned she was a Research Student, should be considered only a private level certification without administrative will."

In the fall of 1970, Imanishi enrolled for a PhD at the University of Helsinki. The university had some old-fashioned rules. Foreign students who had no more than a bachelor's degree were considered still to be undergraduates. To be eligible for doctoral study, the student must have already possessed a master's degree. Imanishi's private-level certification—the significance of its August 1970 date becomes clear—apparently did the trick. She joined the laboratory of Valto Eero Olavi (Olli) Mäkelä, an immunologist of quiet reputation and stern rectitude, a rather specialized figure in the brilliant, far-flung international community that established modern immunology in the nineteen-fifties and sixties. She and Mäkelä worked on a focussed problem, the genetics of certain antibodies produced by a strain of mice when they were repeatedly immunized against a certain simple chemical; their methods were those of classical serology, meaning that they tested samples of the creatures' blood serum, which would be rich in antibodies of all sorts, to determine their activity against solutions of the chemical.

The University of Helsinki had another unusual rule, that four published papers, submitted as a set, sufficed as a doctoral dissertation. (The practice is rare but not unknown in other European countries.) Yet a new-minted PhD scientist must be someone proven ready to work independently: therefore, the papers comprising the dissertation had to include at least one of which the candidate was sole author. Mäkelä explained these oddities in a conversation in the lobby of his hotel in Budapest during the triennial World Congress of Immunology that was meeting there in the summer of 1992. What they had done, which he implied was not unusual, was to arrange to publish one of the papers they worked on together under Imanishi's name alone. Mäkelä was reluctant to talk about Imanishi much further, though he said that once she got the degree she left one of their joint papers not quite finished. Her insouciant lapse had evidently angered him. He thought it "unfair." If the paper had been important he would have pressed harder to get it done. She received her doctorate in 1974. In Finland that August she married an architect named Markka Tapani Kari. (They later divorced.)

In 1975, after a further year in Mäkelä's lab, Imanishi-Kari took a big step closer to membership in the community, moving to the Institute of Genetics at the University of Cologne. She was invited by Klaus Rajewsky, another of modern immunology's founders. Cologne's Institute of Genetics is unusual in Germany in being part of the university, American-style. It was set up that way after the war by Max Delbrück, when he responded to a call to come back on an extended visit to help revitalize German biology. In a conversation with Rajewsky at his laboratory in Cologne on 2 April 1991, and another in New York over coffee at the Waldorf-Astoria on June 18, he said something of the five years Imanishi-Kari spent in Cologne. He is a tall man, soft-spoken with an attractive German accent, patient and shrewdly circumspect. He had met Imanishi-Kari in Finland when she was still in Mäkelä's lab, and he knew of the work the two had done. "It's an interesting history. And that's also how Thereza came into the whole game," he said. "They discovered one of these genetically determined immune responses." He went on, "It was like a mendelian gene. They could cross the mice." The system suggested experiments. "And then I asked Thereza, because Thereza had finished her postdoc in Olli's lab, and I said, 'Wouldn't you like to come to Köln? Because I would like to work on that system.' I wanted to do that experiment."

In the event, they didn't do that experiment. "She was a very enthusiastic, lively, charming person. Had done very good work," Rajewsky said. "She continued to work on the characterization of that response," the genetics of that heritable response. "Serological classification, basically." Imanishi-Kari spent five years in Rajewsky's group at Cologne, the first year in a postdoctoral fellowship funded by the North Atlantic Treaty Organization. Before two years were up, Rajewsky said, "Thereza became an independent group leader in my institute. So she was not under my command anymore, direction anymore. And she took that very seriously. She actually separated herself very much from me. And there were awful tensions." Her group numbered four, including an American postdoc and a German technician. They had bench space in a room crowded with others. "There is sort of a syndrome which many young scientists have and some have it more and some have it less, and Thereza had it to an extreme extent, I must say. And that was that she thought somehow everybody is going, is likely to steal her property away from her," Rajewsky said. "That was a very severe problem with her, actually," he said. "And so that even when she went to America afterwards, I had very little contact with her, because of that."

Mäkelä's dissatisfaction with what he did not explicitly describe as her opportunism, Rajewsky's tactful vagueness about the tensions, about Imanishi-Kari's failure to do the work for which he had recruited her: through the reticence we begin to see the pattern of a temperament, a way of working with colleagues. This provides independent support for descriptions of her behavior later.

The day before my first conversation with Rajewsky, in Cologne, I met Claudia Berek, a German immunologist, in her office at the Charité, the once-celebrated teaching hospital founded by Frederick the Great, in the eastern part of Berlin. Much in decline before the reunification of Germany and the city, the Charité was being reconstructed both physically and intellectually. She had known Imanishi-Kari in Cologne, but said that her husband, Robert Jack, had known her better, professionally—and he was at Rajewsky's Institute of Genetics. Accordingly, after Rajewsky I talked with Jack. He turned out to be a small dark Scot with a pronounced accent. "Thereza then was young. Pretty. Vivacious. Full of energy, full of life. And just a joy to be around. Absolutely. You can't imagine how much zest for life she had at that time. And, ah—very ambitious, very hard working." As for her quality as a scientist, "Good, solid, like most ninety, ninety-five per cent of scientists. She wasn't special." Early in her time in

Cologne, they worked together for a year or so and wrote a paper. That went well, until the end. "We fell out at the end," Jack said. "Ah, yeah. I guess it happens all the time. We, ah—over the question of whose name goes first." On the one paper published together? "Yeah. I don't think it's worth rehearsing the argument, because by this time it doesn't matter." He added, "It was a very simple, straightforward, workmanlike piece of serology. I'm afraid that it is not a paper that set anything alight! Looking back at it, it was simply that I thought I had done most of the work and suddenly she comes in and wants to be first author. I could not understand it. It seemed to me beyond the bounds—"

Rajewsky offered a shrewd observation. "She somehow became caught in the serology," he said. "And when she went to MIT, for a while she continued this kind of work. And that was not very, ah—rewarding. By 'caught' I mean, well, you know, what shall I say? A bit like the Talmud. I mean, some scientists have this problem, that they discover complexities, and then they build complexities on complexities, and somehow the work becomes more complex all the time, and it is difficult to make a cut and say, you know, That's enough!" Classical serology was rapidly becoming outmoded, anyway. In 1975, César Milstein and Georges Köhler had invented a way to make antibodies of a single specificity and in bulk; called monoclonal antibodies, these have wide application in molecular biology, and in particular made immunological work far more precise and controllable.

In 1981, Imanishi-Kari went to MIT as an assistant professor, with a laboratory, two rooms, in the cancer center. Director of the center was Herman Eisen, another of the founding generation of modern immunology but by this time past his scientific prime. She brought with her a narrow repertoire of serological techniques and a variety of experimental materials. "I was thinking about how she would do, in this environment," Rajewsky said, meaning MIT. "As I told you before, she was somehow not very well organized. For example, it was very difficult for her to write something. Because she had these piles of data everywhere, that she liked to work on, rather than to sit and put things together and then write a manuscript. And that is actually a big disadvantage, particularly in a situation like MIT. Because, you know, when you come there you have to publish. If you don't publish something within the first few years, something important, you have a hard time."

Rajewsky's concern was warranted. Despite any tardiness in writing up results, Imanishi-Kari had been moderately productive up to the time of this move, with her name on some twenty-two papers (including several that saw print after she got to MIT). In the next three-and-a-half years, though, to the end of January of 1985, according to her curriculum vitae she published just four—though she claimed another seven, not in press but "in preparation."

São Paulo, Kyoto, Helsinki, then Cologne: Thereza Imanishi-Kari's trajectory began on the periphery and had been no straight line but a spiral, though now reaching to the center. One must conjecture that she had not been effectively inducted into one of the great lineages of science. Nor had she followed the example or bent to the rule of a strong mentor.

To set up her laboratory at MIT, Imanishi-Kari brought with her a German technician, Gertrud (Trudy) Giels, who had been working in another research group. She also recruited an American scientist at the institute in Cologne. Almost on arrival, she was assigned her first graduate student, Charles Maplethorpe.

At that time in that place, Imanishi-Kari was at the collision point of two research programs. Immunology, her serology, was encountering the new generation of molecular biology, the thrust into development and differentiation of higher organisms. Molecular biologists working on development had needed to find appropriate experimental creatures. Fruit flies, whose genetics had been studied almost since the rediscovery of mendelian genetics at the turn of the century, provided one, yeasts another, while still another is the nematode *Caenorabditis elegans,* a tiny worm. For mammals, mice proved close to ideal, once inbred strains of genetically identical mice could be created, adapted, and maintained. (The Jackson Laboratory, in Bar Harbor, Maine, maintains active breeding populations of some twelve hundred different strains of such purebred mice, ready for purchase, and keeps another eight hundred strains as embryos frozen in containers in liquid nitrogen.) And they needed genetical systems to work with. Here, the mammalian immune system offers a complex process of development—many genes and gene families that can be manipulated, while in the laboratory setting most mutations will not be lethal.

Even as molecular biologists were bringing their methods to immunology, originally in order to study questions of development, immunologists began to realize that they needed to learn molecular methods in order to study their own questions, all the interactions among the many types of components of the immune system. From the beginning a tension has persisted between molecular biologists and immunologists. Though monoclonals where applicable have added precision, classical molecular biologists are likely even today to look on immunology as imprecise, murky, tricky, and dubious. But Baltimore had become seriously interested in immunology in the early nineteen-eighties. Of twenty-one papers published in 1981 of which Baltimore was a co-author, ten were immunological. Nineteen-eighty-two saw seven more. Genes, genes, genes: again and again, the questions these papers addressed were genetical and molecular, using the immune system for analysis of development.

As we turn to the science we run up against sharply opposing statements of the fundamental issues in the Baltimore affair. Although the methods reported in the *Cell* paper were intricate, its scientific conclusions could be stated simply enough, as could the reasons for supposing that certain conclusions were not supported by any original data. Yet one view maintains that one does not really need to get into the detail of the paper, because that's not what's most important. What matters is the behavior of those involved. For those who have seen the controversy in this way it is about scientific ethics. They defend what they say is the classic obligation and desire of the scientist to be his own first and most watchful critic.

Baltimore has asserted an alternative view. In a conversation on 31 July 1991, at his office at Rockefeller University, he said emphatically that beside the scientific claims of the paper nothing else matters. "And since basically everything in the paper now has been supported by other data, then there's absolutely no reason to doubt it." A moment later he added, "And so if Thereza, in fact, fabricated some data and I can't say whether she did or she didn't, it did not seem to have produced a paper which is misleading." Questions about the data or about the responses at MIT and Tufts to the problems O'Toole had raised were beside the point: the paper was fine because later work at other laboratories, not all of it published yet, but coming soon, would confirm it, piecemeal. He came back to the assertion repeatedly in the next months.

Weaver et al. made two tightly related claims. They are the crux of the matter. Narrowly, the paper said that the mice with the transgene, compared to the original strain without it, produced a different repertoire, or set, of antibodies. Secondly, the data as published in the paper purported to show that cells in the transgenic mice produced other antibodies similar in specificity—the technical term is "idiotypy"—to the transgenic antibody; that they did so copiously, all the time, without the presence of antigen; and, crucially, that these antibodies were not transgenic but endogenous, an idiotypic mimicry.

The paper leapt to a third claim, a radical extension of the second—that these effects of the transgene were evidence for the Jerne network. Thus, the abstract, which began the paper, introduced the narrow claim, yet culminated in a sentence urging that in these mice the turning-on of new endogenous antibodies suggested that a transgene "can activate powerful regulatory influences." The introductory discussion opened at once with the possibility of regulation by "internal network interactions," citing Jerne 1974. Without this radical claim, despite Baltimore, Weaver et al. was insignificant.

From the time of her first experiments indicating an odd effect of the transgene, Imanishi-Kari began talking enthusiastically of her discovery of possible support of network theory. She gave seminars at MIT and elsewhere about the finding. At the end of January of 1985, five MIT laboratories, led by the director of the cancer center, Herman Eisen, submitted a massive joint grant application to the National Cancer Institute. Into her portion, Imanishi-Kari incorporated her interesting discovery, proposing to pursue it further. Baltimore, too, was excited by the discovery of support for Jerne. In later years, he repeatedly denied that he gave credence to the idea or that it was central to the paper. "We just put those sentences in to please Thereza," he once told me. Again, "Network is not at all— It's not a conclusion of the paper. It's a suggestion in the discussion, that that might be an explanation. That's wrong. I thought that was wrong in the first place. As I think I told you. I never believed in the Jerne networks; I didn't think that either our data or anybody else's data necessitated thinking about networks."

Yet several biologists at MIT remember various occasions back then when Baltimore talked up this first experimental evidence of the network. Nancy Hopkins, professor of molecular and cellular biology at MIT, attended a big formal dinner where she was seated at a table with Susumu Tonegawa, a Nobel-prize immunologist. Baltimore was working the crowd. When he came to that table, he told them of the important new finding, first support for the Jerne network. Tonegawa did not remember that dinner. "But what I remember is probably something earlier than that," he said in August of 1992. "I confirm what she says with respect of how David was. It's just different occasions." Tonegawa was working with Baltimore and another colleague, Philip Sharp, one day when a seminar on some current topic was scheduled for a main auditorium at MIT. "The three of us go to hear the seminar," Tonegawa said. Baltimore told them of Imanishi-Kari's discovery. "You know Thereza was right in this building. So I knew about that, but as far as I remember, that was the first time Baltimore told me and Phil that there's this new evidence for the network. In the Jerne network sense." He added, "And when he said that, I was skeptical." Why? "I just didn't think— I felt it's too extraordinary to accept."

In the summer of 1985, that big collaborative grant application from MIT was reviewed by the National Cancer Institute. On June 4, a team of eight scientists representing the cancer institute visited MIT, attended presentations by each of the five laboratories, and talked to individuals. On July 30 and 31, the cancer institute's program project review committee met. The site visitors and the committee were enthusiastic about the project. They recommended full funding—except for the work that was to have been done by Imanishi-Kari and her laboratory, which they said should get no money at all. The review-committee report noted that she proposed next to test a network hypothesis:

> Unfortunately, it was clear from both the written and oral presentation that her research program lacks focus: she did not present a well-designed experimental approach and it was not apparent that key experiments will be performed in a timely fashion. In addition it is not clear that the experiments proposed would adequately test the hypothesis. . . . Thus, without additional experiments one would not be able to ascertain whether there is any perturbation of the . . . network or not. The second specific problem is one of cause and effect. If indeed the 17.2.25 VH gene [the transgene] does induce a cascade of cellular interaction then it would seem pertinent to examine the early events in both endogenous and exogenous gene expression, not the later events. . . . Instead she outlines a line of investigation that is derivative of the elegant work of Rajewsky. It is difficult to see how the applicant will be able to improve upon his findings.

Reasoned, blunt, and she was knocked out of the project. None the less, Weaver et al., reporting the original anomalous result and floating the network claim, was submitted to *Cell* on 13 December 1985. The journal's editor sent it to three scientists for anonymous review.

Cell's editor was Benjamin Lewin. He has a doctorate in genetics from Cambridge and is the author of a well-regarded advanced textbook of molecular genetics, but had long been a journalist and editor. Lewin was also owner and publisher of *Cell*—an exceptional arrangement, perhaps unique—until he sold his company for a reputed 100 million dollars in 2001. *Cell* comes out fortnightly, and is known as a hot-news journal, one that strives to secure papers with exciting findings and publish them quickly. This in itself is not unusual. Lewin took the policy to an extreme. He was known to solicit manuscripts from top scientists. Many refused. Paul Berg, whose Nobel prize was for founding discoveries in the techniques of recombinant DNA, or genetic engineering, spoke of Lewin's appeals with amused contempt. And many were offended by the blatant conflict of interest. Lewin pressed authors to keep papers short and snappy, not to cloud their central finding with data not directly relevant. He has admitted to overriding referees sometimes, publishing papers they recommended be turned down; in any case, he like any journal editor would usually know the predilections of the referees picked to send manuscripts to. Many younger scientists in fields the journal covers believe that to publish in *Cell* is vital to their careers. James Darnell, a senior scientist at Rockefeller University, said to me ten years ago, "A lot of young biologists think their careers will be ruined if they don't have a paper in *Cell*." As for himself, Darnell said he would never submit a paper to Lewin, calling him "that dreadful little martinet."

The third of the referees of Weaver et al. was Rajewsky. The paper was interesting, he wrote in his review, but "at the same time not more than a first step." He raised two problems, and both went to the heart of the paper, his mild phrases hardly concealing their potentially devastating relevance. He questioned the methods: "One aspect of the work (and in fact a crucial one) is in my opinion insufficiently documented and discussed," namely, the techniques to identify and distinguish the antibodies derived from the endogenous gene. He questioned the interpretation: a simple slowing down in the rate of development of the immune systems of the immature transgenic mice, "as in fact suggested by the data," could explain all the results "without network selection."

Baltimore shrugged off the comments. They modified the final wording to take briefest note of the suggested alternative, but Imanishi-Kari carried out no new experiments. The last page of the paper stated a decided preference for a network explanation of its data. *Cell* published it in the issue of 25 April 1986.

A senior immunologist who is genuinely disinterested in the Baltimore affair is hard to find. One such was Alfred Nisonoff, at Brandeis University. (He died in 2001.) He was a man of old-fashioned principle and classical training. "In those days, when you came to a new lab, the first thing you did was calibrate the weights." Nisonoff had read the paper and detested the developing scandal from the middle distance, uttering not a word in public even when appealed to directly. His work made him thoroughly familiar with Imanishi-Kari's field. In a conversation in June of 1992, he said that he had never had regular contact with Baltimore and knew Imanishi-Kari only by sight at one or two conferences. When he had first read Weaver et al., he had decided at once that her methods were the wrong ones to demonstrate the anomalous genetic expression. But what about the larger claim that they had found support for the Jerne network theory—particularly as Baltimore was now denying that that was important? "Nonsense. That was obviously the whole point of the paper."

Imanishi-Kari's behavior was chronically volatile. "The first year or so in Cologne went very well," said one who worked in her laboratory. "Then I guess I started to get hints that the working relationship was beginning to fall apart." That was echoed and re-echoed. Imanishi-Kari was alternately sweetly charming and a screaming, suspicious termagant. Sooner or later she drove co-workers away. About Imanishi-Kari's approach to science the observations were also unanimous. She was inventive, ingenious, brimming with interesting ideas and approaches, but in the execution slapdash. Plans shifted almost from day to day. Her bench in Cologne, her desk at MIT, were jumbled. She would neglect to run controls. She took notes on scraps of paper. Philip Cohen, an MIT undergraduate, worked in her lab the summer O'Toole arrived: "You know, she'd, like, calculate things off the top of her head and write them down, and I was supposed to use that number. And I remember calculating it myself, and seeing she was off, by an order of magnitude, or something like that." Again, one who worked closely with her for two years: "Someone else in the group would start to do some experiments, she didn't necessarily trust the result; sometimes she would think that they were trying to do *her* experiments—and I remember in one case she didn't trust the result and rather than talk to the person she secretly went about repeating the experiment, which didn't go over too well. She was not a collaborator. She wanted everything for herself." But more: "One characteristic, and I expect you would have heard this from Bob Jack, but Thereza had quite a reputation of doing experiments but never pulling things together." Although O'Toole came to realize this only slowly, her experience was characteristic.

We have independent testimony, difficult to impeach. The committee of the National Cancer Institute that scrutinized the big MIT collaborative research proposal from Herman Eisen and others, in their report at the end of July 1985, had more to say about Imanishi-Kari's part.

> [T]here was little evidence presented at the site visit that significant interaction occurred between Dr. Imanishi-Kari and the rest of the group. . . . Her work is not integrated with other members of the program. Moreover, she does not appear to collaborate significantly with other members in the Cancer Center group. This is an unfortunate circumstance, as it is clear that she could bene-

> fit technically and intellectually from the expertise of the group as a whole.

For such a report, extraordinary language. What the site visit in June had turned up in a day was well known to Eisen, Baltimore, and the MIT administration. Imanishi-Kari was not getting tenure at MIT. Wortis at Tufts arranged that she move there, representing her as a windfall Tufts should seize without the lengthy job search normal for an academic appointment.

The seventeen pages O'Toole discovered in the mouse studbook showed that weeks before she joined the laboratory the previous summer, Imanishi-Kari with Reis's help had got raw data that she presented in the paper just published—but that were not correctly stated there. O'Toole's first glance told her this, she has testified repeatedly, because the data in this raw form actually agreed with the results she had struggled with all winter. The data also suggested that the paper had serious unacknowledged defects. The paper stated that cells producing antibody had been subjected to two crucial experimental procedures that the seventeen pages show, O'Toole says, had not in fact been used. The paper further stated that these cells, with few exceptions, were not expressing transgenic antibodies. These data were represented in a table, Table 2, as fractions of the total number of cells, but the raw numbers in the seventeen pages showed that most were indeed expressing the transgene, but had been ruled out by setting a cut-off point that was conveniently high. Other data in the seventeen pages brought other parts of the paper, notably its Table 3, into question. O'Toole's fortuitous discovery set off the process that wound through the next ten years.

She took her concerns to people she had every reason to expect would determine the facts and act on them correctly. Note that at no time during this initial phase did O'Toole charge fraud, even when directly invited to do so. She asked simply and repeatedly that Weaver et al. be corrected or retracted—that other original data be produced or the paper corrected or withdrawn. Her account of events was circumstantial and unwavering. The response over those years was of three kinds: that all this was mere quibbling about interpretation; that the errors, if any, were insignificant; that data newly produced by Imanishi-Kari on several occasions proved that there were no errors.

The critical stage in dealing with any charge of misconduct is the response at the institution in the first hours and days. The recent cases we have examined and scores of others demonstrate that in most places where science is done response to a fresh accusation is disastrously inadequate. In this respect, senior scientists are no different from corporate executives or government administrators: typically, the first impulse is to minimize the problem, blame it on personality clashes, cover it up, pass it elsewhere. Almost inevitably, senior people close to the problem are caught in conflicts of interest—at the very least, between responsibility to individuals and the need to shield the institution. Some institutions, burned once, do better the next time; others notoriously seem never to learn. In the Baltimore affair, certainly, everyone at Tufts and MIT who got involved behaved unwisely, at best. What follows is only the briefest sketch.

Two days after finding the seventeen pages, O'Toole brought her distress to a scientist she knew at Tufts, Brigitte Huber. She was told she must take the problem to Wortis, who had directed her PhD and was bringing Imanishi-Kari to Tufts. Wortis, Huber, and a third scientist, Robert Woodland, held a meeting with Imanishi-Kari. O'Toole was not present; she was then told that nothing would be done to correct the paper. When she brought the matter to the chairman of Wortis's department, Martin Flax, he told her this was MIT's problem. Huber and Wortis then met with Imanishi-Kari and O'Toole. Imanishi-Kari produced two pages of data. O'Toole says that when she questioned this, Wortis told her, "You will deal with what's presented." In a conversation with me, Wortis refused to go into details of this meeting; elsewhere he has said he does not recall this exchange. (A year later, in May of 1987, Wortis signed a memorandum saying the dispute was resolved; the memo was backdated to the day after that meeting.)

Meanwhile, on May 29 O'Toole met with the dean of science at MIT, Gene Brown. He told her she must either make a formal charge of fraud or drop the matter entirely. (He has since denied this.) Then he phoned Eisen and told him to get in touch with her. On May 30, she drove to Woods Hole, where Eisen was staying, and showed him the data and discrepancies. She has said that when Eisen saw the seventeen pages, his first response was, "That's fraud!" (He has since said he does not remember this, and that she was not coherent.) He asked her to detail the issues in a memorandum. (This was the last day of her year's fellowship; she was no longer a working scientist.) She did so, in a five-page memorandum dated and signed 6 June 1986; on June 9, she sent a copy to Brown. In mid-June, Eisen arranged that he, Baltimore, Weaver, Imanishi-Kari, and O'Toole meet. She was not allowed to bring anyone with her. Imanishi-Kari produced no relevant data. Baltimore, O'Toole has testified, did not ask to see any, but was satisfied that nothing had to be done. After the meeting, Eisen wrote a memorandum to Gene Brown and others, dating it June 17 but filing it away without transmitting it; on December 30 he did send a two-page memorandum to Maury Fox, head of the biology department.

The events of those eight weeks have regularly been referred to as "two investigations." The people at Tufts—Wortis, Huber, Woodland—have been called the Wortis Committee. Such designations are misleading. Wortis's group was self-appointed, had no institutional standing, and met but twice. Flax, Wortis's chairman, has said he did not even know of it until afterwards. Imanishi-Kari's move to Tufts was days away, and Wortis had sponsored her. He is a canny man. We cannot presume to know his motives. But correction or retraction of her one big discovery would obviously have brought his judgement into question and might even have upset the appointment. Eisen acted as a committee of one. Imanishi-Kari's exit had so far been smooth; he is widely known to admire Baltimore. (To his credit, for a while that summer he tried to broker publication of a correction to the paper—to no avail.) Today when scientific misconduct is alleged, the institution's intramural examination of the case is termed an "inquiry," and is an official proceeding, while "investigation" is reserved for the next and most serious stage, where the Office of Research Integrity itself assumes oversight of the case. But even in everyday loose parlance, what was done at Tufts and MIT does not resemble investigation.

O'Toole thought she had dropped the matter entirely. It took on a life of its own. In that pregnant spring of 1986, Philip Boffey, a reporter at *The New York Times,* had written up Walter Stewart and Ned Feder's paper on the Darsee collaborators and their difficulty in getting it published. This of course got the attention of many scientists. One who saw it was Charles Maplethorpe. He had received his PhD the previous September, from MIT, where he had been a graduate student in Imanishi-Kari's laboratory. He had overlapped the three summer months of 1985 there with Margot O'Toole. She had found him intelligent and helpful, but knew that he had quarrelled bitterly with Imanishi-Kari and despised her, and she him. Two friends of his independently called Maplethorpe's attention to Boffey's story. He read it, and filed away the fact that Stewart and Feder were interested in pursuing scientific fraud. During the winter of 1985–1986, Maplethorpe stopped by the laboratory occasionally, and then learned of O'Toole's difficulties. Early in the summer of 1986, he got in touch with Stewart and Feder and told them about the case. Stewart and Feder knew that Dingell and his staff were planning to go after scientific fraud but were waiting to find a robust case and witnesses.

Until the Republican takeover of 1994, Dingell ran the Committee on Energy and Commerce, building it into one of the most powerful and wide-ranging in the House, and also chaired its Subcommittee on Oversight and Investigations. He operated on the principle that where Congress supplies the money it must make sure the money is spent honestly and prudently. He had made a fearsome reputation investigating fraud, waste, and gross mismanagement in government, by defense contractors, in the nation's blood supply, in the securities industry. Dingell's were the hearings on the Environmental Protection Agency early in the Reagan presidency that forced the resignation of the administrator, Anne Gorsuch Burford, and led to conviction for perjury of her assistant, Rita Lavelle; his were the hearings that brought a former Reagan aide, Michael Deaver, a perjury conviction, too.

On 12 April 1988 Dingell held a first hearing on "fraud in NIH grant programs." O'Toole testified. So did Maplethorpe, Stewart and Feder, and others.

Maplethorpe has claimed, in interviews and in testimony before Dingell's subcommittee, that on an evening in June of 1985 he was at work in the larger room of the laboratory; Imanishi-Kari was nearby and David Weaver and a friend came in. At the subcommittee hearing of 12 April 1988, he responded to questions from congressman Ron Wyden, of Oregon, saying in part:

> MR MAPLETHORPE. . . . I was working in the lab at that time trying to finish for my thesis defense in August. But nevertheless, I was paying attention to the conversation and, at that time, I

heard Dr. Imanishi-Kari tell Dr. Weaver that she had some problems with this reagent called Bet-1, which is the reagent that Dr. O'Toole had problems with later.

And what I heard Dr. Imanishi-Kari tell Dr. Weaver was that she was obtaining the same results that Dr. O'Toole subsequently obtained.

MR WYDEN. And how were these issues resolved prior to publication by Dr. Baltimore, Dr. Imanishi-Kari, and the others?

MR MAPLETHORPE. Well, the issues that I heard them discuss were not resolved.

Weaver and Imanishi-Kari have denied any memory of such a conversation; she protests that Maplethorpe is not a disinterested witness; but he made a record of the conversation at the time. The subcommittee heard nobody from Tufts or MIT that day; many found this to be evidence of bias.

In July of 1988, in response to a congressional subpoena, Imanishi-Kari delivered to the National Institutes of Health a bulky loose-leaf notebook, saying that it contained original data, which she had organized into this form. (This came to be called the I-1 notebook.) O'Toole and others determined that some pages of this notebook are palpably faked. At this point for the first time she charged fraud.

Meantime, early in 1988 James Wyngaarden, then director of the National Institutes of Health, had appointed a panel of three scientists to look into the controversy over Weaver et al., chaired by Joseph Davie, an immunologist, who was chief of research at the pharmaceutical manufacturer GD Searle. The Davie report was released by the NIH at the end of January of 1989. It found serious flaws in Weaver et al., but stigmatized nothing as misconduct. O'Toole objected. She said that certain data relied upon by the Davie panel had not existed when she first challenged the paper; she asked that Imanishi-Kari produce the data to back one of the tables. In preëmptive moves at the end of April, Wyngaarden reopened the case and said a new panel would be appointed. He also set up an Office of Scientific Integrity attached directly to his office. (This was precursor to the Office of Research Integrity.)

The explosion came on 4 May 1989, at a new hearing Dingell held. The afternoon session began with three witnesses from the Forensic Services Division of the United States Secret Service. They had examined the I-1 notebook and others, and offered results of tests of several kinds. The forensic team concluded that much of Imanishi-Kari's notebook could not be contemporaneous with the period in which she said the research had been done. Baltimore was last to testify. In a remarkable display, late in the afternoon at the end of his testimony and even as the chairman raised the gavel, Baltimore blew up in anger at Dingell and the methods of his investigative staff.

Impulsive, even foolhardy, the outburst seemed to those who watched it—and a lot of fellow scientists had been recruited to attend. A fortnight earlier, however, Philip Sharp, at MIT, a close friend of Baltimore's, mailed to several score of scientists around the country a form letter—"Dear Colleague"—saying among much else that new hearings were scheduled and that Dingell "has decided to continue to hassle David and the other authors and this has serious implications for all of us." Sharp asked recipients to write to their congressmen and members of the subcommittee, to offer op-ed articles to their local papers, and to enlist other colleagues. He appended a draft letter—"please don't use my sample exactly"—and a page of brief paragraphs making related points. One of these:

> The accusations were immediately investigated by highly qualified immunologists on review panels at the Massachusetts Institute of Technology and at Tufts University and reviewed by the authors. The panels found no sign of fraud or misrepresentation.

Across the country scientists responded to the fear of a Dingell witch hunt, to the call to resist governmental interference. Many did write letters; op-ed pieces appeared in *The New York Times* and other papers. Meanwhile, even as controversy bubbled, the trustees of Rockefeller University offered Baltimore the presidency there. Many of the senior faculty fought the appointment; the trustees forced it on them.

The second NIH panel was assembled that fall, with Joseph Davie, the other two original members, and two new ones. The new Office of Scientific Integrity was the investigative arm. Not until 14 March 1991, however, did the OSI complete and provide to the principals for comment a confidential draft of a report. The second Davie panel found that Imanishi-Kari had indeed committed fraud at the time of the original paper and again when she supplied what she said were her data. The draft report included the details. It leaked. On March 22, an account of it, by Philip Hilts, made the front page of *The New York Times*: BIOLOGIST WHO DISPUTED A STUDY PAID DEARLY.

From spring through summer and fall of 1991 the scandal raged. Baltimore was forced into an abject apology and commendation of O'Toole, which John Maddox, editor of *Nature,* printed on May 9 with a lead editorial about the "End of the Baltimore saga." A week later, Maddox published O'Toole's reply, which itemized charges she insisted Baltimore had failed to answer. The next day, *Cell* published a retraction of Weaver et al., signed by Baltimore and the others—except Reis and Imanishi-Kari. The retraction, too, was obviously forced, and within weeks Baltimore was again insisting the paper was basically OK.

What drove the controversy now was a defense of the norms of science. Over the preceding months a group at Harvard—the elderly, saintly physiological chemist John Edsall; the molecular biologists Mark Ptashne, Walter Gilbert, John Cairns; and Paul Doty, a biochemist, a very senior scientist at Harvard, an eminence—had begun to scrutinize the paper, to talk to O'Toole, and to attend to her analysis of what was wrong. They got the attention of others at Harvard and elsewhere. (In view of Baltimore's later characterization of the Harvard group, it's important to point out that Ptashne had long defended him. For instance, without anyone recruiting him, Ptashne had attended the Dingell hearing in May of 1989 where Baltimore testified.) Their collective conviction grew that Baltimore was wrong, refusing to face the science and attempting to foreclose the issue by the sheer weight of his personality, his prestige, his connections throughout the scientific establishment. Week after week the letters in *Nature* hammered back and forth, and a widening circle of scientists were drawn in. Maddox was the only one having fun: he was on to a journalistic bonanza, publishing nineteen pieces by the various combatants in twenty-three weeks. Then Baltimore lost his temper again.

On July 18, Paul Doty filled five columns in *Nature* with a comprehensive dressing-down not only of Baltimore but of the scientific community for tolerating his behavior.

> So far, attention has been focused primarily on the validity of the data reported and the various means taken to investigate possible misconduct. This has overshadowed the question of whether the account given by the authors themselves represented a departure from the normal standards of research, which puts [*sic*] the uncompromising search for truth first and, if so, whether this reflects a more general lowering of the standards by which research should be carried out and reported.

Doty itemized Baltimore's repeated "lapses from what have been the traditional standards of science." He wrote, "This pattern of behaviour stands in deep contrast to the traditional view that authors of scientific papers have a special obligation to be responsive to criticism and to test their work from every possible angle." He noted the strains created by the growth of science, by "the near cut-throat competition for grants, the possible corruption, on occasion, of peer review, the growing number of cases of deception in scientific papers." He concluded:

> As a result, the scientific community may already be experiencing a gradual departure from the traditional scientific standards; this could be abetted by condoning the behaviour seen in this present case. In this way we risk sliding down toward the standards of some other professions where the validity of action is decided by whether one can get away with it.

Friends urged Baltimore to let the matter drop. He could not. On September 5, he replied in anger, "An open letter to Paul Doty." Amidst indignant assertions of his honor and integrity, he wrote that "there is much published evidence and more coming that support the paper's results in remarkable detail," and he cited six. On October 10, Doty in turn replied, and of this claim in particular he wrote:

> [A]n examination of this evidence shows that it is not at all evident that the six papers referenced provide any support at all. But there is a deeper criticism of Baltimore's claim that all is well if his conclusions are ultimately shown to be correct. The scientific literature would become irredeemably corrupted if this became accepted practice. . . . [T]he reporting of research would be reduced to a lottery.

With that, Maddox shut the exchanges down, and in an editorial for the first time took a stand. "[W]hat are the responsibilities of the authors of a published research report?" he wrote, and went on, in part:

> The issue is simply stated. Dr. David Baltimore, the most celebrated although not the principal author of a disputed paper (*Cell* 45, 247; 1986) has from the outset taken the view that it is for the scientific community at large, and for others working in the field concerned, eventually to demonstrate the validity or otherwise of the disputed data and the conclusions drawn from them. It is a point of view, but hardly a defensible one, especially when the authenticity of the data on which the disputed papers were supposedly based has been sharply questioned. . . .
>
> The plain truth is that the authors of all published research reports have a personal responsibility for their aftercare. . . .
>
> So much has hitherto been generally accepted. Were it otherwise, science itself would be undermined. For the presumption would be falsified that what appears in the literature can be regarded, at least provisionally, as authentic. People seeking to make use of published data would then have to repeat the experiments before carrying on with their own work. Can that make sense?

The events at Rockefeller University in the fall of 1991 would make engrossing chapters in a different kind of book: a tale of principle and expediency, high drama, and byzantine intrigue, the politics of executive offices, laboratories, and boardroom at an august and rightly celebrated institution, a diamond in the diadem of the American scientific establishment. His exchange with Doty activated Baltimore's opponents among the senior faculty at Rockefeller University, and they were now joined by several who had been his supporters, even his good friends. He had enemies. Some were leaving. Anthony Cerami had risen to full professor since he had been a graduate student in the same years as Baltimore; at the end of August, Cerami became convinced the atmosphere was hostile and took a small team with him to another research center. Gerald Edelman, brilliant, a strong writer, elegant in dress, fastidiously incisive in speech, with a Nobel prize for work in immunology, had switched to neurobiology and now besides his own laboratory at the university ran a Neurosciences Institute there, with independent funding; on October 8, he announced that he was moving all that, some twenty-five scientific staff, to the Scripps Research Institute, in La Jolla, California.

Edelman had fought Baltimore's coming to the university. They were notoriously hostile to each other. In a conversation at his office a fortnight after announcing his move, Edelman said, dripping contempt, "So let me summarize what I believe, which is the following. David Baltimore is not a scientist, period. And the reason he's not a scientist is evident in his own acts. A scientist repeats the experiment when challenged, period. There's nothing more to say. And all of us are becoming trapped in all this other human side. We're being fooled, we're fools if we miss that, because that's the *only* issue from the standpoint of science. Remember, I didn't say that there weren't public responsibilities, in what a scientist does. But to the extent that he substituted maneuver, public statements, manipulation, lawyers, personal revenge, all that is irrelevant." Edelman said that, without meaning to sound pompous, he looked on science as the closest he came to a religion. "And what we understand in our religion is this, if you're an experimentalist repeat the experiment. If you cannot repeat the experiment, you ask an associate if you have one, a friend, who can repeat the experiment. You don't argue, you don't hire lawyers, you don't call names, and you don't put the burden off on other scientists' work. You don't."

Sick of the controversy, many of the senior faculty used the fact that Baltimore had brought the affair with him, that he could not let it go, and the charge that it was actively hampering fund-raising and recruitment of faculty, to force him out. The crisis came on Thursday, October 17—a day of lowering clouds, high winds, and drenching rain all up and down the east coast from Maine to Virginia and beyond. That morning, David Rockefeller, who has made the university a life-long personal responsibility, announced that he was giving it twenty million dollars. In point of fact, he had first planned the gift nearly a year earlier, but on this occasion he didn't say so: of course it looked like a thumping endorsement of Baltimore. At two that afternoon, the leading trustees of the university met in one of the smaller libraries: Richard Furlaud, chairman of the board and chairman of Bristol-Myer Squibb, David Rockefeller with his personal assistant discreetly taking notes, the executive committee of the board—which is made up of the chairmen of all the other committees—plus the full membership of the committee on scientific affairs, headed by Roy Vagelos, a research physician and the chairman and president of Merck and Company. Rockefeller, Furlaud, and Vagelos were effectively the board's ruling triumvirate. Baltimore of course was present. The group numbered thirteen. Throughout the afternoon, Furlaud took notes in a big bound book.

At the end of this executive-committee meeting, a group of faculty was scheduled to come in, representing the strong views about David Baltimore continuing as president. At three o'clock, Rockefeller and Baltimore went out to a press conference about the twenty-million-dollar gift. Rockefeller returned without Baltimore. The faculty group came in. From that point, about four o'clock, the meeting ran without a break until half past six.

Norton Zinder led off. In a thick and gravel-voiced New York accent, he said, as he later told me, "These stories keep appearing in the newspapers. We still have these things in *Nature.* The place is rife with rumor." And then he surveyed the trustees up and down the table and said, slowly, very loudly, *"Please! Make. It. STOP!"*

Shocked silence. Paul Berg, a senior molecular biologist and friend whom Baltimore had put on the board, had flown in from Stanford. He remembered: "All he said was, 'The plaintive cry around here is, make it stop!'" In a conversation with Richard Furlaud at the end of the next February, he leafed through his big bound book and pointed to the page. "This is where it starts, right here," he said. The note read: "Zinder: the place is rife with rumor. MAKE IT STOP!" Furlaud had scribbled the three words in large letters with exclamation points and a big circle around them. In Furlaud's note, Zinder continued: "You open the pages of all these periodicals and you see all this. Most people don't know David Baltimore. This whole thing has gotten out of control."

Zinder had seized the attention of the meeting and established the central concern—the effects of the endless controversy on the university, in particular on faculty and student morale, which he said was miserable, on recruitment, and on fund-raising. In a telephone conversation six weeks later, on December 3, Zinder said, "I sort of opened the meeting with a description of what I thought the university events were like, and that we were living in a place which was rife with rumor, and lives in turmoil, and looking over your shoulder each day to find out what's happening next, and arriving in the morning to students asking you, 'Well, did this happen, did that happen?' I tried to frame the issue so the discussion would be in terms of the university at now, present, its current status. That determined the—I believe determined the way the discussion went."

The next day, at Woods Hole, I attended a small meeting of biology editors where Baltimore was to give a talk. The announcement of the twenty million was headline news in *The New York Times.* Baltimore was unaccountably morose. Over the next six weeks, he fought back with his usual tenacity, but on Thanksgiving Day of 1991, by telephone from his country house in Woods Hole, he told Furlaud he was quitting.

Public attention died down. Dingell tied the scalp to his shield with the others. Events had been set in motion, however. The Office of Scientific Integrity was moved out of the NIH itself, and with a new name and some new staff ground on with the Imanishi-Kari case. Eventually, at the end of October 1994, the Office of Research Integrity issued a report, two hundred and thirty-one pages plus appendices, which found, in yet more detail than before, that she had committed misconduct. They set out the charges in a letter to her: eighteen charges, narrowly specific and technical, about the data in the paper and the data she had offered later in its support. The proposed penalty was to bar her for ten years from receiving federal grant money. She appealed to the Departmental Appeals Board of the Department of Health and Human Services. A three-person panel was appointed to hear the appeal.

Consider Baltimore's tactics. He is a great scientist. He is also a great purveyor of the mystique of science and of the authority conferred by the Nobel prize. When questionings first arose, Baltimore could have stopped controversy cold by scrutinizing the disputed data and announcing that he was reconsidering the paper. This he refused to do. Those who disagreed about Weaver et al. he berated as incapable of understanding. From the beginning, as in his only meeting with O'Toole, he sought to substitute his authority for consideration of evidence of the problems. Repeatedly, he redefined what Weaver et al. was about. First to go was its only important claim, evidence for the Jerne network theory. Repeatedly, he defended the paper by saying that the longer-term scientific outcome is the only issue: flaws in data or in conduct are overridden by later independent confirmation. In response to Dingell, Baltimore created a climate of emergency in which he was defending science against governmental interference. In response to the group centered on Harvard, he converted a scientific dispute into a personal combat within the community. In sum, from the beginning he worked to control the controversy—whether by conscious intent or not, and largely with success—by means of a series of shifts of the subject. This was far subtler than what the Harvard group was doing, which focussed, naïvely, on the science and the conduct. Baltimore's intransigence was what kept the Imanishi-Kari case in the public eye.

To comprehend the decision and the hearings before the appeals-board panel, one needs to know that the procedures of the Office of Research Integrity and of the board fall under neither criminal nor civil law but are denizens, rather, of that idiosyncratic domain, administrative law. The ORI is an administrative-law mongrel. It investigated, it charged, it adjudicated and passed sentence—a merging of functions that left plenty of scope for the protest, in this as in other cases, that scientists it targeted were not accorded due process. To show that Imanishi-Kari was guilty of scientific misconduct, the ORI had to prove its case before the appeals board by a preponderance of the evidence, that is, by the standard that in law courts applies in civil suits, and not beyond a reasonable doubt as is more rigorously demanded in criminal trials.

Strangely, given that the panel is in effect a quasi-judicial court of appeal, it hears misconduct appeals *de novo,* like trials, with opposing lawyers, witnesses, exhibits, experts, cross-examinations, objections, the whole boiling. The adversarial apparatus transforms any case appealed to them. As part of this, at the hearings the Office of Research Integrity could not introduce their massive Investigation Report completed in the fall of 1994 and sent to Imanishi-Kari, although it was the basis for the cover letter that went with it summarizing the charges and detailing the sanctions to be imposed. Similarly inadmissible were years of previous investigative reports and all the testimony (taken under oath) before hearings of the Dingell subcommittee. The charge letter, only that, was the finding from which she appealed.

Late in the afternoon of Friday, 21 June 1996, the panel made its decision public. The decision is one hundred and ninety-one pages of thick-textured prose, but the first two sentences made its burden plain: "The Office of Research Integrity (ORI) did not prove its charges by a preponderance of the evidence" and no action was to be taken against Imanishi-Kari.

For Thereza Imanishi-Kari herself, the panel's decision meant that she was once more qualified to apply for grants. After the decision, Tufts University School of Medicine restored her to the faculty position from which she had been suspended at the end of 1994. These results went some way, her defenders said, to redeeming her reputation. Yet they could hardly make whole a scientific career shattered by a decade of bitterest controversy. Moreover, even a cursory reading of the decision brings one up against disturbing ambiguities. Most peculiar is that the panel repeatedly scorned the Office of Research Integrity, for irrelevancy, for slovenly presentation, for violations of fairness and due process. Contempt is not too strong a word, and so sharp was it that several observers remarked that for Imanishi-Kari the decision hardly amounted to "not guilty" but rather to that dry Scottish verdict "not proven."

Coverage of the appeals-board decision in the popular press as well as in scientific journals was immediate and wide-spread. Imanishi-Kari's exoneration is what the press reported, what editorial writers and columnists and op-ed contributors took the decision to mean, what most scientists suppose to this day that she received. Furthermore, although the Office of Research Integrity and the appeals board were not dealing with misconduct charges against David Baltimore, many of the reports and commentaries—in *The Boston Globe, The New York Times, The Washington Post, The Wall Street Journal,* and others—said that the decision also vindicated *his* conduct during the ten years.

Many responding to the appeals-board decision took it as a chance to excoriate Congressman Dingell. Among those who had been grilled before his subcommittee who now hailed the Imanishi-Kari decision was Donald Kennedy, a former president of Stanford University. Kennedy had no connection whatsoever to the Imanishi-Kari case, but had been forced out of office by Dingell's investigation in 1991 of large-scale improprieties in universities' billing under federal grants for research. He wrote of the appeals-board decision as though it exonerated *him.* Another vengeful onetime witness was Bernadine Healey, director of the National Institutes of Health in the latter days of the administration of President George Bush *père.* Dingell had brought her before the subcommittee on 1 August 1991, the last day before a congressional recess, to explain her handling of misconduct charges first at the Cleveland Clinic (where she had been chairman of research at the time the Bush administration recruited her) and subsequently at the NIH. Healey had made a combative and obstructive witness. She succeeded in prolonging the argument over the Cleveland Clinic matter until no time was left to take up the far more serious questions about her treatment of investigations by the Office of Research Integrity, then ongoing. Rejoicing in the appeals-board decision, Healey wrote in *The New York Times* op-ed page of "how due process was trampled in science inquiries" and charged that the Dingell subcommittee and its staff had bullied her.

Those glad of the appeals-board decision also included many senior scientists from all sorts of fields, not particularly interested in Imanishi-Kari, not necessarily admirers of Baltimore nor versed in the arcana of immunology, who none the less saw in the affair—especially in the actions of the Office of Research Integrity and of the Dingell subcommittee—a grave threat to the freedom and integrity of science itself. These scientists comprise what in a loose-knit way amounts to the American scientific establishment; their opinions are ultimately the most important. Few of them will have followed the case closely; almost none will have read the decision. Science is self-correcting, anyway, they said again and again and say today—a phrase soothing and unexamined. And, they believe, science as an enterprise is and must remain self-governing, the community setting the standards of quality and conduct.

Thus for scientists, and for those who run universities and other institutions where scientists work, the Baltimore affair was profoundly disturbing. It upset complacency, forced confrontation of abhorrent possibilities. To many scientists the appeals-board decision came as a relief: everything's OK. Some remained uneasy.

Was the panel of the Departmental Appeals Board impartial? From the beginning, the lawyers of the Office of Research Integrity feared it was not. The panel was made up of three. Cecilia Sparks Ford presided. Her hearing-room manner was disdainfully correct. Ms

Ford is a lawyer. Judith Ballard appeared retiring, even shy. Her questions to witnesses were few and sometimes struck me as uncomprehending. Ms Ballard is a lawyer. In the summer of 1993, Ford and Ballard had been two of the three on a panel that heard an appeal in a previous major case, which the ORI prosecuted and also lost. That time, too, the panel's decision had scolded the ORI severely. After that, criticism had been raised that the appeals-board panel did not have the scientific understanding to assess such cases responsibly. For the Imanishi-Kari appeal, the two lawyers were joined by Julius Youngner, emeritus professor of microbiology at the University of Pittsburgh School of Medicine. His training, with his D.Sc. in 1944, predated the era of molecular biology, although of course in the decades he climbed the academic ladder to department chairman he will have picked up the new science. He had been a virologist, not an immunologist. He was attentive, patient with the *longeurs* of a quasi-judicial proceeding. During the course of the appeal, Youngner turned seventy-five.

The impartiality of the appeals-board panel is to be judged by the text of their decision. However, I must report an incident that I observed during the hearings. Testimony began on Monday, 12 June 1995. The Office of Research Integrity, prosecuting, led off. Their first witness was John Dahlberg, a senior scientist investigator at the ORI, who had received his PhD in 1968 in microbiology, and who had worked for sixteen years at the National Cancer Institute, most of that time in immunological research similar in methods to those of Weaver et al. The hearing room was set up with seats for spectators in back, with tables to the right (from the viewpoint of spectators) for the ORI's lawyers and others to the left for Imanishi-Kari's. Lead attorney for the ORI was Marcus Christ, a ponderous man in build and manner, not intellectually swift. For Imanishi-Kari, Joseph Onek led, an aggressive, experienced courtroom lawyer. Dan Greenberg, journalist, science critic, and at that time proprietor of *Science and Government Report,* a respected Washington newsletter, described him some weeks later as "of the gouge and stomp school of legal practice," which Onek surely took as a compliment. At the front of the room was the bench, raised slightly, and to the right the chair for witnesses—diagonally across from Onek. Youngner sat closest to the witnesses, Ford upright in the center, and at left, leaning back, Ballard.

Somewhat past three in the afternoon of Wednesday, June 14, Onek began cross-examination, his first in the hearing. As he leaned across the table—everyone concentrating on him and the witness—Ballard sat forward, face shining, eyes wide and fixed on Onek. She raised her hands to the level of her head and twice, silently, pumped her fists in the air, *mm mm.* Although other interpretations may be possible, it certainly seemed to me a pantomimed cheer. Another person in the room told me, at the next break, that she had seen it, too.

Was the Office of Research Integrity incompetent? The Departmental Appeals Board certainly thought so, and in many instances they were right. Their first and most general complaint was that during the hearings, and in the thousands of pages of material submitted afterwards, the ORI sought to expand its charges beyond those stated in the charge letter, to encompass conduct that would not amount to intentional and deliberate falsification or fabrication but which the ORI or its witnesses considered bad practice or bad interpretations or judgements of data. This raised questions of fundamental fairness.

The decision cited instances throughout—some involving serious accusations, of which Imanishi-Kari may even have been guilty, but which the board ruled out of consideration, summarily and correctly.

Again, I must report an instance I observed. Baltimore was in the witness chair. Marcus Christ, for the ORI, reminded him of the meeting he had held in his office at the Whitehead Institute—the meeting of 16 June 1986, with Eisen, O'Toole, and Imanishi-Kari. The transcript:

> Q [Christ] Now, at that meeting— After that meeting, you came to some conclusions that there were no serious problems with the Cell 45 paper, is that right?
>
> A Yes.
>
> Q But at that meeting you never really looked at the data, did you?
>
> A At that meeting we looked at some data and I, as I said, don't have a clear recollection what we saw and what we didn't see.

Christ did not press the witness. I winced at Baltimore's vagueness; it seemed to me that Christ had missed an opportunity. He dropped that line entirely, and after a brief break, when the transcript goes on the record again Christ jumped forward to another meeting, 18 May 1988.

On two crucial issues, where the facts are in dispute about who may have admitted what in 1985 and 1986, witnesses for the ORI said that tape recordings had been made that backed their versions—but the ORI neither produced the recordings nor explained why it had not. In each instance, reasonably enough, the panel inferred that the tape recording "either did not exist or did not support ORI's allegations," and was thus enabled to disregard the testimony altogether. The recordings in fact exist and back the testimony. However, although the circumstances were different they had both been made clandestinely. If for this or any other reason the ORI was unable or unwilling to produce them, their blunder was in allowing them to be mentioned at all. Without them, the testimony had considerable force.

Witnesses from the Secret Service testified, as they had done at the Dingell hearing. The appeals-board decision stated: "The case presented before us turns on the forensic attack on the authenticity of data recorded in the I-1 notebook." The Secret Service analyses were important, to be sure, yet other evidence was abundant; but the panel then systematically attacked the forensic testimony.

Forensic evidence about the I-1 notebook was of several kinds: analysis of the sources of paper used in the notebook; analysis of the inks from the ball-point pens used to annotate some of those pages; the finding of altered dates on some pages; detection by a specialized instrument of the faint indentations, mostly invisible to the eye, left on pages by what had been written on sheets originally lying above them. Crucial are the strips of paper tape—like adding-machine tape—on which were printed out radiation-counter readings that Imanishi-Kari said she had obtained in 1984 and 1985 in experiments including ones in dispute in Weaver et al. In assembling the loose-leaf binder, she had attached these tapes to notebook paper. Some of these pages are distinctly odd. One, for example, presents three pieces of tape. Two of these are of moderate length, one mounted directly below the other, while in a space of less than an eighth of an inch between them is the third piece—just one line of print-out.

The panel made a general assertion that deserves examination: "The Secret Service did not have any information about the scientific significance of any of the materials they reviewed and hence attributed unwarranted suspicion to characteristics of the data for which there were scientific explanations." They came back to this point several times. Yet it has no relevance to forensic analyses of inks, paper, and strips of radiation-counter tapes. Similarly, one of the specialists, John Hargett, remarked from the witness chair that the I-1 notebook was created in 1984, at the time the experiments were supposedly being done. Everyone but Hargett knew that I-1 had been assembled in 1988. The appeals-board panel concluded that Hargett's error destroyed the validity of the forensic analyses. Again, the crucial analysis is of the year of origin of the tapes themselves, not of the pages on which Imanishi-Kari later mounted them. Hargett's misapprehension had no possible relevance to that analysis. But the Office of Research Integrity failed to clarify what should have been minor confusions.

In addition to the items of raw data, marching down the tape, radiation counters periodically print in the margins what's called a register number. These serve to identify the experiment and the order in which data was gathered; they advance in sequence and do not start over with each new experiment. Imanishi-Kari had trimmed the register numbers off the tapes in the I-1 notebook—but not off every one. When still there, the register numbers were interesting. An egregious example: a pair of tapes that Imanishi-Kari represented as produced in one experiment spanning two consecutive days bore register numbers so widely different that they must have been generated more than three months apart. The panel dismissed the discrepancies as errors she had made in dating the work, ignoring the fact that she had presented the tapes as recording a single, continuous experiment.

Treatment of the other forensic evidence was similar. And it was piecemeal: the panel did not acknowledge the way in which the many kinds of evidence, forensic and other, point to the same conclusions. Some statements in the decision are patently false. In several places, it said that scientists have not found Imanishi-Kari guilty of misconduct. The strongest version: "All of the scientists who looked at the questions raised about the *Cell* paper over the preceding decade (at Tufts, MIT, and on the NIH scientific panel) found no evidence that scientific misconduct occurred." The NIH scientific panel referred to is the first, three-person group headed by Joseph Davie. Although their published report, in January of 1989, did not find misconduct, an earlier draft had done so. Further, when Davie was questioned at the hearing of the Dingell subcommittee on 4 May 1989, he conceded that certain claims in Weaver et al. were not backed by data. In one instance, an experiment Imanishi-Kari reported in the paper had not, in fact, been performed. He was asked, Is that misconduct? "I believe it is misconduct." A moment later, "Yes, it's misconduct. Any time one describes something that's not accurate, that's not right. That's misconduct. But we called it a serious error. It is a subtlety. I'm not sure I can defend it at this juncture."

The government's investigations of Weaver et al. were full reviews carried out by scientists, some of them employees of the OSI or its successor, the ORI, others working at the National Cancer Institute. They found misconduct. Independently, the second Davie panel, expanded with two additional scientists, concurred fully with the OSI and ORI findings of misconduct.

Considering one after another of the particular instances where Imanishi-Kari was charged with making data up, the appeals-board panel found that she was without motive. They took no cognizance of her situation. In 1985 and 1986, she was moving from MIT to Tufts, and had laid claim to what would have been by far the most important discovery of her career. She was committed to it. Two years later, during a meeting she had with the first Davie panel and investigators of the NIH, the reliability of data reported in a table—Table 2—in the paper came under question: the next day Imanishi-Kari returned to the panel with data that, she said, came from experiments done back then that she had never mentioned before and that bore out Table 2. One can hardly escape the vertiginous feeling that the panel's reasoning reduces to an assertion that if Imanishi-Kari had not faked data then she had no reason to do so.

The decision offered many more instances, whole categories of evidence that the panel treated in broadly this same fashion. A notable feature is its acceptance and extension of the sloppiness defense—that Imanishi-Kari was, notoriously, so chaotic in her handling of data that one must ascribe inaccuracies and misstatements and missing data to incompetence not misconduct. In a twist on this argument, the panel invented a kind of defense not envisaged even by Imanishi-Kari's lawyers. Repeatedly, they speculated about what real fraud ought to look like, and then determined that Imanishi-Kari's anomalous data do not meet that standard. For example:

> Many of the questioned notebook pages contain material that is not helpful in supporting the conclusions of the Cell paper. . . . [I]n many cases the results included on the questioned pages were conflicting or bizarre in ways more likely to raise than to resolve scientific questions. Even if Dr. Imanishi-Kari were cleverly avoiding too perfect a look, there is no demonstrated reason that she would create bizarre and conflicting results which could only call attention to problems rather than mask them.

A decision in favor of Imanishi-Kari had not only to deal with the evidence but with the witnesses against her, above all, of course, with O'Toole. Imanishi-Kari, Wortis, Eisen, Huber, Baltimore—perusal of their statements and testimony over ten years reveals many changes and self-contradictions. O'Toole alone has been constant and consistent. The panel turned this against her. "[W]e question the accuracy of Dr. O'Toole's memory and her increasing commitment to a partisan stand." Or, with unpleasant innuendo:

> Dr. O'Toole worked extensively with the investigators in this case, "indexing" the experiments in the Imanishi-Kari and Reis notebooks (although she said she was not paid for this work).... We are also concerned about the implications of involving a whistleblower too heavily in an investigation. Such involvement can compromise both the ability of the investigators to maintain objectivity and the ability of the whistleblower to avoid becoming too vested in the outcome. We think that happened here.

What is to be done? The remaining chapters offer pathways and suggestions.

Howard Temin died in February of 1994. He had lung cancer—one of the small percentage of cases not related in any way to smoking—and it had spread to the brain. Temin's laboratory, at the University of Wisconsin, was small, with few graduate students or postdocs at any time; every Friday, each of them came to his office, sat on a hard wooden chair by his desk, and explained everything done that week, showing Temin all the data. Never did Temin serve as a consultant or director of a biotechnology or pharmaceutical company, nor did he start a company of his own. He explained that for himself he thought it wrong to put what he discovered with public funding to the making of a private fortune. He was no prig, though, and an active member of the American scientific establishment, friends with many whose conduct of science was different from his. He and Baltimore had known each other since student days, and Baltimore turned to him several times during the controversy over Weaver et al.; Temin never said what they talked about. He gave interviews rarely, and in them was not forthcoming.

In September of 1992, Temin learned the nature of his illness. Now he wanted to be interviewed about his scientific career. He wanted to be debriefed. The discussions took place in his office on 15 and 16 March 1993, in three sessions totalling nearly six hours. Near the end of the last session, as the tape cassette was about to be turned over, he said he wanted to talk more about misconduct. He spoke of several cases then current. He said, "The strength of our society is its plurality. I value that extremely highly—and so I also value that in science. I will allow other styles." He mentioned several scientists, some also controversial. Baltimore's case was simpler, though. "With regard to David, there are not so many layers. There is an important and still unresolved question about the paper itself," meaning the relation of raw data, or its absence, to the published paper. "And the Secret Service claims," he said. "Right, I'm saying— Those are really not related to what was David's error in that—to David's misconduct.

"David's misconduct was— When an experiment is challenged no matter who it is challenged by, it's your responsibility to check. That is an ironclad rule of science, that when you publish something you are responsible for it. And one of the great strengths of American science, as opposed to Russian and German and Japanese science, is that even the most senior professor"—his voice grew firmer, his enunciation flatter—"if challenged by the lowliest technician or graduate student, is required to treat them seriously and to consider their criticisms.

"It is one of the *most fundamental* aspects of science in America."

Jerne's network theory of regulation of the immune system has dropped from sight. Asked about this a while ago, Klaus Rajewsky said, "People don't pay any attention to it now." And he said, "This is not a thing that can be tested and with an experiment proved wrong. It never became a major paradigmatic thing where people could work on it." And Weaver et al.? "I must say, I really never understood the enormous fuss that came about that paper," Rajewsky said. "I don't think the Imanishi-Kari paper is seriously considered. Not by anybody I know."

6

THE PROBLEMS OF PEER REVIEW

So when it is claimed that inquiry must be free, what seems to be intended is that moral, political, and religious judgments should not enter into two important contexts of decision: the formulation of projects for scientific inquiry and the appraisal of evidence for conclusions.

—*Philip Kitcher,* Science, Truth, and Democracy, *2001*

Peer review might disappear because its defects are so much clearer than its benefits. It is slow, expensive, profligate of academic time, highly selective, prone to bias, easily abused, poor at detecting gross defects, and almost useless for detecting fraud.

—*Richard Smith, "The future of peer review," 1999*

The notebooks of Leonardo da Vinci are filled with observations, discoveries, inventions, often far ahead of what others of his time could have conceived: when the notebooks were read and understood, the claim arose that he was not only a great artist but a scientist, a great one. Not so. Leonardo never published. Without publication, the most brilliant understanding of the ways of the world cannot be science. Publication is the distinguishing act of the scientific process. It marks the midpoint in the process: upon publication, the character of scientific work changes from closed to open, as the effort of a relatively small number of individuals becomes the property—indeed, the raw material—of the entire community of science. When a novel or a poem is published, a painting or a statue exhibited, a movie released, that is effectively an end point. Music, dance, drama—those forms that must be remade each time to exist more than potentially—do extend creation into collaborative interaction. But in the sciences publication is but a new beginning, for the collaboration and competition that ensues is what makes the sciences unique. With publication begins the extension, elaboration, correction, adaptation that weave the new into the fabric of the known.

The work of the sciences from inception to publication, since the second world war, has been framed, conditioned, by a heterogeneous set of practices collectively called peer review. These are now in flux, changing more radically than even those most closely involved have fully understood. The changes affect all the sciences, and worldwide. They are structural, which is to say, they are altering the conditions, the building plan, the institutional relationships that underlie and shape the way sciences are done. Structural transformations are probably impossible to resist. Their consequences can be hard to predict. One sure prediction: pick some date safely in the future but not too far, say, 2015: by then, although the term *peer review* will persist vestigially in scientific research and publication, the practices denoted will be all but unrecognizably different from yesterday's.

In November of 1993, in Chicago at the Second International Congress on Peer Review in Biomedical Publication, organized by Drummond Rennie, I gave one of the two opening, plenary talks. Not a physician, not a scientist, not an editor, I was asked to speak as an outside observer—and to shake the hundreds of participants, journal editors and staffs, out of their complacency. Accordingly, I titled the talk "The Structural Transformation of the Sciences and the End of Peer Review." There I set out certain internal factors, closely interlocked, driving peer review into decay.

To understand the coming changes we must begin with the observation that peer review and refereeing are inherently threatened by corruption. Obviously neither is corrupt in any high proportion of cases. Yet they are inescapably under pressure—springing from the basic contradiction that makes peer review possible at all. For the fact is, of course, that the persons most qualified to judge the worth of your grant proposal or the merit of your submitted research paper are precisely those who are your closest competitors. The fact is obvious. Its implications must be confronted.

The vulnerability of peer-review practices to corruption is compounded by the overwhelming growth of the scientific enterprise and the consequent confusion and fatigue; by the concomitant decline of the great tradition by which the elders by their example inducted the new generation into the norms of science; by the relentless pressure on funding and the consequent raising of the threshold below which research projects don't get funded; and by the intrusion of political influences. For many scientists, I said then, the cumulative effect is a sense of futility bordering on despair. The organizers thanked me, the assembled editors by and large remained complacent, and the decay of the peer-review systems accelerated.

Thereza Imanishi-Kari suffered two episodes of peer review related to the contested paper. That single term, though, in its everyday use encompasses two different forms of evaluation of scientists' work. She was hit by both. The first form of peer review—but this is common knowledge—is the screening and ranking of applications that scientists submit to get grants. Thus in the summer of 1985, as a direct result of the site visit by scientists representing the National Cancer Institute, she was knocked out of the big, multi-laboratory grant that was otherwise largely approved for the cancer center at the Massachusetts Institute of Technology. In this instance, we can say that the peer-review system functioned as it is supposed to do. The second activity falling under the term is the critical scrutiny of papers scientists submit to journals for publication. Thus after David Baltimore sent Weaver et al. to *Cell,* the then editor, Benjamin Lewin, sent it out to several scientists to review, and one, Klaus Rajewsky, put his finger on the problematic passage—yet the possibility of fraud did not cross his mind. The system is not designed to catch fraud, and could be expected to do so only rarely. In this instance, however, the cautions Rajewsky raised in his report were brushed off by the editor and authors.

Those two events epitomize the strengths and problems of the peer-review system. The problems begin with the inescapable awkwardness of nomenclature—two divergent activities covered by the single term. Critical review of journal submissions is also called refereeing, a label that avoids confusion. This is more than a quibble, for the functions, methods, and histories of the two forms, though conceptually related, are distinct. Yet the dual usage is too deeply embedded in the way scientists and editors think and talk to be corrected. For example, a young scientist's career is crucially dependent on getting papers published in what are called peer-reviewed journals.

We cannot consider the two modes of peer review independently of each other: in basic purpose they are similar, and both are intrinsically part of the system of science we now practice. Both are locked into the way we distribute recognition, money, hierarchical position, and power in science in the United States and Europe. Crucial to that system, both are methods that have evolved to protect the autonomy and self-regulation of the sciences. Peer review and refereeing reserve judgement of scientists' work to other scientists, acting in the name of the community. In review of grant applications, scientists evaluate plans of other scientists prospectively. In refereeing, the view is retrospective, scientists appraising completed work of other scientists. The two modes stand at two ends of the scientific process prior to publication.

From their strategic positions—prospective, retrospective—peer review and refereeing are institutions of vital importance: they are thought to be, or ought to be, the means to control the process and at the same time to validate the results. They have been characterized as "gatekeeping." The review machinery for grant applications shields the choice of projects or groups to support from direct pressures from administrators, politicians, and the public who are outside the system of science. Review and refereeing also serve the prophylactic function of protecting from charges of bias those who dispense the money and those who publish the results. Their significance goes deeper: because of their day-to-day role in the self-governance of the sciences, most scientists defend them as essential to the foundation of their enterprise. As such, they seem to most to be immovable, eternal, blocks of stone.

Of course, they are not. Although peer review and refereeing seem rational, indispensable, and immutable, their histories demonstrate that they are institutions of recent date. They are not laws of nature, nor of epistemology. They have changed and evolved. At the first of the congresses on editorial peer review, Elizabeth Knoll, then science editor at the University of California Press, spoke tartly of "a remarkably uncritical faith in the peer-review system." She went on:

> In only a generation, editorial peer review has become a powerful social system. In the process, formal peer review has taken on some of the supposed objectivity of research that the peer review process is meant to judge. Institutionally and individually, we tend to forget that just because peer review reviews scientific work does not mean that it is itself a scientific process.

Knoll is right. Though she was speaking of refereeing of journal articles, what she said applies with equal force to review of grant applications.

To be sure, in the past two decades an immense and stupefying literature on peer review has grown up, scores of books, many hundreds of papers, attacking, defending, testing, analyzing. A recent, careful survey of studies of peer review of grant applications, fourteen pages long, cited one hundred and twenty different publications—and the grants process has received far less attention, by a factor of ten or more, than refereeing of articles. Yet questions are hard to formulate, data hard to come by, conclusions cloudy. Good studies are few. The best are indecisive. Faith is strong and eloquent, trustworthy evidence for or against the efficacy of peer review is vanishingly rare.

Peer review and refereeing became dominant only after the second world war, and despite their similarities their earlier histories differ greatly.

*

In the United States before the second world war, little scientific research was paid for by the federal government. The great exception was agricultural research, which was carried out at Department of Agriculture field stations and the land-grant colleges, a system which originated at the time of the Civil War—the southern states having opposed anything of the kind—and which by the beginning of the twentieth century had grown into a network with activities in every state, politically untouchable. The United States Geological Survey also spent federal dollars. Money for other scientific research was provided by a few large private foundations, chief among them those established early in the twentieth century by Andrew Carnegie, steel and railroads, and John D. Rockefeller, Sr, oil and banking.

Preaching what Carnegie called the gospel of wealth, practicing a new economic order so ruthless as to give Darwinism a bad name, in an age of unprecedented private fortunes these two along with the Nizam of Hyderabad were the richest men in the world—in constant dollars, worth several times the richest of today's parvenus like William Gates III or Warren Buffett. Of necessity they shared broad capitalist interests and attitudes. Yet they differed greatly in personal styles, Carnegie ebullient and secular, Rockefeller dour, tight-fisted, devoutly Baptist. By the turn of the century they had become eleemosynary rivals. Steel was planning the Carnegie Institution of Washington, which he endowed with an initial ten million dollars in 1902—an amount comparable to some two hundred million a century later. Not to be preëmpted, in March of 1901 Oil announced the founding of the Rockefeller Institute for Medical Research, in New York, though only in 1906 did the institute open its first laboratories. (The priority is an irritant to this day.) John D. Rockefeller's trusted, all-powerful lieutenant had long been Frederick T. Gates, originally a Baptist minister, a man of imaginative and rhetorical flair; by the end of the old century he was joined by John D. Rockefeller, Jr, as architects of the Rockefeller charities. They acquired the East River site of the institute in 1903 and began recruiting senior scientists.

Money guaranteed these benefactors control. Independence from government was a first condition. The model available to them was the Institut Pasteur. Louis Pasteur had founded it in 1888 on the wave of public enthusiasm and donations following his demonstration of immunization against rabies: it was free of control by the French state, otherwise notoriously *dirigiste,* and was eminently successful in medical research. Independence from the established universities was another desideratum, which allowed management to become centralized and professional, increasingly in the new corporate style, while research was undiluted by teaching. Further, idiosyncratically, the elder Rockefeller had a rooted objection to alliance with any of the great medical schools, for he was a believer in homeopathy.

In the years just before the first world war, a number of new, large, general-purpose foundations were established. Once again, Carnegie and the Rockefellers were by far the most munificent.

Carnegie endowed the Carnegie Corporation in 1911 with one hundred and twenty-five million dollars for starters—which would be some seven hundred million today. Initially, in large part, it functioned as a consolidation of the man's remarkably diverse philanthropic ventures: establishing public libraries, ultimately 2,509 of them throughout the English-speaking world; organs for churches; a large additional grant to the Carnegie Institution; an unfocussed scattering of small grants to small colleges and universities. After Carnegie's death, in August of 1919, the corporation's board and staff rationalized its interests under the broad slogan "the advancement and diffusion of knowledge and understanding" (which they're still using). Some support went to the natural sciences, though this diminished in the nineteen-twenties and ended by 1931.

The Rockefellers and Gates established the Rockefeller Foundation in 1913. Always distinct from the Rockefeller Institute, the foundation awarded grants for a variety of purposes, including science. By 1927, Rockefeller had given the foundation one hundred and eighty-three million dollars—upwards of a billion, today. Between the world wars, it was the most generous and powerful supporter of basic research, especially in biology. From 1932 to 1955, the director of the foundation's division of natural sciences was Warren Weaver.

Weaver's way of funding science stands in radical contrast to what we do now. For that very reason, his example can still exert a powerful attraction. In September of 2002, a dozen people came to a dinner meeting in a private room in one of the more distinguished clubs in midtown Manhattan. Most of those present were science journalists; I was there as a journalist with pretensions to be a historian. We had been brought together by the retired vice president of one of the country's rich foundations. The purpose was to give advice and suggestions to the president of another, among the richest of all—so rich that it needs new areas to fund, and he was looking first at the sciences. He introduced Warren Weaver's name early in the conversation. Towards the end, I suggested we come back to Weaver's way, because he had two lessons to offer.

A mathematician by training, Weaver didn't wait for grant proposals but actively sought out scientists to support, choosing them in order to shape the direction of research in one field or another. For example, he selected Boris Ephrussi, a French geneticist (born and educated in Russia) to develop modern genetics in France. Ephrussi had already spent a year at the California Institute of Technology with Thomas Hunt Morgan, the leading mendelian in the United States. As a first step, Ephrussi asked to return to Caltech for another year, and to bring along an assistant to be trained. The year was 1936, the assistant the young Jacques Monod, and the experience and contacts formed Monod's interests and style—leading twenty years later to his crucial contributions to the golden decade of molecular biology.

That term itself Weaver coined, and he backed the field's earliest development. In the foundation's report for 1938, he wrote of aiding explorations in "those borderline areas in which physics and chemistry merge with biology" and said that "gradually there is coming into being a new branch of science—molecular biology—which is beginning to uncover many secrets concerning the ultimate units of the living cell." Far-sighted, brilliant—yet the principles apply in the present day. The first: go hunting for promising people. The second, more subtle and more difficult: find those borderline areas where fields of research previously not closely related are beginning to coalesce.

Indeed, I have long believed that the coming together of previously disparate elements is a general and powerful generator of important new work. The new flourishes in the cracks between the paving stones of the old. It happens in the arts. Think of Joseph Conrad, Polish writing in English. Vladimir Nabokov, Russian writing in English.

Günter Grass, growing up in Danzig, Gdańsk, at the intersection of three language cultures, German, Polish, and the dialect, Kashubian, of a local people, his mother's tongue. The list is very long. Think of Handel in music, Balanchine in dance. Think of the migration to the west of Greek artists, artisans, writers, and philosophers, and their books and tools, after the fall of Constantinople to the Turks in 1453, and the tumult that followed, called the Renaissance.

In the sciences, a simple instance has been the confluence of molecular biology and immunology, which began in the nineteen-seventies. Molecular biologists moved into immunology at first because they thought they could tweak the immune systems of, say, mice, to see how alteration or addition of one or a few genes would affect the creatures' development. David Baltimore was a leader in this. Immunologists wanted to bring the methods of molecular biology, the new precision and clarity, to bear on some of their recalcitrant classical questions. Part of Thereza Imanishi-Kari's problem was that she never really absorbed the molecular approach. The bitter fact, then, is that the paper Weaver et al. was a product of the confluence gone wrong. That was an anomaly. Present-day research in immunology, including such urgent practical problems as AIDS and malaria, is the ongoing product of the confluence.

Or consider plate tectonics. The idea that the continents have changed positions over many hundreds of millions of years, though obvious to you as a schoolchild when you compared the east coast of South America with the west coast of Africa, was first seriously proposed by Alfred Wegener in 1915. Professional geologists scoffed. Then in the mid–nineteen-sixties physicists developed new methods to determine the ages of relatively young rocks, using the rates at which radioactive isotopes of certain elements decay. At that same time, geologists were finding that down the middle of the Atlantic oceans runs a rift from which molten rock has emerged and cooled, leaving bands of rock parallel to the rift. When these two sets of discoveries from previously unrelated fields came together, the bands of rock could be dated. They proved to have the same series of ages as others all over the world: the sea floors were spreading, driving great plates on which the continents, or portions of them, ride. The confluence gave birth in less than a decade to the new science of plate tectonics, which unifies many things, from the deepest oceans to earthquake zones to the uprearing of mountains—and, yes, the relation of the west coast of Africa to the east coast of South America.

Mathematicians, several times in the past fifty years, have broken ancient difficult problems by new-fangled methods that brought together fields never before thought to bear on each other. One such was the proof, in 1993, by Andrew Wiles, a British mathematician at Princeton, of Fermat's Last Theorem, a famously baffling conjecture from the seventeenth century. Wiles did it in two hundred dense pages that only a few thousand people in the world are equipped to follow, but he made front-page headlines in *The New York Times.* The coming together of unrelated fields can yield remarkable results at the individual level, too. The most spectacular example is the move in the fall of 1951 of the young James Watson to the Cavendish Laboratory, in Cambridge, England, where he met Francis Crick. Watson was trained in the genetics of microörganisms. Crick was a physicist by background, working with x-ray crystallography to determine the three-dimensional structures of large molecules of biological interest. Each taught the other. Their collaboration got the structure of DNA, the molecule of heredity, the most portentous discovery in twentieth-century biology.

Search for individuals of promise, find disparate fields on collision course: these are principles for productive support of important new science. Federal funding of science does not work that way. Probably it cannot. A great private foundation can. The function of such a foundation should not be to try to fill in the bits missing from what the National Institutes of Health or the National Science Foundation already finance. The best method, the best justification, for support of sciences by great private foundations lies here, in the right appraisal of the success of Warren Weaver at the Rockefeller Foundation in the nineteen-thirties, finding the big ones.

In the history of the world, wars have often been followed by revolutions. The second world war led to the revolution in the federal funding of American science, whose prophet was Vannevar Bush and whose leading institutions are the National Institutes of Health and the National Science Foundation. Although their procedures for awarding grants differ in details, many of the underlying problems are the same. The way the National Institutes of Health go about it will make clear the difficulties and suggest the faults. Decisions at the NIH determine the spending of billions every year—just over twenty-seven billion dollars in fiscal 2003, twenty-eight billion in fiscal 2004 (which began 1 October 2003). They potentially affect the health of millions of people—and the health of the research enterprise.

The charge sheet against the system of peer review of grant applications carries three principal headings: that it is unfair, loaded with biases; that it is unreliable, not effective in choosing the best work to support; and that it is costly. In the quotidian details lies the potential for decline and corruption and for reform. Scientists apply to the NIH either for new lines of research or to renew current work. A successful lab chief may prepare several grant proposals a year—they speak of "writing a grant"—or perhaps as few as one every two years or so. A grant must be renewed every five years. The heart of the proposal gives the details in no more than twenty-five pages. Single-spaced, with a rigidly enforced minimum font size, that can run to nearly ten thousand words. The components and their sequence are fixed: the specific aim of the research, its background and significance, preliminary results or a progress report, and the research plan. The twenty-five pages are preceded by an abstract and followed by a budget, detailed down to individual items for the first year, more cursory for the full term. Ancillary materials include curricula vitae for all principal research workers. To write a grant takes from a fortnight to a full month of long exclusive days.

The process devours time. All told, the NIH receive some forty thousand applications a year. At, say, three weeks' preparation each, that bulks up to one hundred twenty thousand weeks, or twenty-four hundred scientist-years of work.

Proposals reach the NIH by several deadlines across the course of the year. A National Institute of Health, under the Public Health Service, was first authorized by Congress in 1930, but the act required that the money come from private donations. In 1937, Congress established a National Cancer Institute, financed initially at seven hundred thousand dollars a year. Springtime 2004, the NIH comprised twenty institutes, ranging from aging, alcoholism, and allergy, to human genome research, neurological disorders and stroke, and nursing, and including the National Library of Medicine. Atop sits the office of the director—its holder a highly politicized presidential appointment—but the foci of power are the directors of the individual institutes. The earliest, after the cancer institute, were four established in 1948; the latest, the National Institute of Biomedical Imaging and Bioengineering, was set up in 2000. To these, add seven centers of disparate function, and a number of offices concerned for example with AIDS or with women's health. New institutes or reorganizations of established ones are authorized by Congress, which appropriates the money annually, each by each. Scientists submit their proposals electronically to the individual institutes for evaluation, and they go first to an initial review group, usually called a study section or panel.

In the quarter-century after the second world war, peer review of grant applications functioned pretty well. Robert Pollack, an eminent cell biologist at Columbia University, told me some years ago that when he first served on study panels these were exciting and instructive occasions—offering several days of "discussions with seriousness and scientific depth." He went on, "In those days, to be at a meeting of a study section was an invaluable experience—in effect, a three-day seminar with top people on the most advanced work going on in your field." But the quality has degenerated, he said. "The system has virtually collapsed." Above all, the dedication and enthusiasm of panel members have sadly waned. The underlying reason, he said, was that as an ever-smaller proportion of grants gets funded and as the applications themselves, in the top quartile, are more difficult to put in any reliable rank order, politics became overt. "In the Reagan-Bush years, peer review became politicized: things like sex, racial, and geographical balance of the panels became important, rather than simply their quality." Rivalries between scientists, laboratories, schools of thought became palpable factors in the review process. What began as a means of keeping external pressures at arm's length has, to some extent at least, turned into a cockpit in which the internal politics of the sciences are sourly bickered over.

Listen now to the experience of S, an immunologist who prefers to remain anonymous. "I served a four-year term in the nineteen-nineties." His panel worked only with a subspecialty of applications—those for research in immunology—to the National Institute of Allergy and Infectious Diseases. "A study section numbered some twenty people, with the terms staggered." In the nineteen-sixties and seventies, he believes, the study sections, or his at least, had been dominated by physicians. "By the early nineties, though, our section drew membership from laboratory scientists, these being young associate professors, perhaps five years into their independent research careers, reasonably well known—that is, with established credibility—and yet with time free from, say, administrative responsibilities." A few panelists ranked higher. "The chairman of the section"—he named his—"would be more senior." Each study section had an executive secretary, an NIH person, and staff to do the work of duplicating, distributing, and tracking applications.

"The panel met three or four times a year, for three days. They met, for example, at the Bethesda Holiday Inn, or at some soulless hotel in Chevy Chase. Nothing to do but work. Sitting long days in a big room.

"The twenty panelists got ten grants each, to review. These included renewals and new grants. They arrived three weeks before the meeting. Each was twenty-five pages. Each got read by two panelists who wrote critiques—and these were ten to twenty double-spaced pages—and by a third panelist, a reader, who did not write a formal critique" but who stood ready to add an opinion in case of disagreement. "The reader was a new function. Critiques were turned in the first meeting day to the executive secretary and the chairman" and distributed promptly to the panelists. "The work was onerous." Indeed, S said, a panelist had to put in the equivalent of nearly two full-time months a year.

Formal studies confirm this. One published fifteen years ago said that NIH panelists gave thirty to forty days a year to reviewing. At a minimum, then, ten applications in thirty days by each of two reviewers: this suggests a figure per application of at least six days of reviewer time. With the number of grant proposals at present-day levels, this works out to at least two hundred and forty thousand days, or forty thousand six-day weeks, or some eight hundred reviewer-years, per year. Again: reviewers are for the most part established younger scientists who should be at their peak.

"Often the grants were from well-known people. Of your ten proposals, you'd know your best couple," S said. The meeting took the proposals up one by one. Panelists argued the science in each. "After the science discussion, you'd suggest a score—the two readers." The scores ran from 1, the highest, down to 5, in fractions. "When the scores of the two reviewers were widely different, then there was protracted discussion—a lot of soul-searching between the reviewers. This was very effective." Bargaining, S called it. "The value was that it brought out differences in understanding and evaluation of the science. It also exposed personal prejudices. Then when the two readers could agree on a rating, everybody voted."

A real though rarely noticed defect: "The discussions lacked a higher-order structure. They stuck at the level, for instance, of 'What's he going to find out?' or 'Has he made the knock-out mouse yet?'" He said, "The panelists were concerned about fairness, due to the budgetary constraints. Very few of the applications were obviously no good." The constraints were stringent. "It used to be that a score of 1.6 was OK—but soon even a 1.4 was the kiss of death." Therefore, he said, "The work was horribly depressing. While I was on, there were two levels at which the stringencies affected us. First, the payout line got so high that grants were funded only down to the ninety-third percentile. The situation was demoralizing, for all those on the panel. And also it was because everybody on the panel got concerned about how his own grants would fare—that they couldn't pass muster, either. We had an impossible task."

Triage has been called the most dread word on the battlefield. Originally meaning nothing more ominous than to sort materials into groups, typically three, by quality—as with newly shorn wool or newly picked coffee berries, two early instances of the word in English—it soon took on the connotation of being triaged *out,* as at a field hospital with some soldiers desperately wounded whom the medics would set aside, so sure to die that treatment would be a waste of time and resources. In the work of the study sections, "Triage came in in the mid-nineties," S said. "It took place, but informally. Then it became formal, and you would know before the meeting what had been triaged out. But anyone on the panel could call an application back." The rating was not the end. After it was agreed upon, the panel moved to the requested budget—line by line, arguing it down.

"One served four years, then tried not to serve again. Each year, people rotated off and on," S said. "Previously, the study section was made up largely of classical immunologists of the old school." But in the time of his chairman, and before and after, "the panel was largely reshaped." They dealt only with immunology. "But in those years applications in this category were becoming mostly to do with molecular biology, molecular genetics." The study section, S said, felt the fact that as early as the mid–nineteen-eighties, immunologists, especially of the newer generation, had begun taking up the common methods of molecular biology, while at the same period molecular biologists were turning to immunology to find model systems in which they could attack their own problems, in particular the stupefyingly complex nature of development and differentiation, or how the single fertilized egg becomes the multicellular, multifunctional adult creature. "David Baltimore expressly went into it as a developmental system." In any case, "The newer panelists were versed in those methods. Sometimes, the study section brought in ad hoc panelists for certain applications. The shift took place over a period longer than my tenure: maybe ten or fifteen years." Disappointed applicants complained. "'What's happening to the study section? What do those young people think they're doing?,' or 'The damned panel is too molecular.'"

At the same time, S said—just what Robert Pollack had told me ten years earlier—the work of study sections was becoming political.

"There was pressure from the NIH to diversify, ethnically, by gender, and geographically. Geographic diversification was hardest: molecular biology is a bicoastal science, with few good centers in the Midwest, such as Washington University in St Louis, maybe the University of Chicago. This pressure emanated from the director's office, but ultimately it was to satisfy Congress."

The pressure to diversify affected recruitment of panelists, but it operated most invidiously in the selection of institutions whose laboratories got grants. If they were awarded strictly on the evident quality of the work performed and proposed, the result could not help but favor top people at top places, the like of Harvard, Stanford, Johns Hopkins. "Elitism" became a battle cry, even within study sections. Yes, a plausible argument can be made that such a bias for demonstrated quality deprives able young scientists at provincial institutions of the chance to prove their work just as good. Yet those who press for even a modicum of diversification, on these grounds, are unable to offer, within the present review system, a reliable alternative method of trying to pick winners.

A further effect of such strictures has been to keep study sections focussed on research results in the short term. "There has to be something wrong with the system that doesn't look at an applicant's lifetime achievement," S said.

Applications passed by a study section must then be approved at two more levels, the institute's council and its director. By law, each institute is required to have a national advisory council, made up of twelve or more members, some scientists, some laymen. In principle, study sections are supposed to rate proposals strictly on scientific merit, while the councils consider wider questions of policy, such as a proposal's place in that institute's overall direction, balance, and budget. In practice, councils very rarely change the priority ranking of individual applications. Advisory councils' recommendations are anyway just that—advisory, not binding. Having passed the council, proposals go to the director. The institute directors make the final decisions, and they have considerable leeway, particularly, of course, because proposals approved perennially outrun funds appropriated.

Harold Varmus took the directorship of the National Institutes of Health in 1993, welcomed by the community as successor to Bernadine Healey. Immediately, he took up the problems of grant peer review, from the point of view of the practicing scientist—albeit one who had never had an application rejected. He summed up the devastating effects on morale of "the low success rate, the high number of resubmitted applications, the unwillingness of talented people to serve on study sections. They're making distinctions between grants that are equally excellent. It's very, very demoralizing to do that kind of reviewing."

At the National Science Foundation, review procedures are broadly similar. The NSF is more centralized than the NIH. Rather than quasi-independent institutes, it has a hierarchy. At the top are six directorates, according to subjects—mathematics and physical sciences, biological sciences (but "not health related"), geological sciences, engineering, social sciences and economics, and education—plus the Office of Polar Programs, which is concerned chiefly with research in Antarctica in all relevant subjects. Congress appropriates—five and a third billion dollars in total for fiscal year 2003—and specifies the amount for each directorate. Within each directorate are several divisions, within divisions programs. Each of these is headed by a program officer. Applications are considered in two cycles, with filing dates of February 1 or August 1; they are submitted electronically, in a format much like that required by the NIH and enforced just as rigorously. Review is a hybrid: the program officer sends a proposal to as many as six outside reviewers, who file written opinions; an advisory panel meets to consider those in conjunction with their own views. Each year, the NSF boasts, some fifty thousand outside reviewers will submit a quarter-million separate reviews. They are, of course, unpaid.

No doubt some of these are informed, reasoned, detailed, scrupulous, comparable in quality to the ten- to twenty-page critiques written by NIH panelists. Some are not. I recently submitted an application to the science-and-technology-studies program (that being modish academic jargon for sociology and history of science), in the directorate for social sciences and economics. Six outside reviewers commented. Several were sensible, helpful. None was as long as five hundred words. One, of two hundred and sixty-eight words rather hastily typed, began:

> Regretfully, I am doing so badly with grants or other job applications (e.g. Harvard decided not to hire anyone) that I really cannot devote a lot of attention to over haf a million garnts when others with comparable expertise in molecular biology have had to beg NSF over 3 rounds for a mere 25K.
>
> There is soemthing very wrong with this idea of paying a post-doc 34K and an accomplished scholar a mere $25. . . .

Digesting the external reviews, scrutinizing the proposal, the NSF panel comes up with its own critique and a rating: should be funded, could be funded, should be turned down but revised in light of the critiques and resubmitted, or should be rejected flat. The program officer then determines how far the available funds can be stretched: many proposals judged meritorious lose out at this stage, and the applicant has the hollow satisfaction of a letter that says We liked it but—

Unfair, unreliable, and costly: the full range of criticisms of peer review of grant applications must lead to the question, Can it be fixed, can it work at all? First, the charges of bias. Most often raised are four: institutional, with premier research centers unduly favored; age, with the more established scientists more likely to get grants; sex, with women doing less well, proportionally, than men; and cronyism, or institutional nepotism, with reviewers favoring applicants with whom they are personally affiliated. The first two are surely plausible, yet although practicing scientists are concerned about them, some deeply so, what studies have been done of institutional or age bias fail to reach the experience of panelists—subtler, more vague. They offer no convincing support or disproof.

Gender bias is a different matter. The most damning study was carried out by two Swedish immunologists, Christine Wennerås and Agnes Wold. Throughout the world, although women in increasing numbers have been obtaining doctorates in the sciences, men still dominate the upper levels of academic research. In Sweden, for example, Wennerås and Wold noted, "women are awarded 44 per cent of biomedical PhDs but hold a mere 25 per cent of the postdoctoral positions and only 7 per cent of professorial positions." Dwelling on such numbers, in 1995 they got a grant from the Swedish Ministry of Education "to investigate whether the peer-review system of the Swedish Medical Research Council (MRC), one of the main funding agencies for biomedical research in Sweden, evaluates women and men on an equal basis" in choosing which research proposals to support among applicants for postdoctoral positions.

In 1995, twenty postdoctoral fellowships were offered; one hundred and fourteen scientists applied, sixty-two men and fifty-two women, with a mean age of thirty-six. Applications go to one of eleven different review committees, each covering a particular research field, each with five or more reviewers. Invoking the Swedish Freedom of the Press Act, comparable to the American Freedom of Information Act but far stronger, Wennerås and Wold obtained the primary documents, the reviewers' work sheets for each applicant. Three factors were supposed to count. Two of these judged the applicant's research proposal—its "relevance" and the quality of its proposed methodology. The third was the applicant's "scientific competence." In practice, scientific competence meant productivity, the number and quality of original papers published.

Their first observation was that reviewers gave women lower scores on all three factors, judging them most damagingly on scientific competence. Scores on the three factors combined left female applicants, on average, nineteen per cent behind the men. "That year, four women and 16 men were awarded postdoctoral fellowships." Judgements of relevance and methodology are unavoidably to some degree subjective; competence, or productivity, the women's apparent weakest factor, the reviewers also judged subjectively—but can be quantified objectively. So the first question was, Were the women indeed less productive? Wennerås and Wold constructed a measure using the number of publications weighted by the applicant's rank among the authors of each paper and by the quality of the journals where published. Rank was simply whether the applicant was a paper's lead author or not, relying on the convention that begins the byline with the scientist contributing most importantly to the research. For a journal's quality they took its impact factor, which is computed, for more than eighty-five hundred journals publishing tens of thousands of papers a year, by the Institute for Scientific Information, in Philadelphia, as the number of times an average paper appearing in a given journal is cited by papers elsewhere during a year. Top specialist journals—the authors offered examples, including *Gut, Neuroscience,* and *Radiology*—have impact factors around 3. *Nature* has an impact factor of 25, *Science* 22.

The methodology was rigorous, the presentation relentless, the conclusions unequivocal, their impact high indeed. The report appeared in *Nature* in May of 1997. The first conclusion was that "a female candidate had to be 2.5 times more productive than the average male candidate to receive the same competence score as he." And, they said, "This represents approximately three extra papers in *Nature* or *Science* . . . or 20 extra papers in a journal with an impact factor of around 3." Wennerås and Wold considered a variety of other factors that one might think could influence ratings. Statistical analysis ruled all but one of these out. "The applicant's nationality, education, field of research or postdoctoral experience did not influence competence scores."

The exception: their analysis of sex bias yielded a bonus, strong evidence of cronyism. A reviewer for the Swedish Medical Research Council who is personally or professionally associated with a candidate "is not allowed to participate in the scoring of that applicant." Yet the statistics demonstrated that "this rule was insufficient, as the 'neutral' committee members compensated by raising their scores when judging applicants affiliated with one of their peers." The effect of this was about the same for men and women, and was as powerful an influence on an applicant's overall rating as was gender bias. Thus the second conclusion was that their analysis, the first ever "based on actual peer-reviewer scores," provided "direct evidence that the peer-review system is subject to nepotism, as has already been suggested anecdotally."

(The last line of the last page: "*Nature* adds: this article was peer-reviewed by three males.")

Irrespective of biases, critics—and not just those whose own proposals have been turned down—assert that grant peer review is not effective in choosing the best work to support. Can a panel, scoring the best of the proposals before it, predict which of those near the top will produce the best research? The most perplexing variant of this that scientists raise is that the system favors safe research. Neither applicants nor panel members are likely to take risks. Unorthodox ideas, science that might turn out to be radically new, is not going to be funded through peer review. The point is plausible on its face and has been made repeatedly, vehemently, by eminent scientists (David Baltimore among them) from their personal experience. The NIH acknowledges the criticism. They have a Center for Scientific Review, whose director until the summer of 2003 was Elvira (Ellie) Ehrenfeld. In a recent conversation she said, "There is a tendency to give very good scores to projects that are guaranteed to work," and went on, "In general, I think NIH peer review works quite well. But there is a fraction of stuff—impossible to say how much—that will never get past peer review," and not that it's substandard but because it's too unusual for the present system to evaluate and to risk. In sum, whether the system is in fact an unreliable way to pick winners is peculiarly hard to get at. How can one compare the outcomes of science funded and not funded?

The charges most relevant to fraud are two. First, many believe that grant review facilitates plagiarism. The anonymity of the process is sacrosanct, yet is another potential contributor to decline and corruption. Panelists would be useless if not involved with the specialties they are dealing with. But the better the applications before them, the more they may be able to learn about new ideas and methods relevant to their own research, in that seminar with top people on the most advanced work which Robert Pollack described. This should be a strength, on balance—yet at the extreme it can lead to what in this context is euphemistically called leakage. Tales and grumbles abound. Hard data would be intrinsically hard to come by. None exist. The matter is quite otherwise in the refereeing of journal articles, as shown by a number of recent cases of such fraud and discussed below.

The second issue for those concerned about fraud is a further variant on the bias towards success. Rarely do scientists even bother to point out that research that repeats others' work has no chance to get funded. So there goes the hope that science can be self-correcting.

The greatest and most indisputable problem with the present system of peer review of grant applications, the one charge that can be documented irrefutably, is the inordinate expense of research time in writing grant applications and reviewing them. It is expensive in money and extravagantly so in time stolen from productive research—and as with S and his colleagues, this cost falls for the most part on scientists in what should be their most productive years.

The question remains: What's the alternative?

Periodically—and whenever the National Institutes of Health get a new director—they announce reforms and revisions in procedures for grant applications. Typically, these involve efforts to cut down on the detail and hence the time that writing a grant requires, for example, by simplifying applications to renew a grant already in operation. Thus, the NIH offers a number of Merit Awards, given to scientists whose applications have regularly scored well: they allow the scientist to bypass full peer review for several years to continue a particular line of work. The most interesting recent change in NIH funding methods are the Director's Innovator Awards, intended to respond to the charge that high-risk, potentially high-payout research can't get grants. These awards go not to projects but to outstanding individual scientists, to enable them to follow their instincts. They began in the 2004 fiscal year. Ehrenfeld now runs this program. She calls these "people awards." She says that over the next several years they may be complemented (if Congress agrees to fund them) by two more new ventures. One she calls "special projects," awards "that will be reviewed separately in a risky-but-potentially-ground-breaking context." The other she describes as "top down," or officially "grand challenges": the NIH will itself designate new areas, problems, ripe for development and seek the right scientists to take them on. Yet the scale of all these ventures is modest, which only emphasizes the greater problem. They will be pasted onto the present system, and by no means represent fundamental reform.

Alternatives exist. In Britain, the Medical Research Council has long followed the practice of finding a young scientist of promising talent in a promising field, giving him a small grant and following his first successes with further, larger grants, like starting a campfire with twigs and feeding it larger and larger sticks and chunks of fuel. The most conspicuous instance was the support the MRC gave in 1947 to a young x-ray crystallographer at the Cavendish Laboratory in Cambridge, Max Perutz, who was attempting to get the structure of hemoglobin, a large protein molecule. That grew into the MRC unit of four scientists including Francis Crick that James Watson joined in September of 1951. By 1962, when Perutz with his colleague John Kendrew shared the Nobel prize in chemistry and Watson and Crick shared (with Maurice Wilkins, from King's College London) the prize in physiology or medicine, that unit had grown so large that the MRC spun it off from the Cavendish into a separate building of its own. The Medical Research Council Laboratory of Molecular Biology has been one of the world's premier biological institutions, with nine Nobel prizes so far, more than France. And when Perutz retired as director, in 1979 as he was required to do at sixty-five, the MRC followed its standard practice and completely reorganized the laboratory. (He kept working and publishing; when he died in 2002, aged eighty-eight, he left yet two more papers in press.)

Fund the scientist rather than the project. This is what Warren Weaver did in the 1930s. It does happen in the United States today. Rosalind Yalow worked for decades at the Veterans Administration Bronx Medical Center, never had to write a grant proposal, was always adequately financed, and won the Nobel prize in physiology or medicine in 1977 for inventing a method called radioimmunoassay for detecting minute quantities of target substances in blood or other liquids. She has called repeatedly for the system that supported her to be extended. The Howard Hughes Medical Institute is America's second-richest foundation (after the Bill and Melinda Gates Foundation). In fiscal year 2003, starting 1 September 2002, it spent four hundred and seventy-nine million dollars in direct support of its own investigators in its own laboratories, which are set up at a number of leading American research institutions. (It also has a division that awards grants in the more usual way, one hundred and seven million in 2003, but these go for science education with a small amount given individually in support of a number of international scientists.) Its vice president for grants and special programs, Gerald Rubin, said recently, "HHMI has done this experiment. It works." The Lucille P. Markey Charitable Trust was established in 1983 as a spend-down trust, meaning that its founder had directed that it give out all her money within fifteen years of her death and then close its doors. That was well over half a billion dollars for basic biomedical research. In 1984, the trust launched the Markey Scholars program, supporting selected young scientists for seven years, the last three of their postdoctoral years and their first four years on their own, so that in this critical period in establishing their careers they were free of the pressure to write grant proposals. Five years ago, shortly after the Markey Trust closed, the W. M. Keck Foundation began a similar program though on a more niggardly scale.

Confounding all, defeating all, depressing all, has been the fact that so long as only the most highly rated applications can be funded, to suppose that a study section can make reliable and just decisions is ridiculous. Compounding that problem, even as scientists have themselves become more skillful at presenting their work and plans, the past decade has seen the emergence of a new specialty, the professional grant-writer. Thus, when an application must be compared to several others any one of which could be in the eighty-eighth or just as reasonably in the ninety-second percentile—and the cutoff is at the ninetieth—peer review becomes a lottery. At every meeting, panelists are made bleakly aware of this. As are scientists nationwide when awards are announced.

The sum of these influences is a grim problem of morale. The overloading of the system, the volume of applications and the passage of decades, the increasing sense of unease, even futility, can only discourage top people from continuing to play. True, funding has grown. In a series of increases in the five years through fiscal year 2003, Congress fully doubled the total appropriated to the NIH. (Fiscal-year 2004 over 2003, and 2005 over 2004, show much smaller increases, barely matching the inflation rate.) Yet the structural problems of the research-grants system will not go away.

*

"Refereeing is expensive, time-consuming, inaccurate, subject to bias or worse, doesn't catch errors let alone fraud—and there's no objective evidence that it works better than the independent judgement of a qualified editor." Thus emphatically Richard Smith, editor of the *BMJ*, in a conversation at his office, in London in the summer of 2002.

A change of seismic proportions is now under way in the publication of scientific papers. As one element in this revolution, refereeing of papers is being transformed. Smith and his august journal are among the leaders of change—which shows how serious it is, how far it has come.

A myth is told of the origins of peer review of scientific papers. Like other origin myths it is devoutly believed by many; like other origin myths it is false, bearing little relation to the world of today. But it makes a pretty story. The conventional history has it that refereeing dates back to the mid-eighteenth century and the Royal Society of London. The society itself had been founded a century earlier. Looking into its beginnings, one is struck by the contagious excitement, the sheer energy of the founders' interest in what they called "the *New Philosophy* or *Experimental Philosophy*." A number of natural philosophers had been meeting weekly in London as early as 1645, during the civil wars, and in Oxford as well by 1648. They were "our invisible college or philosophical college"—the phrase grown famous that the physicist Robert Boyle used in letters then. The monarchy, in the person of Charles II, was restored in the spring of 1660. On Wednesday 28 November of that year, twelve men gathered at their usual meeting place, Gresham College, in London, to hear a lecture by Christopher Wren, Gresham's professor of astronomy (and architect and polymath, who after the great fire, September 1666, built the new St Paul's Cathedral and fifty other churches in the City of London). In discussions after Wren's lecture, "something was offered about a designe of founding a Colledge for the promoting of Physico-Mathematicall Experimentall Learning," according to the memorandum of the meeting in the journal book they started that day. They appointed a chairman, drew up a list of forty-one suitable candidates for membership, and agreed to meet every Wednesday afternoon.

The following week, said the entry for the day, one of their number brought word that "the King had been acquainted with the design of this meeting. And he did well approve of it, and would be ready to give encouragement of it. It was ordered that Mr. Wren be desired to prepare against the next meeting for the pendulum experiment." Eighty-five signed up. In October of the next year, Charles himself offered to become a member; in July of 1662, the first charter of the Royal Society passed the Great Seal. In a warrant of 1663, the king ordered a silver mace to be presented to "the Royal Society for the improving of Natural Knowledge by experiments." At the weekly meetings, experiments were demonstrated; the other matter discussed was the correspondence, vigorously pursued, with like-minded natural philosophers on the Continent, and this emerged as the most important work of the early society.

On the first of March of 1664–1665* the society "Ordered, that the *Philosophical Transactions,* to be composed by Mr Oldenburg, be printed the first Monday of every month, if he have sufficient matter for it; and that the tract be licensed under the charter by the Council of the Society being first reviewed by some of the members of the same." Henry Oldenburg was the notable one of two secretaries first appointed by the society. He brought out the premiere issue a week later. Oddly, in Paris less than two months earlier the *Journal des Sçavans* was launched—not the least of the many simultaneities in the history of the sciences. Though by a few weeks not the earliest of all scientific journals, *Philosophical Transactions* has been in continuous publication to the present day.

Its charter called, thus, for peer review of a sort. But the essential features were informality and openness, while responsibility for selection of material remained with Oldenburg. Secrecy was not unknown to the society: from 1663, debates over nominations for membership were closed. Yet the world of science was tiny. Everyone knew what

*Britain did not adopt the Gregorian calendar, with the year beginning on January 1, until 1752; interestingly, the resetting of the calendar was supervised by the Royal Society—an early instance of the government calling on the society for advice and aid in scientific matters of national moment. By that time, the old-style calendar because of the way it had used leap years was ten days behind the new. Rioters protested the loss of the ten days.

everyone else was doing. The correspondence and the demonstrations for the journal had already been discussed at the meetings.

A century later, in 1752, the society reorganized publication of *Philosophical Transactions.* They set up a Committee on Papers to review all articles. The committee could call on "any other members of the Society who are knowing and well skilled in that particular branch of Science that shall happen to be the subject matter of any paper which shall be then to come under their deliberations." This has often been marked as the birth of refereeing. John Ziman, an English physicist turned commentator on the sciences, has called it "the key event in the history of modern science."

Despite this hallowed story, refereeing of journal articles is a creation of the twentieth century. Its development is little explored, the territory rough, the evidence scant and scattered. As the sciences in the nineteenth century though vigorous were less specialized and far less populous than today, one might suppose that journals were then few. On the contrary, they multiplied, particularly medical journals. But we would hardly recognize them, just as we are bemused by newspapers of those days, and for some of the same reasons. Most were founded and run by individuals; some of the most celebrated came out under the names of their editors, an early example being *Hoppe-Seylers medicinisch-chemischen Untersuchungen* (founded by Felix Hoppe-Seyler in 1877 as *Zeitschrift für physiologische Chemie*), the first journal of biochemistry. Many of the minor journals, particularly in medicine, began as vehicles for editorial opinion rather than evaluated research. Many editors rather than picking and choosing had actively to seek out material to fill their columns. Yet among the best journals professional responsibility and especially the increasing specialization of the sciences led, though sporadically, to change. Thus in 1893 Ernest Hart, editor of the *British Medical Journal,* told an audience of American medical editors:

> The only system that seems to me adequate to the real needs of professional readers is that in which every unsigned editorial paragraph is written by a specially selected expert. That is the principle on which I have modelled the journal I have the honor to conduct. Every letter received, every paragraph, every cutting editorially dealt with, is referred to an expert having special knowledge and being a recognized authority in the matter.

This was a manifesto. He went on:

> It is a laborious and difficult method, involving heavy daily correspondence and constant vigilance to guard against personal eccentricity or prejudice or—that bugbear of journalism—unjustifiable censure. But that method may, I venture to think, be recommended as one that gives authoritative accuracy, reality, and trustworthiness to journalism. A medical journal, in order to rise to the height of extended usefulness, needs to be written from end to end by experts; and so far as the [*British Medical Journal*] may be considered to have been a success, that success has, I believe, largely been due to the fact that no pains or necessary outlay has been spared to provide that every line in every department shall be written by persons who are themselves trustworthy experts.*

No better justification and standard for refereeing could be written today. But the practice of referring to experts did not spread. Well into the twentieth century, journal editors by and large were like Oldenburg in the young years of the Royal Society, in that they worked single-handedly and judged submissions on their own authority.

*I owe the reference to the historian John Burnham, who, though perceptibly frustrated by the paucity of evidence, in several reports has bagged a wide-ranging assortment of materials about the prehistory of journal refereeing.

They could do so: the sciences were not necessarily simpler then, but laboratories and their papers were so very much fewer that with experience and plentiful contacts an editor could be fully cognizant of the field his journal covered. Refereeing of journals spread idiosyncratically, piecemeal. As with peer review of grant applications, only after the second world war did it become standard.

Speaking to the assembled biomedical editors in Chicago in 1993, I said that the refereeing process is inherently fragile, and that its decline is closely parallel to what has happened in peer review of grant applications. It, too, is under pressure because the enterprise of the sciences has grown so. The half-century has seen not only the vast increase in the number of scientists but the multiplication of specialties and subspecialties—and notoriously the proliferation of journals.

Estimates vary, but upwards of fifty thousand peer-reviewed scientific journals are now publishing, perhaps ten thousand in biomedical fields alone. Publishing printed journals can be immensely profitable. Cold economic calculation rules. When the price of subscriptions passes a certain point, individuals can no longer afford their own and none will buy but science libraries that feel a compelling necessity. At this point, demand becomes inelastic. That is, the publisher can kite the price far higher with little further loss of circulation. Indeed, the most specialized have circulations in the low thousands, while subscriptions to them may run from two thousand to as high as five thousand dollars a year. Those equations have a corollary. Added up, from the moment a manuscript arrives at an editorial office to the moment the first subscriber reads the resulting paper, for the most specialized journals the costs of printing, production, and distribution, plus the costs of the refereeing process, plus the other routine in-house editorial costs, push the average cost of papers up to three to five thousand dollars, sometimes higher, for each one actually published. In still another dimension, across those thousands of journals the total cost of the refereeing process in time and money is vast—and this despite the fact that referees serve without pay. This, of course, parallels the problem in peer review of grant applications.

Who pays? The science libraries that subscribe write the purchase orders. But their budgets come almost entirely from what are called *overheads* in government research grants. These are the percentages that are added on to the basic, direct-to-the-laboratory amounts to cover what are loosely termed administrative costs, from the salaries of secretaries to the travel of university presidents, from copying machines and telephone calls to the stipends paid graduate students and postdoctoral fellows in the sciences to the accountants who determine these numbers—and to libraries. A research university negotiates with the federal agencies, every several years, to set its overhead percentage, justifying it with itemizations of costs. It then applies this to the budget for each grant application. Typically, a university's overheads for the NIH or the NSF will run to fifty per cent, more or less. Thus, in a grant application of my own to the National Science Foundation, for two years of a research project at George Washington University in history of molecular biology, the $449,085 budget for direct costs was inflated by overheads to $605,379. (I was turned down, largely because of the expense.)

Some rates are startlingly high. In the mid–nineteen-eighties, Stanford University after several rounds of negotiation with the Office of Naval Research had driven their rate up to seventy-nine per cent. When congressman John Dingell's investigative subcommittee looked into the expenses the university had itemized, and discovered they included maintenance of a yacht, flowers for entertainment at president Donald Kennedy's house and sheets for his beds, Dingell hauled Kennedy before the committee and denounced the scandal so vigorously that the university repaid considerable amounts. A number of other universities quietly lined up to repay excessive overheads. Six months later, Kennedy had to resign. (But the network takes care of its own: Kennedy retained his professorship and in June of 2000 took over as editor of *Science.*)

Overheads of course come out of congressional appropriations for research. In a telling contrast, private foundations allow much lower overheads, often none at all. The federal figures are curiously hard to come by; what follows is an order-of-magnitude estimate, for no one appears to compile the numbers this way. The National Institutes of Health and the National Science Foundation are the chief, though by no means the only, federal agencies funding research. The NIH does considerable research intramurally, mostly at the vast laboratory, hospital, and administrative campus in Bethesda, Maryland. This part of the NIH appropriation amounts to about ten per cent; leaving that out, in fiscal year 2003 the NIH gave 24.5 billion dollars in extramural grants. Approximately one third of that went to indirect costs, which indeed means an average overhead allowance of about fifty per cent—amounting to 8.2 billion dollars. The National Science Foundation gives far less money. Leaving out grants for educational activities as well as the foundation's administrative costs, in 2003 about 3.9 billion went to research, all extramural. Questioned, a spokesman came up with an estimate that in 2003 the foundation allowed institutions overheads at an average of about nineteen per cent, which would yield a total of six hundred million dollars. In sum, the two agencies paid out some 8.8 billion dollars in overheads that year. There's still space on the back of this envelope for an educated guess of the overheads other federal agencies pay out: we can say that by this route the federal government subsidizes American research universities at the rate of about ten billion dollars a year. Without this, many would be hard pressed. In short, as overheads support libraries, not only do print journals exploit referees, they profit by exploiting the entire system of government-sponsored research—ultimately paid for, or course, by the taxpayer.

Of all those thousands of journals, a very few are high impact, meaning as we saw that papers they publish get cited far more frequently than do those appearing elsewhere; their content is relatively general, interesting to readers of many specialties, and their circulation is large. You are to suppose that here's the important stuff—and so high impact is self-fulfilling. In biomedical publishing, high-impact journals are *The New England Journal of Medicine, JAMA, BMJ,* and *The Lancet.* Others include *Physical Review Letters, Proceedings of the National Academy of Sciences* of the United States, and *Annales de l'Institut Pasteur.*

But by far the highest-impact journals are *Science, Nature,* and *Cell. Science* is published weekly by the American Association for the Advancement of Science and a subscription comes with membership; its circulation is about one hundred and forty thousand. *Nature,* also weekly, is owned by Macmillan in London, and is a richly money-making enterprise, with a worldwide circulation of some sixty-seven thousand, sixty per cent in the United States. The two compete for the big papers. In an egregious instance, the two not-quite-complete sequences of the human genome were published in the same week in February of 2001. The one produced by the public, international consortium of more than twenty laboratories, coördinated by Francis Collins, at the NIH, and John Sulston, at the Sanger Institute, outside of Cambridge, England, appeared in *Nature. Science* carried the sequence generated by Celera, a private company founded and run by Craig Venter, a venture scientist in hostile competition with Collins and with the intention of profiting from the sale or licensing of genome information. *Cell,* biweekly and more specialized, has a circulation in print of about ten thousand but heavy readership on line; personal and institutional subscribers download copies of upwards of half a million articles a month.

These three provide the most telling example of the effects of competition among journals and their editors and publishers. Scientists view them with ambivalence. They scorn them as hot-news journals—"the tabloids," they call them. Yet to have a paper accepted by one of them is highly rewarding in readership and prestige. (It's true: I have published commentaries and book reviews in *Nature* and *Cell*, and the telephone rings, the e-mail notes fly.)

The effects of the competition among the tabloids have been invidious. David Botstein, an expert in the molecular genetics of yeast, observed to me a while ago that he or any of his colleagues used to be able to pick up a newly published paper in their field, read it through, and judge with some confidence its credibility and worth. This, he said, is now rarely possible. The reason is only in part the increasing complexity of the subjects—cell biology, say, compared to molecular genetics. More important is the pressure passed from publishers to editors and on to would-be contributors to keep papers short, assertive, punchy. Papers get condensed. Discussions and conclusions are simplified. Qualifications, cautions, alternative interpretations are abbreviated or pencilled out. The use multiplies of that indicator "(data not shown)"—a temptation to sloppiness or worse. The general scientific reader, the idealized other for whom the editor ought to imagine the journal is publishing, is baffled, as my informant suggested, and is even to some degree misled.

To be sure, what by the mid–nineteen-nineties had become an acute problem in the tabloid journals may be present in a milder but chronic and concealed form in papers published without such overt editorial pressures. In a subtle study, Richard Horton, editor of *The*

Lancet, raised the question "whether the views expressed in a research paper are accurate representations of contributors' opinions about the research being reported." He selected ten papers from his journal, with multiple authors and of varied subject and method, requested permission from the lead author of each to write to the others sharing the byline, and asked them all to answer six open-ended questions. Of the fifty-four contributors, two-thirds replied. The questions seemed innocuous: "In your own words, how would you: 1. Summarize the results of your study? . . . 3. Define the weaknesses of your study? . . . 5. Assess the implications of your results? . . ." The conclusions were far from innocuous.

> The results of this qualitative study show that a research paper rarely represents the full range of opinions of those scientists whose work it claims to report. I have found evidence of [self-] censored criticism; obscured views about the meaning of research findings; incomplete, confused, and sometimes biased assessment of the implications of a study; and frequent failure to indicate directions for further research. . . . What was striking was the inconsistency in publishing evaluations, especially regarding weaknesses. . . .
>
> A scientific paper is an exercise in rhetoric; that is, the paper is designed to persuade or at least to convey to the reader a particular point of view. When one probes beneath the surface of the published report, one will find a hidden research paper that reveals the true diversity of opinion among contributors about the meaning of their research findings. For both readers and editors, the views expressed in a research paper are governed by forces that are clear to nobody, perhaps not even to the contributors themselves.

(In the light of these strictures, I should note that Horton discussed his work's limitations and what might next be done.)

In the tabloids, the editorial forces are open and often fierce. Consider, then, that the first independent readers to be affected by these practices are, after all, the referees—which leads to the paradox that even as the general reader finds himself dependent more than ever on the assurance seemingly offered when some particular paper appears in a refereed journal, the referees themselves are being deprived of the means to be confident in their judgements. This is a problem that is hidden but not trivial. Note the parallel with the review of grant applications: in each mode, though in different ways, the grounds for decision have become opaque.

As with review of grant applications, the first difficulty that evaluation of journal refereeing presents is opinion in abundance and the paucity of hard evidence. Hundreds of studies have attempted to evaluate the effectiveness of refereeing. At the Fourth International Congress on Peer Review, this one in Barcelona in September of 2001, Tom Jefferson and three colleagues, medical scientists all, presented an analysis of such studies. They had located one hundred and thirty-five "comparative studies assessing the effects of any stage of the peer-review process that made some attempt to control" for confounding factors. Of them, only nineteen approached the criteria they had set for quality. To these they applied methods that have been developed over the past two decades—rigorous, proven methods for rational evaluation of claimed scientific research into the efficacy of clinical procedures. Though they didn't put the matter quite this bluntly, even the nineteen best studies failed. "We discovered remarkably few well-designed studies assessing the effects of this process," they said.

> Given the widespread use of peer review and its importance, it is surprising that so little is known of its effects. However, the re-

> search needed to address these questions would require a well-funded and coordinated effort involving several sectors of the scientific community as well as the cooperation of large numbers of authors and editors, and the methodological issues involved in conducting proper studies of the subject are daunting.

Their curt conclusion: "Until such research is undertaken, peer review should be regarded as an untested process with uncertain outcomes."

Yet perhaps we are not altogether reduced to anecdote. As with peer review of grants, the underlying question is, Can it be expected to work at all? One difference: unlike unfunded grant proposals, papers rejected by a high-impact journal are not necessarily dead. Stephen Lock summed the evidence a while ago. "A high proportion of articles submitted to top journals, but rejected, get published elsewhere, often without revision." At that rate, refereeing at best may help top journals screen out poor papers, but does little to protect the integrity of the scientific record over all. But no studies rigorously constructed have demonstrated that the system even at the top picks better papers more reliably than editors would do without refereeing. Nor are referees likely to catch authors who publish essentially the same material more than once. Lock again: "Every editor can tell of examples where neither he nor the referee has detected that exactly or almost exactly the same work has already been published, without any mention of this by the author." Worse still, refereeing cannot help but create conflicts of interest between author and reviewer over research methods and results: plagiarism is a built-in temptation. And a temptation yielded to: recall the convoluted comedy of the Soman–Felig–Wachslicht-Rodbard case.

Why does refereeing persist? Scientists are innately conservative, to be sure. Academic standing seems to them to depend on extensive publication in peer-reviewed journals. Yet that smells like an evasion.

We must put such individual prejudices in their setting. Elizabeth Knoll, whom I quoted earlier, advanced an explanation that reaches to the simple sociology of present-day science. "In many respects editorial peer review is now a bureaucratic rather than a collegial process; it is used for many reasons, only one of which is trying to discern the strength of an article's argument. Journals use peer review because it is the way the game is played—not simply because the editors need consultation," she said, and went on:

> New journals spring up virtually every month, sometimes to cover remarkably tiny subspecialties, and their announcements emphasize that the journals will be peer-reviewed. In cases like this, the use of peer review is more a claim to professional status than an intellectual necessity, since the editors of such a highly specialized journal are surely able to judge most of the specialized material they receive.

Anonymity for referees is all but universal among scientific journals, defended as essential. Of the many ideas for saving the system that have been urged from time to time, the most recurrent and interesting is that this anonymity be scrapped. A strong plea for this has been made and reiterated by Drummond Rennie. In 1994, he wrote, "The only ethically justifiable systems of peer review are either completely closed (with no one but an editorial assistant knowing the identity of the authors and only the editor knowing the identity of the reviewer) or completely open." Yet a closed system cannot succeed: various trials, randomized and controlled, have found that when authors' identities were kept back reviewers none the less correctly identified the authors of a quarter to nearly half the papers.

The case for open reviewing has been made repeatedly, accepted rarely—with recent significant exceptions. The most eloquent statement of the case has been made by Fiona Godlee, who is well placed to judge whether a change to open refereeing is not only right but feasible. She was an editor at *BMJ,* which in 2000 made open refereeing the policy; she went to BioMed Central, perhaps the most successful of on-line publishing ventures, to be medicine editor, and introduced open refereeing there; she moved back to the BMJ Publishing Group, where she is head of a unit called BMJ Knowledge. In a piece in *JAMA,* she concluded, "I look forward to a time when open commentary and review replace the current, flawed system of closed prepublication peer review and its false assurances about the reliability of what is published." She gave four reasons: that open refereeing is ethically superior, that it has no serious adverse effects on the quality of reviews, that it is feasible, and that while it imposes greater accountability on reviewers it offers public recognition for the work they do. Her simplest yet most compelling reason was the flat assertion of ethical superiority: quoting Rennie in a statement from 1998, she wrote, "It is not open review that should have to justify itself but the 'ethically unequal and inconsistent system' practiced by most journals, in which authors are identified and peer reviewers are not."

Should we be surprised that the ethical argument appears to have practical force? A leader here, again, is Richard Smith and the *BMJ.* Because refereeing has not been studied objectively, he said in our conversation, he has moved towards open reviewing in incremental steps and by way of trials, as objective as they could make them. They have some five thousand scientists and clinicians they call on, occasionally or frequently, to serve. A preliminary survey found that a high proportion of these objected to open refereeing. None the less, in 1999 they tried an experiment, a randomized trial. They put submitted papers, as yet unpublished, into two categories, A or B, arbitrarily in the order received. Papers of one of these groups went to reviewers assured of their usual anonymity, while those of the other group went to referees who were told their names would be made known to the authors. They developed a protocol or instrument that objectively measured the quality of each review. Open reviewing seemed to have no negative effects. So the next year they converted entirely. When it came to it, of the five thousand referees only some forty refused to relinquish anonymity. Then in the summer of 2002 Smith moved the *BMJ* to electronic submission of manuscripts—and, of course, of referees' reports. The next step, after suitable trials, he said, will be to make the open fully open, putting every manuscript and its referees' reports on a web site for anyone to read. "This will transform the refereeing process, moving it towards opening a scientific discourse."

Extravagant words like "revolution" become devalued by overuse, worn thin. But now, very now, a revolution is indeed under way in publication of scientific papers, a revolution as great as anything since the invention of the printing press—for it amounts to abandonment of the printing press. The move to open reviewing is one small part of this.

7

AUTHORSHIP, OWNERSHIP

PROBLEMS OF CREDIT, PLAGIARISM, AND INTELLECTUAL PROPERTY

Authorship no longer carries the responsibilities it should—the ability to justify intellectually the entire contents of an article.

—*Stephen Lock,* British Medical Journal, *1988*

In the social system of the sciences, nothing matters more than priority. Recall Robert Merton's formulation of the dynamic: the community enforces its norms, insofar as it does, by a pair of rewards. These are to be recognized for originality, which puts high value on priority, and to be held in esteem by other scientists. Priority is the main driver of esteem. To have priority ignored, denied, stolen—nothing else can generate such bitterness, such anger.

The Royal Society of London has occupied since 1967 a magnificent Georgian town house, an aristocratic town palace in Carlton House Terrace. The Fellows' Room looks out over the Mall and St James's Park, with Admiralty Arch near at the eastern end, Buckingham Palace away to the west. Portraits of dead great scientist fellows make heavy the walls. Nearest the windows on the left hangs Gottfried Wilhelm Leibniz, a German mathematician of genius, elected fellow in 1673. Leibniz peers across to the opposite wall, where Isaac Newton is suspended, elected president of the society in 1703. Long of face, long of nose, pale, Newton stares haughtily out the window. The two men and their partisans quarrelled rancorously, messily, for years, over priority for the calculus. Newton had invented the calculus in the seventeen-sixties, and had written it up but not published it. Leibniz independently developed the same methods with a different notation ten years later, let them be known to the Royal Society in 1676, and published in 1684. Controversy erupted. An international scandal blew up; at its height, in highly publicized correspondence, each was calling the other a fraud and a plagiarist.

History burps—repeats itself—as comedy. On Monday, 6 October 2003, in Stockholm, Hans Jornvell, secretary to the Nobel committee of the Karolinska Institutet, announced the year's prize in physiology or medicine, split between Paul Lauterbur, of the University of Illinois, and Sir Peter Mansfield, of the University of Nottingham, in England, for the invention of magnetic-resonance imaging, or MRI, a diagnostic technology widely used in medicine. Controversy erupted. Full-page advertisements appeared in *The Washington Post* on Wednesday, *The New York Times* on Thursday, and in the *Los Angeles Times* and the Swedish paper *Dagens Nyheter.* Headlined THE SHAMEFUL WRONG THAT MUST BE RIGHTED, black with massed ranks of angry type, the ads charged at length that the Nobel committee had "ignored the truth" and "is attempting to rewrite history." In what way? By failing to include Raymond V. Damadian, a physician and inventor who has long asserted that it was he who made the initial discovery of MRI and that Lauterbur and Mansfield merely refined the technology. The protest, so bizarre, attracted the attention of the *Times*'s Nicholas Wade, who covered it in a story on that Friday, where he estimated the cost of the ads at about eighty thousand dollars in the *Post* and one hundred thousand in the *Times*. The advertisements were placed by the Fonar Corporation, of which Damadian is president. Then ten days later the *Times*'s op-ed page ran a commentary I wrote comparing Damadian's conduct with that of a score of other scientists who over the years have been left out of Nobel prizes, and chiding him for his unseemly whining. Not recognizing the counter-productive effects, over the next two months Damadian ran a half-dozen more of those full-page advertisements, each more black with outrage than the last. On December 9, the day before the climax of the Nobel ceremonies in Stockholm, Damadian sprang for a full two-page spread in the *Times*, this time with splashes of muddy color. Total cost of the campaign, Damadian has acknowledged, was one million, two hundred thousand dollars.

While the fact is that Damadian took an early first step in MRI and holds a basic patent, many scientists in a position to know maintain that the two to whom the prize went were indeed the ones whose work led most directly to the present technology. But putting aside any merits of Damadian's claim and the extravagant language and cost of his advertisements, consider how the Nobel committees have dealt with priority, in particular with precursor discoveries.

Non-scientists are often surprised to realize how frequently Nobel prizes have been awarded to inventors of new technologies. In 1926, Theodor Svedberg got the chemistry prize for the ultracentrifuge. In 1948 and again in 1952, the chemistry prize went to inventors of electrophoresis and chromatography, methods by which vanishingly small quantities of very similar substances can be separated from each other in solution, isolated, and identified. Frederick Sanger, one of only three scientists ever to receive two Nobel prizes in science (Marie Curie, John Bardeen), got the first for demonstrating how to sequence the amino-acid chains that make up proteins and the second for how to sequence DNA. Prizes have also been given for the use of radioisotopes as chemical tracers and latterly for invention of the polymerase chain reaction, by which a few molecules of DNA containing a gene of interest can be amplified a billion-fold so they can be sequenced. All of these were precursors of multiple discoveries: they enabled scientists to ask whole new categories of questions.

Predecessors have been included a number of times in Nobels awarded for later work. In 1945, Ernst Boris Chain and Howard Florey got the prize in physiology or medicine for penicillin, but shared it with Alexander Fleming, who indeed first observed the antibiotic properties of the mold that secretes the stuff but never developed it. In the most gratifying instance, in 1986 Gerd Binnig and Heinrich Rohrer received the physics prize for the scanning tunneling electron microscope—and Ernst Ruska was included, who had invented its remote ancestor, the first electron microscope, in 1933, and was fortunately still alive. Such cases would seem relevant to Damadian's complaint.

Longevity can make the difference, for the prizes are never awarded posthumously. Barbara McClintock, an American geneticist who did important work in the 1930s and 1940s, won the prize in physiology or medicine in 1983 at the age of eighty-nine—and although nobody doubts it was deserved, the oddity here is that according to her biographer the particular discovery cited in the text of the award was not original with her.

Deserving scientists have all too frequently been skipped over. Lise Meitner, the physicist first to recognize that experiments reported by two former colleagues in Berlin meant that atoms had been split, never got a prize, even though one of those colleagues did, Otto Hahn in 1944. Arne Tiselius, himself a laureate (chemistry, for electrophoresis, 1948) and a big man in Nobel deliberations at the Karolinska, admitted that the most conspicuous omission was Oswald Avery, the first to establish that genes are made not of protein but of DNA. Rosalind Franklin, whose data James Watson and Francis Crick used in 1953 to elucidate the structure of DNA, died before the prize in physiology or medicine went to them in 1962. The third scientist in that prize was Maurice Wilkins, a colleague of Franklin's, who indeed spent seven years refining the structure—after it was published—but whose crucial contribution to the discovery itself was to show an x-ray-diffraction photograph Franklin had made of a form of DNA to Watson, from which he learned facts essential to the model.

Many predecessors have been left out. In 1923, the prize in physiology or medicine went to Frederick Banting and John Macleod for the discovery of insulin just a year earlier. Macleod was head of the department at the University of Toronto but had no hand in the work. That was done by Banting and a medical student, Charles Best, together with a biochemist, James Collip—and so upsetting was the omission that Banting insisted on splitting his share of the prize money with Best, while Macleod split with Collip. In 1974, when Martin Ryle and Antony Hewish were awarded the physics prize for the discovery of pulsars, the cosmologist Fred Hoyle, an irascible man, rose in wrath at the omission of Jocelyn Bell, the student astronomer who had made the initial radio-telescope observation that led to the discovery. Bell (now Dr Jocelyn Bell Burnell) has serenely ignored the fuss. Erwin Chargaff was not so quiet. In the late 1940s, he had investigated the biochemistry of DNA and published certain facts that he did not himself interpret but that were explained by the Watson-Crick structure. Chargaff entered into a voluminous correspondence with scientists all over the world about his exclusion from the prize and became a bitter polemicist against molecular biology.

Others have been too proud to raise the issue and too aware of the spectacle they'd make. I know two, well enough to be sure of the facts and of their attitude. One is a scientist who first discovered a biological phenomenon, wrote it up—and had publication blocked by a doubting editor. Meanwhile, another made the same precursor discovery, wrote a paper that, in fact, acknowledged receiving materials and technical help from my acquaintance, and got into print. Later, two other scientists developed the phenomenon into a powerful tool for genetic engineering. When they were awarded a Nobel prize, the third scientist included was that second discoverer. I will not identify the scientists or the phenomenon, for my acquaintance has never spoken publicly of the matter and in private, even to close friends, only in answer to a direct question. The second instance is of a scientist long an ornament to his field, whose methods and insights were directly relevant to a pair who emerged with a Nobel prize. Many in his field or at his university, though they don't resent the winners, will say that my acquaintance should have been included. I believe he has never mentioned the matter. The behavior of these two, like that of Jocelyn Bell Burnell, stands in elegant contrast to the spendthrift petulance of Raymond Damadian.

*

Defending his handling of the Imanishi-Kari affair, David Baltimore said among so much else, "One has to trust one's collaborators." That statement in this case is a perversion of the idea of trust among scientists. A countervailing principle has long maintained: that every author of a scientific report is responsible for all of its content. We have seen that the actual behavior of many a scientist has ignored that principle. While the junior scientists elevated unknowingly to authorship by Robert Slutsky can hardly be blamed, recall the forty-seven co-authors on one or another of the hundred and nine papers by John Darsee. Many of them if they had read with even minimal care a Darsee paper to which they were attached would have tripped over absurdities. The dereliction is evident. Yet such individual behavior is regularly, predictably, part of such cases. The causes are general—once again, structural. They are built in to the traditional approach to authorship as it has failed to respond to growth and change. And, once again, fraud casts a lurid, revealing light on the accepted practices of science.

Trust in a rightful sense is so obviously essential to the enterprise of the sciences that it is hardly ever mentioned, yet it lies at the foundation—trust among co-workers, co-authors, trust that is more generally required of those who build on others' findings, and most broadly the citizenry's trust of the evidence and ideas of science, once lost difficult to regain. Any notion of norms of science is meaningless without it. Trust is in trouble when scientists grow reluctant to discuss their work in progress, fearing their competitors' ambitions. It is threatened when fraud is perceived to be widespread—particularly plagiarism. It comes in question when the motives of scientists appear to be mixed, as when fashionable academics declare the content of sciences to be negotiated under the pressures of power relationships, or when the public understands that some scientists, exploiting work they have done with government grants, are making personal fortunes.

Trust in the sense Baltimore parodied is indispensable, for example, among co-workers in carrying out the immensely complicated yet finicky experiments of present-day high-energy physics. A single research report from SLAC, the Stanford Linear Accelerator Center, or from its European rival, the vast installation outside Geneva of the Conseil Européen pour la Recherche Nucléaire, or CERN, may list a hundred or more authors, physicists theoretical and experimental, computer experts, armies of variously specialized technicians. In no way can the physicist whose hopes are being tested, or the experimentalists who designed the test, check up on the work, say, of the technicians who maintain the radically hard vacuum in the accelerator tunnel or those who calibrate the beam and man the detectors. Those hundred authors must each live with the knowledge of the others' reliance on them. This is the basis of morale, of *esprit.* "Team science" connotes more than merely science done in large groups.

A paper in *Nature* in July of 2003, L. W. Hillier et al., "The DNA sequence of human chromosome 7," listed one hundred and seven authors from eight different research facilities. Biology, particularly when following out the consequences of the genome projects, has become sometimes as massive and almost as diversely specialized as is much physics. But the contexts are different. In that kind of physics, the pattern of interactions was foreshadowed from the start, with the Manhattan Project during the second world war. Team physics grew up across half a century as the major installations evolved. Division of function and responsibility created institutional frameworks, formal but also tacit—cadres—and these inducted recruits into the ongoing enterprise. Visitors, too, who have been allotted beam time, a precious commodity, could not function without those frameworks. Nor is this pattern limited to particle physics. It is found, for example, though on a smaller scale, in the teams running the big telescopes, conspicuously the Hubble space telescope but also the network of great earth-based optical astronomical observatories. At the other extreme lie fields of physics that are chiefly theoretical, highly specialized, recondite, and not directly allied with massive technology. Such a field may well number no more than a few hundred scientists, dispersed worldwide—described to me by one such scientist as elitist in relation to outsiders, but among themselves democratic and egalitarian. Here, trust is guaranteed by the fact that every member of that anarchic community is competent to judge the work of any other.

Reflecting on the contrast between physics and biology, in July of 2003—a week before the hundred-and-seven-author paper—the editor of *Nature* wrote that in biology teams are emerging within a tradition where individual investigators set the problems, pose the hypotheses. The new order, he said, "requires that biologists, chemists, physicists and engineers mingle and even merge their disciplinary cultures and languages—sometimes an extremely tall order." In biological research, team science has sprung up recently and precipitately, without much established institutional ethos, without strong cadres.

In an antithesis of teamwork, the early 1960s saw a frantic competitive scramble to elucidate the genetic code, which is the set of biochemical relationships by which the genes, stored in the cells as strings of DNA, come to be expressed as protein chains. As DNA, a gene is a long sequence of four different chemical entities, the bases adenine, thymine, guanine, and cytosine, while proteins are made up of twenty different amino acids. A four-letter alphabet must specify a sequence of twenty different objects. Two-letter words, any of the four bases in the first position followed by any of the four in the second, could encode no more than sixteen amino acids. Three in a row can make sixty-four combinations, which meant that breaking the code was complicated by a lot of redundancy. Each triplet is called a codon. The technique for identifying what each codon specifies was discovered by Heinrich Matthaei and Marshall Nirenberg, working at the National Institutes of Health: Nirenberg announced the first codon at an international congress of biochemistry in Moscow in August of 1960. Competition became intense, bitter, and messy, with three major laboratories, several lesser players, and a number of hasty, erroneous claims. Francis Crick, not often an experimentalist himself, had to call the field to order.

Massive clinical trials, testing out new drugs or treatment methods, have long been the one type of biomedical team science. Though sometimes vast in scale, with many thousands of subjects' case records tracked over many years at many hospitals, they are rudimentary in organization. They are expensive, decentralized, difficult to supervise, and effectively impossible to replicate. They are predicated on trust: the architects of the over-all study must depend on their scattered colleagues to be accurate and honest in what is for the most part the humdrum reporting of clerical detail. Nor can a true sense of interdependent teamwork develop, with its reinforcing morale. In short, large clinical trials invite sloppiness and cheating.

In the late 1980s, a radical change took place in treatment of breast cancer. Mastectomy had been the standard: when a cancerous lump was discovered in a woman's breast, the entire breast was cut off, and if the disease had spread as far as the lymph nodes in the armpit all that tissue was extirpated as well. Many women were cured of cancer this way; but the procedure was disfiguring to an extent that some found emotionally harrowing. Then came the announcement that in cases where the lymph nodes are not involved, simple excision of the lump itself, conserving the rest of the breast, followed by radiation therapy, is sufficient. Lumpectomy—clunky but unforgettable term—produced survival rates as good as mastectomy.

The change was an outcome of a multi-institution, long-term clinical study organized by the National Surgical Adjuvant Breast and Bowel Project, a network of centers that has carried out scores of clinical trials of cancer treatments complementary or alternative to surgery. Co-founder of the project, its chairman for more than a quarter-century, and organizer of the lumpectomy study was Bernard Fisher, a surgeon at the University of Pittsburgh. The study recruited 2,163 patients, at ninety hospitals and other institutions. The largest contingent was 354, enrolled by Roger Poisson, a physician at St Luc's Hospital in Montreal. The first and definitive report was published by *The New England Journal of Medicine* in 1985; a further report, accompanied by two papers about chemotherapy in breast cancer, appeared there in 1989. Fisher was lead author on all these, Poisson a co-author.

Women facing surgery were of course enormously heartened by the change, though some were apprehensive. Then on 13 March 1994 the front page of the *Chicago Tribune* carried a story by John Crewdson, as ever relentless in pursuit of scientific misconduct, with the headline FRAUD IN BREAST CANCER STUDY. Crewdson had dug up the fact that Poisson had doctored some of the case records at St Luc's, to include patients who did not actually qualify.

The reaction was instantaneous and cataclysmic: as most American women are aware, breast cancer is the leading cause of cancer deaths among them, and the disease has taken on a fearsome significance, indeed becoming for some a feminist issue, a symbol of society's failure adequately to address the needs of women. Crewdson's exposé hit at a time of high public concern about fraud in science, with the Baltimore affair and others repeatedly in the headlines and with congressional hearings harvesting publicity. The National Cancer Institute and the Office of Research Integrity got involved. Congressman John Dingell's subcommittee on investigations scheduled hearings. The University of Pittsburgh forced Fisher to resign as chairman of the breast-and-bowel project.

The case illustrates neatly problems of supervision and trust in large-scale clinical studies, problems that emerge often enough that they must always be considered a risk. But the case also speaks to the dilemmas of institutional response. Although Chapter 9 treats these more generally, here is how the Fisher case played out. Unquestionably, both Poisson and Fisher had been guilty of misconduct. Yet the kind and degree of their culpability differed. What Poisson had done was to misrepresent the dates of the surgery in at least six cases, so that these patients could be included in the study. According to an account in *The Lancet,* he had believed that patients in the study would be followed up more carefully, and so claimed he thought he was acting in their interest. Five weeks after Crewdson's article appeared, Drummond Rennie published a detailed analysis in *JAMA.* He wrote, among much else:

> It is not the fault of Dr. Fisher that one of his many associates, and one in another institution and country, committed fraud. Nor is it the fault of the *New England Journal of Medicine* that it failed to detect falsifications; the review process cannot and does not pretend to do so. If coauthors can be tricked, then scientific journals, removed from the evidence, can be deceived that more easily [*sic*].

Where Fisher was gravely at fault, Rennie said, was in his dilatory, secretive handling of the fraud. In June of 1990, nearly four years before Crewdson's article, Fisher's group had noticed problems with Poisson's data; three months later, a routine audit had found more discrepancies. Fisher himself then reviewed some hundred of Poisson's cases; in February of 1991, he suspended Poisson and for the first time notified the National Cancer Institute.

The institute handed off the problem to the Office of Scientific Integrity (just then being moved out of subordination to the director of the NIH and renamed the Office of Research Integrity). Their team seized the records. They took two years, until February of 1993, to issue a report. There the investigators said they had found "115 well-documented instances of data fabrication or data falsification which constitute scientific misconduct." Poisson asserted he was fully responsible for all this. In March of 1993, he was barred from receiving United States research funds for eight years. By May, he had admitted falsifying results in other studies, and the Food and Drug Administration disqualified him for life from receiving grants.

Yet for the forty-five months Fisher knew data was contaminated neither he nor the National Cancer Institute published any correction. When Crewdson's story broke, the National Surgical Adjuvant Breast and Bowel Project asserted that the National Cancer Institute had ordered that the problem not be made public until the investigation was completed. They said also that they had conducted a statistical reanalysis of the data in the study, first excluding those six Poisson cases, then excluding all 354—and that the reanalysis showed that the fundamental finding of the study was not altered by excluding Poisson's data. The conclusion held that in breast cancers before the lymph nodes are involved, lumpectomy was as efficacious as mastectomy. This would seem crucial for the medical profession and the public to know. If Fisher's group had published the reanalysis, Crewdson would have had no story and women no new cause for anxiety.

The failure to correct the scientific literature was the culminating outrage, in Rennie's view.

> Not the least extraordinary aspect of this sad affair is that the authors seem to have ignored the rights and interests of the journal that had published the fraudulent data, the *New England Journal of Medicine,* and of its readers. Surely this journal had the

> greatest interest in publishing a correction as soon as possible and had the right to be told.

Rennie linked this with an aspect of trust and its justification:

> Meanwhile, Fisher's group has spent a good deal of time reassuring the media that their conclusions are the same when purged of Poisson's patients. Maybe they are. Fisher's accomplishments are so considerable and the group has so much credibility that I'm inclined to believe them. . . . But maybe they aren't. Without the data, how can we really know? We seem to be expected to take the NSABP's promissory note on trust, though their handling of the problem shows just the sort of behavior that most rapidly erodes this trust.

Protection of the integrity of the scientific record and the scientific process: once again, beyond questions of the guilt of individuals, this protection emerges as the prime necessity in the response to scientific fraud.

In one area of biology, team science with the concomitant trust of the sort physicists would recognize has developed fully over the past two decades: the genome projects. As an impulse, a nascent idea, the origins of the human-genome project go back to the early nineteen-eighties. By the late eighties, conferences had built agreement among biologists on the goal, reports had been written, Congress had held hearings, money was beginning to flow, work was starting in the United States, Europe, and Japan, and the National Institutes of Health had established an Office of Human Genome Research, with James Watson its first head. From the start, everyone knew this was going to be big science, costly, multi-centered, international. A vocal minority of biologists opposed the project in its formative days for just that reason, because they claimed it would drain funds from other research but more fundamentally because they feared the reorganization of biology away from the small-scale, individual-investigator-driven version of Vannevar Bush's vision under which they had prospered. Part of that opposition was neutralized when Congress appropriated new money in bulk. At the same time, and crucially, new technologies were automating and speeding sequencing—at a rate comparable, say, to the celebrated growth of computer speeds and memory. For the genome projects are like big physics in this way, too, with the central role of high technology, and not just the gear but the consequent division of labor driving the evolution of the team, of *esprit,* of trust.

The project was sold to Congress with the promise of completion by 2003, with great advances in understanding and treatment of disease sure to follow (the V. Bush doctrine lives). *Nature* for 15 February 2001 was the human-genome issue, book-thick, one hundred and fifty-four pages (not counting advertisements) plus two snazzy full-color, two-sided, fold-out charts, eleven inches by thirty-six, displaying all the chromosomes. The central article, "Initial sequencing and analysis of the human genome," carried a five-word byline, International Human Genome Sequencing Consortium, but directed the reader to the "partial list of authors" on the facing page. These included more than two hundred names, grouped by their research institutions in descending order of the amount of the sequence contributed. The project was led by Eric Lander and thirty-three colleagues at the Center for Genome Research, in the Whitehead Institute for Biomedical Research, affiliated with the Massachusetts Institute of Technology. Next came Jane Rogers and John Sulston with thirty colleagues from the Sanger Institute outside of Cambridge, England;

then seventeen other centers, including three in Germany, two in Japan, one each in China and France. Among other leaders in the consortium were Maynard Olson at the University of Washington in Seattle and Leroy Hood, originally at the California Institute of Technology but by the time of the report also in Seattle, who had been the driving force in development of automated-sequencing technology. Scientific management had been coördinated by Francis Collins, director of what Congress had elevated to the National Human Genome Research Institute, at the NIH. An ocean liner surrounded by tugboats, the report was flanked by thirty short reports, commentaries, articles, and letters, offering news, analyses, and prognostications.

In the issue of February 16, *Science* published "The sequence of the human genome," with J. Craig Venter as lead author. The byline listed two hundred and seventy-five scientists. This was a markedly different kind of group science: all but fourteen authors worked at Celera Genomics, a commercial firm in Rockville, Maryland, hard by the NIH. Venter, a biochemist, had begun doing gene sequences at the NIH, but quit to found Celera, in May of 1998, with a biotech company as partner. His aim was to beat the international consortium to the complete sequence—and exploit it for profit. He indeed moved fast. Celera exploited the rapidly evolving technology of automated sequencing to obtain its sequence fully a year ahead of what had been the target date for the human-genome project. Venter was in a bitter personal rivalry with Collins, and for that matter with Lander at MIT and Sulston at the Sanger Institute. The rivalry drove the consortium to get its sequence out a year early, too. Several leading molecular biologists brokered a deal: the two sequences would be published the same week. In an extraordinary departure from the scientific norm of open publication, Venter negotiated with Donald Kennedy, the editor of *Science*, an agreement that the data in the Celera genome were a proprietary product and could not be used except by paying a fee. (Venter immediately went after the genomes of mouse and rat. But selling the rights to sequences turned out not to be richly profitable. In January of 2002, Venter quit Celera and the firm turned to development of new drugs.)

The sequence in 2001, however much a triumph of team science, still needed considerable work, with gaps and a certain low level of inaccuracies. It amounted to a pretty good rough draft. I must add that, obviously, Congress was bamboozled by that promise that the human-genome project would be completed by 2003. The genome projects will still be expanding in 2053—for the human nucleotide sequence is the barest beginning. It's not a map until labels are put on, genes located and their functions determined, control elements identified and their working explained, and so forth—and biologists have already encountered important features they could not have predicted. Further, the gains in understanding and treatment of disease can only fully be realized when the project accumulates copious knowledge of the variations in sequence among individuals: for the clinician, the map of the genes remains what it has always been, the map of genetic defects. Still further, the genomes of other creatures allow comparisons of great power—power to illuminate the functioning of genes and controls, and ultimately to elucidate precise evolutionary relationships. Yet the human details are steadily being filled in, while biologists have turned avidly to getting out the sequences from viruses, bacteria, and yeast, to worm and fruit fly to mouse and chimpanzee, already a score or more. All is being carried out by teams coöperating across many centers, as demonstrated, indeed, by that report on the DNA sequence of human chromosome 7 by Hiller and a hundred and six colleagues.

Which returns us to the problems of authorship. In 1978, a number of editors of medical journals got together in Vancouver, British Columbia, to set standards for manuscripts submitted to them. They became known as the Vancouver Group. This grew and evolved into what is formally called the International Committee of Medical Journal Editors. It meets every year. They published a first, brief set of uniform requirements in 1979. This has been revised and expanded repeatedly, most recently in 2003, and now addresses ethical issues related to publication, as well. The Vancouver Group first issued guidelines for authorship in 1985. These said that "each author should have participated sufficiently in the work to take public responsibility for the content," and that every author must meet three criteria: having made "substantial contributions" to the conception and design of the work or to analysis of data; to drafting or revising the article; and to "final approval of the version to be published."

Three years after that definition, in 1988, Stephen Lock wrote in the *British Medical Journal,* "Authorship no longer carries the responsibilities it should—the ability to justify intellectually the entire contents of an article." Today that reads as both lament and acknowledgement.

The problems are brought into sharp focus by fraud. The principle that every author is responsible for the entirety of a paper has been urgently defended—pithily, for example, in the correspondence columns of *Science* in December of 1996. The occasion was a case that arose in the laboratory of Francis Collins, at the National Human Genome Research Institute. Collins, although the director, with a budget of one hundred and eighty-nine million dollars that year, was also running other programs including a genetics laboratory of his own. On the front page of the *Chicago Tribune* of 29 October 1996, Crewdson reported that fraud had been uncovered in Collins's lab.

Collins was retracting a number of papers on the genetics of leukemia and on October 1 had sent an open letter to other scientists in the field, though NIH lawyers had told him not to publicize the problem beyond that. A day later, Lawrence Altman of *The New York Times* jumped on the story, filling in details. The next week, Eliot Marshall, a senior correspondent at *Science,* reconstructed the case.

It was simple, as these things go, though it had two unusual and heartening features. The first was the admirable dispatch with which Collins had addressed it. The second was that the initial warning of something amiss came from a referee. In August, Collins had received a call from the editor of *Oncogene,* published in England, who told him that an alert reviewer had noted a discrepancy in a manuscript Collins and several colleagues had submitted. On close inspection, one of the figures had elements that were duplicated, part of one inverted to create the other. It was "absolutely unequivocal once you look," Collins told Marshall. The colleague who had done the work shown in the figure was a graduate student. Hoping this was an isolated incident, Collins with the help of another scientist in the laboratory combed through the student's notebooks and dug into the freezer to test materials he had used for other reported experiments. After two weeks, Collins told Altman, "the significance and the scope of the fabrication in this circumstance, of which I had not the slightest idea, began to be very apparent." Two papers published in 1995 and 1996 would have to be withdrawn, three others substantially corrected. Evidence in hand, he confronted the student, who after three-and-a-half hours confessed both verbally and in writing. Collins did not name the student, so of course the journalists looked at the bylines of the retracted papers. They found one that had just two authors, Collins and Amitov Hajra, who was working for a PhD at the University of Michigan.

The case bears familiar marks. Hajra had come to Collins highly recommended and his work in Collins's lab was impressive. "The data he showed me every week looked absolutely wonderful," Collins told Altman—and in a wry twist, "Some of the things he did to fabricate experiments were quite creative." The affair left Collins devastated. In an e-mail note to me recently, Collins said, "This was the most painful experience of my scientific life." He went on, "Amit Hajra was by all measures a student with remarkable intelligence, dedication, and promise. His command of the scientific literature was exceptional. Of the dozen graduate students I trained from 1984–1995, he was one of the top two or three." I mentioned Efraim Racker's observation that he had treated Mark Spector like a son. Collins replied, "Yes, I think any serious mentor-student relationship includes aspects of the parent-child relationship. That made it all the more devastating to discover that Amit had manufactured large amounts of data for a significant length of time in the lab, presenting it directly to me month after month without blinking."

Crewdson wrote that "questions are likely to be asked about how he could have signed his name as senior author to a half-dozen articles based wholly or in part on fabricated or non-existent data." Marshall raised a related matter, whether a director of a complex national institute could reasonably also run a laboratory. He put the question to Harold Varmus, overall director of the NIH, who was running *his* own laboratory and who told him that it would always be difficult to guard against "someone who's very smart and very determined." He talked to Tina Gunsalus at the University of Illinois, who as associate provost there at that time dealt with misconduct cases. "A committed liar is going to get you every time," she told him. "You just can't afford to write rules in a cooperative community—where the foundation must be trust—for the bad actors."

Marshall's report prompted protest. *Science* on December 6 published a pair of letters. Charles Wooley, at Ohio State University, wrote, in part, "[P]erhaps we could agree on the following. If you haven't done the work, don't put your name on the paper. If you put your name on the paper, then you are stuck with it." Paul de Sa and Ambuj Sagar, at Harvard, wrote, in part, that "co-authorship of a paper is not always indicative of participation in the reported research—or even of knowledge about its content. We feel that co-authors should bear collective responsibility for their publications, sharing blame as well as credit. It is a contradiction to be a co-author but then plead ignorance (and assume victim status) if there is controversy regarding data in the paper."

Six years later, similar questions were prompted by the Schön affair, the case of the physicist at Bell Laboratories who falsified data in seventeen papers that claimed sensational discoveries and that he published with an assortment of twenty colleagues. The Beasley committee, which Bell Labs had appointed, reporting at the end of September of 2002, found Jan Hendrik Schön the only perpetrator, and exonerated all the co-authors. Three weeks later, Donald Kennedy, at *Science,* offered an editorial. *Science* had been prominent among journals that had published Schön papers: Kennedy's first move was to defend their peer-review practices, saying, in part, "We would reiterate that it is asking too much of peer review to expect it to immunize us against clever fraud." Beyond that, though, the Bell Labs committee "raised but left hanging" the problem of authorship.

> In dealing with authorship issues in other institutional roles, I have encountered vigorous arguments on both sides of this question. One claims that given the interdisciplinary nature of science and the coparticipation of people with various specialties in

> a project, each author cannot be expected to take responsibility for the validity of the results. Another asserts that because all coauthors receive professional credit for the entire product, all should share the consequences if it is invalid.

Kennedy concluded: "If the benefits are enjoyed jointly and severally by all authors, then shouldn't the liability be joint and several too? The answer has to come in the form of a decision by the scientific community."

Reform begins with recognizing that the principle that all authors are responsible for the entire report is idealistic and impracticable. It has not been working. Bylines on an article look straightforward. They are not. An understanding prevails that when relatively few scientists are involved, six or eight, say, the first author listed is to be understood as getting substantial credit—university tenure committees add up first authorships—while the last is senior author, in some sense the over-all supervisor, top of the hierarchy. Yet the practice is not necessarily observed by journals nor codified by individual institutions.

The sequence can be important. The strangest instance I know is the paper that appeared in *Nature* in June of 1970 in which Howard Temin and Satoshi Mizutani reported proof of the existence of the enzyme soon christened reverse transcriptase. Though the fundamental idea was Temin's, Mizutani had had at least an equal part in designing the experiments that succeeded and had done the work at the bench. On the typescript they submitted, the authors are Mizutani and Temin. As published a fortnight later, the authors are Temin and Mizutani. I asked Temin, Mizutani, and John Maddox, who had been editor of *Nature* at the time, about the switch. None could account for it. That same issue carried David Baltimore's report of his independent demonstration of the enzyme. The Nobel prize went to

Temin and Baltimore. Mizutani was left out. Third man in the prize was Renato Dulbecco, a fine scientist who had not contributed directly to that work. Did the switch make a difference? I know of no way to resolve the question.

Guests and ghosts: even when a scientific report names few authors, the byline may be deceptive. Robert Slutsky, at the University of California, San Diego, was one of the most brazen at recruiting guest authors to his bylines who actually had little part in the work; John Darsee was another, and Malcolm Pearce in England. The practice has spread. A variant has been called "gift" authorship, where department heads are listed even when they have done nothing—as was nearly universal in Germany until recently, and is still widespread there and elsewhere. James Watson has supervised many graduate students, postdocs, and other colleagues yet has never put his name on a paper where he had no hand in the work. Eugene Braunwald, at the time of the Darsee scandal, had accumulated six-hundred-odd papers, which is hard to believe without gift authorship. Guest authorship is risky for the guests. It is fraudulent in itself, and often compounds fraudulent work. Its inverse is ghost authorship, when persons who have participated don't make it to the byline, perhaps not even to the small-type acknowledgements that precede the endnotes. In December of 1993, an editorial in *The Lancet* reported a practice that had sprung up at certain pharmaceutical companies—a ghost-guest combo. Typically, the trick is to get a journal to agree to run a review article—that is, not original research but an extensive analysis of the literature on some subject, in this scam one "broadly related to the company's product."

> Thus one might envisage a review of mechanisms of peptic ulceration for a company with an anti-ulcer agent, or one on the role of inflammation in asthma to aid the promotion of a new

> inhaled steroid formulation. A staff writer prepares the review to the sponsor's satisfaction,—whereupon the publishing house contacts a doctor with a special interest in the relevant topic to inquire whether he or she would like to be the guest author, subject to approval of the content, for an honorarium. The pinnacle of success, presumably, is to sign up a prominent academic and to achieve publication in a respected peer-reviewed journal.

Six years before Schön, at a conference at the University of Nottingham in June of 1996, Drummond Rennie offered a solution: a simple-sounding yet thorough-going revision of the entire concept of authorship, of the allocation of credit in published work. Authors were to be redefined as contributors, with the precise contribution of each to be specified—and thus the responsibility made clear. He compared the system to the way movies list credits, and made some pungent points. Gift authorship would be exposed: "contribution: head of department." Ghosts would be exorcised, particularly that pernicious, industry-sponsored form that *The Lancet* had condemned. On the other hand, specialists like statisticians, often left off the byline, would get due credit. Contributors would name one or more of their number as "guarantors" of the whole. Rennie then brought the idea to the annual meeting, immediately following, of the Vancouver Group.

Within a week, Fiona Godlee, on the staff of *BMJ,* reported Rennie's suggestion favorably, but said that the Vancouver Group editors were not collectively convinced change was necessary and had put the decision off. A year later, writing in *JAMA* with two colleagues, Veronica Yank and Linda Emanuel, Rennie went into the ramifications. Five sentences in the abstract stated the basic proposal.

> Credit and accountability cannot be assessed unless the contributions of those named as authors are disclosed to readers, so the

> system is flawed. We argue for a radical conceptual and systematic change, to reflect the realities of multiple authorship and to buttress accountability. We propose dropping the outmoded notion of *author* in favor of the more useful and realistic one of *contributor.* This requires disclosure to readers of the contributions made to the research and to the manuscript by the contributors, so that they can accept both credit and responsibility. In addition, certain named contributors take on the role of guarantor for the integrity of the entire work.

They then took six pages plus a page of fine-print references to lay out the multiple defects, the temptations to corruption, of the prevailing system, and the practicality of their alternative. The proposal works, they said: "[T]here are good examples of the contributor, or job-description, approach already in the literature," at least for large clinical trials. They cited a study of heart-attack survival that listed "roughly 2000 'members,' whom we would call contributors, usefully divided by committee (writing, steering, data monitoring, and the like) and by country and hospital, and with the tasks of the overseeing research unit described in detail."

The reform caught on only slowly at first, but in recent years has gained. By 2001, the Vancouver Group had adopted it as its standard. Rennie has pushed it tirelessly. On reading Donald Kennedy's editorial on the Schön affair, he published a letter in *Science,* urging it once again and observing:

> The Beasley committee, in investigating Schön, would have had a far easier time, and been more convincing in their assessment, had they been able to see in print what Schön and his colleagues had asserted they had actually contributed to the work at the times when their joint papers were submitted. Then the

committee, like the readers, would have plainly seen that the coauthors did nothing.

*

To be sure, fabricators and falsifiers in the sciences originate their frauds alone. One rarely finds conspiracies to commit them, as in the great financial scandals of the past fifteen years. (Of course exceptions occur, as with Herrmann and Brach in Germany.) Even when the attempt is made to hush up or to cover up a fraud that has been detected, this can hardly rival the size or audacity of the cover-ups that endured so long at Enron, say, or in the Roman Catholic church. Yet fabricators or falsifiers operate, and for a while successfully, in the social setting of the laboratory: they interact with colleagues, especially with their lab chief, as Darsee did with Braunwald, Spector with Racker, Hajra with Collins. As we've seen, typically they display great charm and freshness, intellectual brilliance, mastery of the literature, mastery at the bench. They are attractive people. In photographs, Schön was youthful, open-faced, clear-eyed—the sort who would leap to help a little old lady cross the street, and whom no one would suspect of lifting her wallet.

Plagiarism differs from other forms of fraud in essential ways. It is a solitary vice; interaction with others can play no part. It knows no boundaries, appearing in fiction, history, the humanities, scholarship of all sorts. Charles Babbage did not include thieves in his typology of trimmers and cooks. But in the sciences plagiarism is the worst enemy of trust, for anyone doing worthwhile work is potentially threatened by it. We first think of plagiarism as the direct copying of language, perhaps data, from someone else's printed—or, now, electronic—work. But it takes many forms. It can strike at almost any stage in the scientific process, from discussions of work planned or in progress, to peer review of grant applications or refereeing of manuscripts, to purloining of published work. Most pathetic is self-plagiarism, a Darsee or a Slutsky retitling and slightly changing their own papers to submit them to a second journal. Most insidious is the misappropriation of ideas, approaches, methods, with no duplication of language or data. In any form, plagiarism attempts to steal credit. Once again: whereas the fabricator or falsifier creates something almost sure to be incorrect, the plagiarist appropriates only what appears to be true.

Scientists, especially younger ones with reputations still to establish, have reason these days to be wary of revealing too much about work in progress. Consider, for example, the Gordon Research Conferences. These are universally known in the community but hardly at all to the general public. They provide a place where scientists can discuss what they are doing and thinking, in total privacy and so in freedom. The rules are strict: scientists only, closed to press and public, and any information offered in any form is a private communication and can never be published or even referred to as coming from a Gordon conference. Often, conferences are designed to bring together workers from different fields. Attendance is by invitation only, at the discretion of the organizer of the particular conference, and for a young scientist is a mark of coming-of-age. The conferences were founded in 1931 at Johns Hopkins University on a small scale, and moved to New Hampshire in 1947, with ten conferences that summer. They take place summer-long on the campuses of private boarding schools, eight at present count. They last four or five days, with papers read and discussion intense and merciless. The year 2003 saw one hundred and seventy-five Gordon conferences, from Aging, Biology of, to Viruses & Cells and X-ray Physics.

Unquestionably, scientists find that Gordon conferences can be richly productive, stimulating, liberating. They enshrine the ideal of free exchange of ideas and information. Yet despite the rules and the reverence with which they are regarded, even here plagiarism is not unknown. One instance is emblematic. I shall not mention the title or the year of this conference, nor the names of the scientists involved: those who know, will know. Young colleagues of two established figures, from different but equally august research institutions, presented solutions of the three-dimensional structure of a particular sort of large biological molecule. This was important progress, eagerly awaited. The structures differed. Afterwards, both labs published—but those who had been at the conference saw at once that one of the scientists had altered his structure to conform to the other's. The ensuing row was all the more vicious for being quiet. A senior personage had to step in.

Although the incidence of plagiarism coming out of peer review or refereeing cannot easily be estimated, by their nature the processes match what Bernard Shaw once said of marriage, that it combines the maximum of temptation with the maximum of opportunity. Recall what Robert Pollack told me, that a meeting of a grant-review study panel used to be an invaluable experience, "in effect, a three-day seminar with top people on the most advanced work going on in your field." A panel's scrutiny of a grant application, though confidential, involves all the people present in that hotel conference room. Outright appropriation of a research plan might eventually be noticed. Yet how could a panel member not be helped in his own work sometimes by a new approach, perhaps a new method that could be adapted to other materials, a slightly different problem?

Referees of journal articles have labored under fewer constraints. To duplicate a manuscript takes but minutes with scanner or copying machine—and is not even necessary as journals switch to electronic refereeing. To write a doubting review can delay publication while the author responds. Meanwhile the referee can employ what he has learned in his own work. It took an extraordinary chain of coincidences to expose Vijay Soman's use of material from the manuscript of Helena Wachslicht-Rodbard's he reviewed—and careers were wrecked. But Soman was a stupid plagiarist, for the duplication was obvious. One can hear the ironical earnestness with which Charles Babbage would have explained that the accomplished thief, having delayed a rival's publication, avoids direct incorporation of language or data while, inspired by new understanding, he uses the methods to generate original data leading to a similar result. This happens.

Plagiarism of published reports is represented in its most straightforward—and most stupid—form by Elias Alsabti, that Iraqi non-MD, non-PhD who copied articles from obscure journals and sent them to other obscure journals, with new titles and his own byline.

A more enterprising variant was uncovered in the fall of 1997. In London, Richard Smith and others of the recently formed COPE, the Committee on Publication Ethics, convened a meeting on September 4 of more than one hundred journal editors to discuss scientific misconduct. "Cases are still exposed mostly by chance, and we worry about the scale of the problem," Smith told a reporter from *Science.* At that meeting, individual cases were presented, one of them by Marek Wroński, an MD-PhD brought up in Poland but working at a hospital in New York City as director of neuro-oncology research. The details are entertaining. Wroński said that a year earlier he had come across a note in the *Danish Medical Bulletin* about a case of plagiarism: a paper by Danish scientists in English in that journal in 1989 had been translated into Polish and published in 1992 as original research in the journal *Przeglad Lekarski.* The London meeting, partly because British libel laws are so stringent, allowed no names to be published. After Wroński returned to the United States, on December 3 he posted a notice to the active Internet list serve SCIFRAUD, or Discussion of Fraud in Science. "Because of British libel law, only here can I reveal the real names and places," he wrote. The culprit was Andrzej Jendryczko, a chemical engineer and former professor at the Medical University of Silesia, in Katowice, Poland. He had recruited four guest co-authors. Denmark has a Committee on Scientific Dishonesty, and when the original authors found out about the Polish paper the committee investigated and in February of 1995 declared Jendryczko guilty.

Wroński looked into the matter. He found that in thirteen years Jendryczko had published some one hundred and forty articles. Scouring through them, Wroński found another twenty-nine papers "where the text was taken nearly verbatim from different medical journals." The originals had appeared in top British and American journals, and in others from Scandinavia, the Netherlands, Germany, and Japan. Jendryczko had published most of the stolen goods in Polish in journals there; a few showed up in English in lesser European journals. They displayed an assortment of Polish co-authors, including professors and chairmen of departments of biochemistry and surgery and of three departments of gynecology. As late as October, as Wroński told SCIFRAUD subscribers, the president of the Silesian medical school was stone-walling, but by the end of November the senate of the school had opened an inquiry and the president of Poland's General Council of Education had joined the call. The case was making headlines in Polish media.

It would never have arisen but for the fact that Wroński was in the right field, reads Polish, and was angered by fraud. At the end of January of 1998, Eliot Marshall at *Science* picked up the story. He had to get Jan Latus, an editor at a New York Polish-language newspaper, *Nowy Dziennik,* to interview Jendryczko and his wife by telephone. Jendryczko was still insisting he was innocent. On 11 March 1998, SCIFRAUD posted a letter in dignified English, dated two days earlier, from Tadeusz Wilczok, Prorector for Science Affairs of the Medical University of Silesia, saying, in part:

> [T]he accusation of plagiarism of the articles published in the international science magazines mentioned by Dr. M. Wronski, had been positively proved. As the consequence I would like to apologize on behalf of The Senate and myself to all the victims—authors of the articles whose work happened to be the subject of plagiarism and also to all the editors of the science magazines in which Dr. A. Jendryczko had published his plagiarism.

The most significant part of the Polish story is the method by which the plagiarized papers were identified. The Internet and an immense archive of medical journals at the National Library of Medicine provided the tools. Medline is a computer service of the National Library that finds medical abstracts, for a fee. One of the Danish authors of the first plagiarized paper had used Medline to search for articles in his field. One paper that came back was his own—and along with it the Polish duplicate. By the time Wroński began digging, Medline had been opened to the general public through a free service called PubMed. Tens of thousands of people use it every day. A tool of PubMed is a button on the screen labelled "find related articles." This activates a powerful program that identifies key words and searches for other articles containing them. It was intended to facilitate literature searches. If you see a paper you suspect was plagiarized PubMed will produce likely candidates for the original. That's how Wroński did it.

The best news about plagiarism is that the Internet and new programs are making most forms detectable and therefore dangerous to attempt. In the early nineteen-nineties, Walter Stewart allowed himself to digress from his harrying of David Baltimore to address the problem of detecting plagiarism. He invented a computer program he called the plagiarism machine. It compares documents to find strings of characters that are substantially the same. It requires, though, that your hard disk carries the materials to be compared: it can't go hunting in the great wide world. Stewart and his colleague Ned Feder took as their target not a scientist but an eminent American historian of the Civil War. (The choice seemed to me lunatic, irrelevant to fraud in science.) With tireless energy they scanned in entire books of his and of other historians. They claimed that the plagiarism machine found multiple passages their man had copied. But the target fought back. With public stature and private connections, and as Stewart and Feder had strayed from pursuit of fraud in science, he got them shut down. Stewart and Feder found themselves locked out of their laboratory, their computers, their banks of filing cabinets where the trays of snails once had stood.

A PubMed search for plagiarism has its own limitations. It doesn't work for an unpublished suspect manuscript. Other programs do. Most of these have been built originally by university faculty in software development responding to the enormous increase in plagiarism by students tempted by the depth and speed of research through the World Wide Web. Some of the new software is correspondingly powerful. Commercial versions offer to take over an instructor's plagiarism search, for a fee: one such, said to be making money, is www.turnitin.com, founded in England in 1996 but operating, of course, the world around.

At the new London South Bank University, Fintan Culwin, head of software and computer engineering, with his students has created a suite of remarkable engines. These are freeware, at his Web site Centre for Interactive Systems Engineering: Plagiarism Prevention and Detection: http://cise.lsbu.ac.uk. One, called OrCheck, employs Google to locate documents resembling the text in question. To use it, you must create an account with Google and get what they call a key; this allows you to load documents in and grants you up to one thousand Google hits a day. Google ranks the documents it finds in order of similarity, and presents you the most likely ten for inspection.

In a recent conversation, Culwin said that, in short, good tools for catching plagiarists already exist. What they won't do, of course, is to catch the smart thief who takes not text or data but ideas and methods.

*

Plagiarism is a subset of the misappropriation of intellectual property. Consider now the case of Carolyn Phinney. She is a research psychologist, with a doctorate from the University of California, Berkeley, in cognitive psychology. At the University of Michigan in 1988, she was completing a postdoctoral fellowship from the Institute of Social Research, while working part-time in a collaboration at the Institute of Gerontology. There the principal investigator on the grant that paid her was Marion Perlmutter, an authority on aging. The director was Richard Adelman. Phinney was attempting to determine whether wisdom can be measured objectively and whether it increases with age—an amorphous topic, perhaps, but the research seemed to go well. Perlmutter suggested that Phinney write an application for a grant to extend her work, first to the National Institutes of Health and then to the National Science Foundation. Perlmutter then decided that *she* should submit the application, as principal investigator, listing Phinney as co-investigator. In the event, Perlmutter took Phinney's draft and a box of her research materials, and in the application presented the work as her own, not including Phinney. Certain other promises were not kept. Phinney complained. Baffled and hesitant at first, through the eight years that followed she learned to be determined and assertive. Adelman started an investigation. Over the next months, four faculty panels considered the case. None found Perlmutter guilty.

So in 1990 Phinney brought suit in the Michigan Court of Claims—an unusual step in academic disputes. She charged Perlmutter with plagiarism and with fraud as well, for making false promises to gain access to her research. She charged Adelman with trying to discredit her, in violation of the state's Whistleblower Protection Act. She also showed that on each of the four investigating panels Adelman had included faculty members who had been colleagues of Perlmutter's on other grant applications. In 1992, the university dropped Phinney's contract. She was unemployed. In May of 1993, a seven-member jury awarded her 1.1 million dollars. At the end of September, the judge confirmed that and added one hundred and twenty-six thousand dollars in interest. At this point, the case made *Science, Nature, The New York Times,* the *Chicago Tribune*: the amount was sensational. The university appealed. In April of 1997, the Michigan Court of Appeals upheld the award. On July 30, the university paid her 1.67 million dollars, including more interest, and returned her research materials. She has not worked in science again, but moved to northern California and became an active campaigner against fraud and for protection of whistle-blowers.

No other form of scientific fraud is so destructive as the appropriation of a subordinate's work by a superior. "I have lost my data on ten years of work," Phinney told Philip Hilts of *The New York Times* after her victory. "I've lost my career. I got very sick. I believe what happened to me was intellectual rape."

In one of the most outrageous of such cases, Heidi Weissmann, highly regarded for her research in nuclear medicine at Montefiore

Medical Center in New York City, won a federal lawsuit against her department chief, Leonard Freeman: he had replaced her name with his own on a chapter she had contributed to a textbook, and, the court said, "actually attempted to pass off the work as his own." Montefiore forced Weissmann out and promoted Freeman. She was unemployed and stuck with more than half a million dollars in legal fees, while Montefiore paid Freeman's and then investigated her for supposed misconduct. Antonia Demas, a PhD student at Cornell, was another such victim— But the pattern is clear. The fifteen meetings of the Commission on Research Integrity, the Ryan Commission, attracted sparse audiences, but always among them were victims of plagiarism of this kind, obsessed, eager to testify, sometimes in tears.

John Crewdson's most conspicuous case was the multi-faceted controversy over Robert Gallo. The central dispute was about priority. Was the human immunodeficiency virus, HIV, the cause of AIDS, first found by Luc Montaigner and colleagues at the Institut Pasteur in Paris, or by Gallo and colleagues at his laboratory at the National Cancer Institute, on the NIH campus in Bethesda, Maryland? The dispute was complicated by genuine difficulties in the virology. It was poisoned by Gallo's aggressive, volatile personality; it was loaded with charges of theft of ideas and even of biological materials; it was enveloped in a cloud of unscrupulous ambition on both sides, by charges, counter-charges, and bitter resentments. The sad part was that to discover the cause of AIDS was science of world importance, and Nobel committees stay away from priority fights. In the midst of all this Crewdson believed he had found evidence of fraud, the misappropriation of intellectual property, and cover-up. Crewdson's culmination came with publication in the *Chicago Tribune* on 19 November 1989 of his fifty-thousand-word account of the discovery and the controversies.

The affair had its tap root in discoveries that Gallo and collaborators had made in the 1970s about a disease-causing agent called human T-cell leukemia virus, or HTLV. This was implicated in a particular form of leukemia, a cancer affecting a type of white blood cells important for the immune system which is itself rare, found chiefly in fishing villages in southern Japan. Its interest lay in the fact that scientists had long wondered whether some cancers might be caused by viruses. Though instances had been known as early as 1910 in other animals, HTLV was about the first to be clearly associated with a human cancer. Further, the virus was of a particular class whose genetic mechanism had been elucidated only a few years earlier, in 1970, with the discovery of reverse transcriptase. This enzyme enables certain viruses to incorporate their genes, written in RNA, into the DNA of their host cells—hence, retroviruses.

Gallo's laboratory betrayed the style of the man. His office when I saw it, at the height of the controversy, was bright, large, with the walls adorned with framed blow-ups of awards and of newspaper and magazine articles about him. Prosperity, publicity, and no-holds-barred competition were the themes. The laboratories were large and busy. Over a span of decades, Gallo had come up with valuable results in cancer virology—and had repeatedly distorted and exaggerated the significance of those results. In the most naked way, he was bucking for a Nobel prize. In conversation he could alternate between ingratiating and ranting; he was in the habit of telephoning opponents late at night and talking at them unstoppably for an hour or more. In the lab, the pressures on and among the junior colleagues were reputedly harshly competitive. Mentoring can be negative.

When AIDS first was recognized, in the summer of 1981, a number of laboratories began working on it. Gallo's was one. Another was Montaigner's. By May of 1983, Gallo was claiming that AIDS was caused by a retrovirus closely related to his T-cell leukemia virus. His chief colleague in this was Mikulas Popovic. Gallo called the new virus HTLV-3, and fought for it in a campaign that grew more political in method than scientific, enlisting collaborators, publishing a stream of papers, organizing scientific meetings, holding press conferences.

Montaigner, however, had come up with an unrelated retrovirus, the one we now call human immunodeficiency virus, or HIV. It was discovered by Françoise Barré-Sinoussi in his laboratory, and was published that summer with her as first author, a number of colleagues, and him as senior author. In May of 1984, in a paper in *Science* with Popovic as one of the co-authors (and the one who did most of the laboratory work), Gallo announced that *his* laboratory had isolated that one from AIDS patients, too. But scandal bloomed when it turned out that what he now claimed to have isolated was Montaigner's virus in literal fact—that sometime in the winter of 1983–1984 a sample from Montaigner's lab had somehow made its way to Bethesda. Whether the transmission was accident or theft has never been quite settled. What is hard to escape is that Gallo was determined to steal the credit from Montaigner.

The uproar that followed looked to have reached a climax in September of 1991 when Crewdson and then other journalists got hold of an unpublished draft report of an eighteen-month investigation of Gallo and Popovic by the Office of Scientific Integrity. The draft asserted that Popovic was guilty of fabrication, falsification, misstatements, and errors in that paper of May 1984. Gallo's actions "do not meet the formal definition of scientific misconduct," the draft said, but "they warrant significant censure." Popovic was formally charged. Whereupon his lawyers appealed to the Departmental Appeals Board. In the summer of 1994, saying that the OSI had failed to prove the case, the board found him not guilty. This was later considered a preview of the outcome of the Baltimore–Imanishi-Kari affair. In despair and disgust, the Office of Scientific Integrity dropped the case against Gallo. His reputation was damaged, though, and his career as well. Though noisy, the Gallo case was narrow.

On the way, the feud between Gallo and Montaignier became an international *cause célèbre,* calmed only by governmental intervention. The two eventually swallowed their bitterness, in public at least. In *The New England Journal of Medicine* in December of 2003, they published a jointly-written brief history, "The Discovery of HIV as the Cause of AIDS." It was predictably anodyne. They went so far as to acknowledge, "The year 1984 was a time of both intense excitement and harsh discussions between members of our two groups." For those who knew the men and their polemical exchanges, one sentence offered a certain ironical amusement: "Many lessons can be drawn from this early intense period, and most suggest that science requires greater modesty."

For every case of fraud this book has discussed, five more of comparable size and interest could be cited. We have examined frauds of many types, and have seen the many ways people and institutions implicated have responded. We have laid out the repeated patterns and settings of frauds, and the exceptions. Implicit in those responses and patterns may be measures that can be taken to change the settings in which fraudulent behavior flourishes and when those measures fail, as of course they sometimes will, some more effective ways of dealing with frauds, at the level of the laboratory, the institution, the supervising bureaucracies of government, and the law—always mindful of the integrity of the scientific process and the scientific record. The Internet is opening the sciences to a transforming scrutiny.

8

THE RISE OF OPEN PUBLICATION ON THE INTERNET

The literature will all be interconnected by citation, author, and keyword/subject links, allowing for unheard-of power and ease of access and navigability. Successive drafts of pre-refereeing preprints will be linked to the official refereed draft, as well as to any subsequent corrections, revisions, updates, comments, responses, and underlying empirical databases, all enhancing the self-correctiveness and interactiveness of scholarly and scientific research and communication in remarkable new ways.

—*Steven Harnad,* "Open Archiving for an Open Society," *July 2000*

In 1991, the next great structural transformation of the sciences was only just beginning: electronic publishing. A dozen years later, the transformation has progressed so far and is proliferating so fast that much of what one can say about it will be outdistanced, superseded, within a year. Yet in the seemingly chaotic richness of the transformation a broad outline is clear—as are its implications for scientific fraud, which though little noticed are powerful.

From the start, publishing of science on the World Wide Web has moved along two lines. These can be characterized as the conservative or main line forced to embrace the new—to be fair, sometimes welcoming it—and the radical forced to acknowledge practicalities.

Conveniently for bringing order out of that seeming chaos, each line has a publishing model and a leading figure, the *BMJ* and its editor, Richard Smith, and a Web entity called arXiv and its creator, Paul Ginsparg. The two approaches are antithetical. Yet they are adapting to each other, converging. Among the many who are driving the convergence, representative figures are Harold Varmus and Vitek Tracz.

Early on, timidly or boldly, print journals began posting part or all of their content—straight up, already peer reviewed, accepted, edited—on Web sites. By now, driven by competition to maintain readership, most print journals are publishing both ways, at the least by duplicating themselves electronically. Some journal editors are pushing deeper, and among these is Smith at *BMJ.* Several years ago, *The Scientist,* a magazine of news and features (you can read it on the Web), ran a profile of him. The writer described Smith as an "ethical populist." On the mark: and though *BMJ* is a journal that would seem to epitomize the biomedical establishment, Smith is a strong and level-headed promoter of electronic publication.

Radically, subversively, scientists in certain fields have organized to bypass journals and refereeing and make papers openly accessible directly on the Internet. Ginsparg, in 1991 a little-known physicist in an esoteric specialty, is prime mover in Internet open publishing—prime in several senses, first to see and act on the possibility offered by the Web, first in his combination of vision and activism, first in the influence on the transformation he has achieved. Dozens of others have been drawn to the development of this program. Two have been most significant, Andrew Odlyzko, a theoretician, and Steven Harnad, a proselytizer. They foresee a rapid evolution to what one of them calls "a global knowledge network."

How quickly we have come to take for granted what the Web can do! Periodicals are appearing electronically far afield of the sciences. Today, we take it for granted that every newspaper or magazine we might wish to scan will be on our screen with a couple of clicks. Want to know what's happening in Pakistan and Afghanistan three years after the destruction of the World Trade Center? Go to the Web sites of the Pakistani newspapers, published in English, frontierpost.com.pk or dawn.com.pk. *The Economist's* Web site gives American readers more stories about Britain than the United States edition carries, but only if you are a subscriber. At the Web page of *The New York Times,* anyone can sign in and read stories of the last seven days, but older articles incur a fee. In London, *The Daily Telegraph* and *The Guardian* offer free access and archival search, allowing you to print hard copy; *The Times* charges a fee. Variations are many.

The changeover came with the swiftness we have learned to expect from the electronic permanent revolution. In the sciences, I said in my talk at the peer-review congress in 1993, the transformation holds remarkable promise—nothing short of a remedy for many of the ills of refereeing, but far beyond that for the deficiencies of print-and-paper publication. Used rightly, I said, electronic publishing should even make fraud more difficult.

From the point of view of the scientist reading the literature, electronic publication offers considerable advantages. In the fall of 1992, Joshua Lederberg gave a talk in Woods Hole to an international conference of science editors. Lederberg had won his Nobel prize years earlier and had recently stepped down as president of Rockefeller University, returning to active research there. His is an elephantine intelligence, supported by massive, systematic reading. "One of my main functions with my own laboratory group is that I try to be its principal reader," he said. "I want to be alert to events that might have a very important bearing on the way we think about our own research, our planning, of the data coming in." Lederberg saw the future and welcomed electronic publication. First, he said, it offers the only possible response to the problem of sorting out from the vast scientific literature all the articles, but only the articles, that are directly relevant to one's own work. Secondly, it will allow one to record, retain, and keep accessible one's own responses—notes, commentary, linkages, inspired ideas—to those articles. Many journals will continue to function best in their present form, on paper, to be read on planes or trains or sitting in the garden. But any journal should be available for electronic browsing and searching.

> Each of us faces the task of selective retrieval from a cosmic domain of stuff that every other eager beaver in the world has been busily putting into the repository. Our present technology offers the researcher an approximation of recovery with reasonable confidence. Keeping track of what you have accumulated on pieces of paper is the frustration. . . . The next step, to integrate that into your own private library of useful knowledge, is simply not achievable with last year's technology.

Electronic publication together with more sophisticated search software will even ease the problem of locating highly specific but elusive matters within the vast literature, what Lederberg called "the exquisite detail needed to take the next step." Beyond helping the individual scientist, he predicted that electronic publication will make possible a running dialogue, "a dialectic," between readers, journals, and authors. He saw this as the greatest long-term gain.

Consider now the viewpoint of scientists publishing papers. Their behavior—although the system takes it for granted—seems distinctly odd: diametrically unlike authors in other modes of organizing publication, they give their work away without charge. Scholarly journals don't pay contributors, they don't pay referees. Scientists tolerate this because they must. Their overriding need is to get their papers read. Electronic publication vastly multiplies potential readership. It follows that the transition pits the interests of scholars against those of traditional publishers. With access the essential desideratum, the scholars' interests must strongly influence the shape of the future.

*

The alternative to ink-on-paper journals is the phenomenon called open publishing. It began about the same time as the open-software movement, which it resembles, and its originators had the same subversive intent. They wanted to loosen the control of established institutions, in this case the scientific journals, radically speed and cheapen the flow of information, and democratize the consideration of it. For some proponents, the entire journal system, starting with refereeing, has been a prime target. They scorn those journals that merely put their print versions onto a Web site as "clones," while proclaiming that the day of the printed journal is past.

In 1991, Paul Ginsparg, a young physicist who had recently moved from Harvard to the Los Alamos National Laboratory and was working in a notably abstruse corner of high-energy particle physics, reflected on a mundane problem bedevilling everyone in his field. For fifteen years or more, their practice had been that whenever they wrote a new paper, before even submitting it to a journal the authors and their institutions would mail out preprints to colleagues all over the world. So doing, they simultaneously staked claims and invited the next step, comment and collaboration.

Ginsparg and I talked about all this in a long conversation early in July of 1998, at Los Alamos. His office, in a non-secret building, was cramped and stacked with papers; he was lanky and hairy—beard, shorts, hiking boots. "At Harvard, we had hard-copy preprint distribution," he said. "The system had essentially already decided that, you know, that peer review was useful— That the journal was useful for the archival structure. But, you know, we needed more rapid distribution for ongoing research." Therefore preprints. "And what we would do is we'd finish a paper, and the lab would run off five hundred, six hundred copies of it. You'd send fifty of them off to, you know, your own personal distribution list, those who were in the preferred loop, and who would then feed it to people lower down in the food chain—graduate students, postdocs at those universities. And then simultaneously there was an institutional mailing list which I believe was around five hundred fifty high-energy physics institutions that—"

I interrupted: This was at Harvard?

"This was every one of those five hundred fifty institutions," Ginsparg said. "They all did it. There was this network, people were getting things already, in this organized network, and it had grown— This five fifty was a number from the late eighties.

"And of course that's completely dead now. I killed it.

"Or, sorry, replaced it, it evolved to something much better," he said.

"So there was this experiment, that I have claimed that I was doing, starting in ninety-one, but it wasn't really an honest experiment, because I already knew the answer." The answer was given by the preprint-distribution network.

Like other revolutions, Ginsparg's had a prehistory. It was made possible by a big leap that had taken place six years earlier, in the software for writing physics papers. By 1985, "We all switched to using TeX as our word processor. T-e-X, but the X is a chi [χ]. So it's pronounced 'Tek.'" TeX had been developed by Donald Knuth, at Stanford University, and immensely facilitated the writing of papers with mathematical equations, for it replaced the secretarial typing and retyping and the repeated writing-in of mathematical expressions by hand—with errors springing like dandelions at every step. "Suddenly, we could produce something that was not only just as good as what the final journal version looked like, in many cases it was actually better. In many cases we could make a better layout and most importantly, our version— We would, in the course of making this, proofread it fifty to a hundred-odd times. Get all the errors out." Only to find, he said, that the journals, setting type, put new errors in. "In 1986, two or three errors per page that they'd inserted, and in important places, in equations and things like that. And I just said— That was *Physical Review D,* and I said OK that's it. I'm not submitting to you any more, because we can do better on our own." At that time, "In the physics community we were about a decade ahead in electronic connectivity," with bigger computers and a prototype international network, predecessor to the Internet.

"So, um, by the late eighties, we would then start e-mailing our TeX manuscripts around—again in a limited circle. And I was always bothered by the unfairness of this, because our notion of the way the discipline should work is equal access, equal dissemination. We're very elitist, but within our elitist group we want equal access; and our elitist group effectively contains all the researchers."

Physicists were still using TeX. "It's obsolete in conception now," Ginsparg said. "I mean it was so brilliantly conceived that we all still use it because—well basically because Bill Gates could employ a hundred thousand computing monkeys typing forever and they'd never reproduce a small fraction of what Knuth's program gives, nor Shakespeare for that matter." (A more structured version also came out in the nineteen-eighties, called latex—not a rubbery joke, but named after Leslie Lambert, who wrote it—and is widely used by computer scientists as well as mathematicians and physicists.)

Thus by 1984 or 1985 physicists were writing papers with colleagues at other labs not by telephone but by computer. That worked far better. Telephone calls are often intrusive at the other end. "And when you do get on the phone, there's a certain amount of socializing beforehand, socializing afterward, catching up on this and that, and it's less efficient." International collaborations required conforming to time zones. Drafting papers back and forth on the network cut all that out. Best, though, was the gain in clarity of thought: "Frequently when you're collaborating with someone, the act of just writing it down, committing it to print, so that you won't be embarrassed by what you're saying, is sufficient to formulate it well enough so suddenly you know the answer."

Unbelievable as it now seems, back then the problem for electronic distribution of full texts of preprints had long been that one was allotted so little space for e-mail on the institution's mainframe that if you went away to a meeting you would return to find your quota filled and who-knows-what messages bounced back to the senders. "And it just occurred to me, clearly we were doing it wrong. It wasn't a self-sustaining system. We were a subcommunity— Back then, we were doing matrix models of two-d gravity, which was a subcommunity of string theory which was a subcommunity of, ah, relativistic quantum field theory, which is a subgroup of, you know, high-energy physics. So we were a really, *really* small subdiscipline. It was a couple of hundred people, at the time, worldwide. And I said, clearly what we should do is set up some automated system where people submit these things. You don't send out the full text to everybody, unsolicited," as the paper preprint system had been doing. "That's insane." In three afternoons of programming he set up a system that processed each incoming electronic preprint, sending out the title, bylines, and abstract, storing the full text. "So we settled on once a day that we sent out notification of the things that, uh, arrived and then the people who want to can retrieve the full text, only if they want to."

That began on 9 August 1991. "My original design estimate was that we'd get one submission every three days. So on the order of a hundred, a hundred twenty, submissions per year, based on the community of a hundred fifty to two hundred people," he said. "I was only planning to keep these things for three months." After that, even by "boat mail," hard-copy preprints would have arrived. But then about a month after the start-up a friend visited, a computer illiterate in Ginsparg's opinion, who pointed out the obvious: by that time, computer memory was becoming cheap, disk space a trivial matter, so why not keep them forever? "That really opened my eyes. If he, a computerphobe, saw this, then the approach was going to cut across disciplines. And not just for the younger scientists brought up with computers, but across ages. And so at that point it became an archive rather than just this temporary expedient and that's where we got the name e-print archive."

He used the address xxx.lanl.gov—lanl.gov designating Los Alamos National Laboratory, xxx for no other reason than that it was handy. "This was pre-Web, so that when people were sending e-mail they needed an address that was easy to remember." The consequences of the address were sometimes comical. "You wouldn't believe— We can look at our reference logs all day, the false hits we have coming through here because they go to the big search engines and they search 'xxx + Japan + young girls, xxx +' uh, you know, 'anal + hard core.' I mean, you see all these things in the logs—and then people send messages, 'I thought this was supposed to be a child porn site, where's the child porn?' And, um, no, I'm not exaggerating." (Late in 1998, he realized he needed to register a "dot-org" domain name, so in homage to Knuth's TeX he changed the address to its present form, home page http://www.arXiv.org.)

The idea took off. "What happened in ninety-one was, um, it grew very quickly and it also instantly expanded its purview. From the original two hundred people, the people who were actively using it grew to a thousand within a few months, by the end of that year. Just this constant expansion. First to all of formal theory—and then by the spring of ninety-two we had separate archives set up in general relativity–quantum cosmology, another for high-energy physics–lattice theory, others for astrophysics, for condensed matter, and one in algebraic geometry— It just, it grew very quickly and it was very uncomfortable for me because that wasn't supposed to be my job. It's still not, and I had no funding to do it."

Bypassing ink and paper in this way reduced the marginal cost of distribution, once a preprint was drafted on-screen, vanishingly close to zero. "At Harvard they were giving us twenty thousand dollars a year, from a National Science Foundation grant, for this hard-copy preprint distribution—postage, photocopying, labor to stuff envelopes. If there are just a hundred such United States institutions that's a couple of million dollars a year. And I was telling them honestly that that's what I was going to save them." The electronic archiving quickly became an integral part of the research, the communications structure, of the community. Yet at Los Alamos, Ginsparg said, "Nobody— They had no idea what was already running here; and the National Science Foundation and DOE"—the

Department of Energy, within which the laboratory sits—"were both telling me, 'Well, we don't support that kind of thing.'"

By the end of summer 1993, some eight thousand physicists the world around were using the e-print archives. On September 27, anyone who sent an e-mail message found the archives shut down. The screen suggested, "If you are in the U.S. and feel strongly about the utility of the system, now is a good time to contact your NSF and DOE program officers, and encourage them to find a way to support it adequately." The response was international, voluminous, anguished. And effective: the Los Alamos administrators were shocked into promising help. Gary Taubes, a skilled journalist who had been following Ginsparg's progress for *Science,* reported the incident a week later. The episode demonstrated, Taubes wrote, "just how much physicists have come to rely on the bulletin boards"—the archives—"and the extent to which they have changed the culture of physics." He quoted Jim Horne, a physicist at Yale, who told him, "It would really be painful to do without them," and that now, "The only thing I use journals for is looking back for papers that came out before the bulletin boards existed."

Ginsparg told Taubes he had shut down only the most visible part of the system, the daily distribution of announcements of new papers. For one thing, September being the start of a new academic year many graduate students and postdocs changed e-mail addresses, and hundreds of the daily postings were bouncing. But he calculated the way he staged the shutdown to evoke a response that would dramatize his need for support. He was still running the system single-handedly. He required a clerk to do administrative chores and comb out the e-mail bounce-backs. He had ideas for improving the system but needed a good programmer to implement them. To his considerable surprise—he was "staggered," he told Taubes—he was promised what he wanted. Within three days, Ginsparg put the system back up.

Running fast and silently the torrent swelled. By the end of 1993, the archives covered nearly one hundred per cent of the various specialties within high-energy physics, and were still expanding. In 1994, the NSF relented, and awarded him a million-dollar grant for three years, with the funds coming from the NSF physics program and from an office recently created to support unusual projects. He hired a full-time professional software assistant. By the spring of 1995, they covered some twenty-five fields of physics and were getting more than forty-five thousand hits a day. In our conversation the summer of 1998 Ginsparg said, "In the last two and a half years, the biggest increase has been in condensed matter and astrophysics—which had never had these organized preprint-distribution systems."

How large was the physics community?

Depending on what you include, he said—"because there's this nether region of applied physics closer to engineering, and stuff like that"—it numbered from fifty to seventy thousand scientists. The American Physical Society is the biggest and most prominent publisher to that community worldwide. "Between *Physical Review A* through *E,* and *Phys Rev Letters,* they publish on the order of ten or eleven thousand papers a year," he said. "And this system here is handling much more than twice their volume, right now." A couple of years earlier, he recalled, "One of my employees—it was a funny thing!—left here to work for the American Physical Society. You know, got there, and they said, Well you're leaving Los Alamos and, uh, that was really diddly volume and now you're coming to us and you'll see what real volume is." Ginsparg smiled innocently. "You know, they thought it was always this marginalized thing, just some tiny high-energy-physics community where all the practitioners know one another and nothing to do with the size or scale of *their* operation." So they compared numbers. "It was like a bolt of lightning. He told them, Well guess what, it's not only already bigger than all of your operation, it's also growing at an accelerated rate."

He spoke of what he saw ahead, not so distant. "It's a stealth operation. If they knew the real goal, which is really to replace— Oh well, we don't know what's going to happen across the board, but in physics certainly we will have a global, unified database of everything, one that transcends all journals and all of these picayune things they're working out about how different print journals can be on line as well. I mean that's all completely irrelevant." The vision came tumbling out. "That's all artifacts of this paper-based mentality. I mean the electronic medium is so different; you can just throw all of that out, um, and do it right and also do the peer review right in the form of these intellectual overlays that take the, you know, the individual submissions which we could call the data or informational building blocks and provide the glue to build into a knowledge network instead of knowledge resources. Um, and in physics we can do it because we've got such a head start and probably in math and computer science. Math is already starting, slowly, but it's moving."

Over into biology at all?

Pause. "Um. Not this system. I think there's been very little movement in the life sciences." Five months later, however, a group of biologists, mostly computer competent, met at the Cold Spring Harbor Laboratory, on Long Island, to discuss assembling and putting on the Web massive data sets on the expression and interaction of genes. Ginsparg spoke there.

Addressing a meeting at UNESCO in Paris in February of 2001, Ginsparg said that the relation of audience to electronic archive or journal can take two forms. In those highly specialized subdisciplines of physics that his original preprint archives served, the pool of potential contributors was essentially the same as their total audience. Everybody is competent to judge each submission. Alternatively, for some types of journals the size of the audience can be a hundred times the number of contributors. Medical and some biology journals are of this kind. Many in this audience may indeed not be good judges. In a conversation in July of 2001, Ginsparg suggested a third category. "Of course, there may be journals where the total number of contributors is significantly larger than the audience. As in the humanities."

In Paul Ginsparg's vision in its purest form, judgement of a paper's quality becomes a function of the entire community of the particular science. Everybody is a peer. The paper shows up on-screen. In the preprint tradition, it is viewed as a draft, anyway, a work in progress. But all science is a work in progress. The electronic paper attracts commentary, questions, corrections. These attach themselves to the original. The process becomes a multiple dialogue, Lederberg's "dialectic."

In 2001, Ginsparg moved to Cornell University, taking arXiv with him.

Ginsparg's archive came first, its growth and achievements dramatic, well publicized among scientists, universally acknowledged. He is by no means the only one who has been pushing for open publication on the Internet—in some form, not necessarily his. Conferences, articles, books about Internet publishing have multiplied, and are of course available electronically. Proponents are a loquacious, disputatious bunch, and do their arguing, of course, through list servers on the Internet. They generate thousands of pages. And in a dozen years, the number of archives of one form or another has continued to expand and the number of papers lodged in them to double and redouble. The revolution is self-sustaining, the phenomena too rich and diverse to be easily chronicled.

Of theorists of the revolution, besides Ginsparg two have been prime movers: Andrew Odlyzko, a mathematician long at Bell Laboratories but now director of the Digital Technology Center, at the University of Minnesota, and Steven Harnad, an enthusiast in the Department of Electronics and Computer Science at the University of Southampton, England. These three and their many interlocutors are conspicuously of that culture that created the Internet and the World Wide Web. They are technologically sophisticated visionaries, and their vision is of a global, decentralized, massively open, democratic and non-hierarchical intellectual world where the presentation of ideas and theories is not constrained by fixed forms but, rather, is fluid and interactive, organic and evolving. Above all, access is to be free. On how to reach that world, they differ.

They agree, and have convinced mainline scientists, publishers, and funding agencies, that drastic change is inevitable and imminent. Odlyzko observed early, in an extended review, that change is driven by two trends, the exponential growth of the scientific literature and the exponential growth of computer power. More than fifty years ago, Derek de Solla Price, a historian of science, asserted that since the dawn of modern science in the seventeenth century the number of scientific papers published has doubled every fifteen years. I have always thought his data scant for the earlier period, but his claim has been all but universally accepted, often quoted. Since the mid-nineteenth century, anyway, increases of that order have certainly maintained. Odlyzko narrowed the analysis just to mathematics, his field. "In 1870 there were only about 840 papers published in mathematics. Today, about 50,000 papers are published annually." He went on, "[A] more careful look at the statistics shows that from the end of World War Two until 1990, the number of papers published has been doubling about every 10 years." Since Thales and Euclid, perhaps a million have been published—and the startling fact follows from the geometric increase that half of them have appeared in the last ten years. That's an ocean of ink, forests of paper, an insupportable weight on library shelves and budgets. Odlyzko again: "Good mathematics libraries spend well over $100,000 per year just for journal subscriptions, and the cost of staff and space is usually at least twice that."

However, the exponent of growth in computer power—information processing, transmission, and storage—has been higher. Computer speeds are still doubling not every ten years but every eighteen months or so. A mathematical paper of average length set in TeX, Odlyzko figured, would occupy some fifty thousand bytes. Storage of a year's papers would require 2.5 gigabytes. The Macintosh laptop on which I'm writing this could hold a couple of decades' worth. Odlyzko:

> We conclude that it is already possible to store all the current mathematical publications at an annual cost much less than that of the subscription to a single journal. What about the papers published over the preceding centuries? Since there are 1,000,000 of them, it would require about 50 GB to store them if they were all in TeX. . . .
>
> This ability will mean a dramatic change in the way we operate. For example, if you can call up any paper on your screen, and after deciding that it looks interesting, print it out on the laser printer on your desktop, will you need your university's library?

Harnad went to Southampton in 1994 as professor of psychology, after some years at Princeton. He has been editor of the print journal *Behavioral and Brain Sciences* and as early as 1990 had started *Psycoloquy,* a peer-reviewed journal published electronically. He, too, has made much of the subversive nature of electronic publishing. He saw an immediate problem in how to get past the initial growth spurt without arousing the opposition of traditional publishers, who could retaliate in a variety of ways, for example by declaring they would refuse to publish papers that had already appeared on the Internet. Harnad proposed an approach that was sly and curiously idealistic—in the way that, say, pacifism is idealistic—in a reliance on individual activism. Ginsparg offered a service. Harnad called for a mass movement. Ginsparg built on the central nodes provided by the various open but specialized archives. Harnad envisioned a system radically dispersed. "If every esoteric"—irritating term: he merely meant writing of limited audience and not for profit—"author in the world this very day established a globally accessible local archive for every piece of esoteric writing from this day forward, the long-heralded transition from paper publication to purely electronic publication (of esoteric research) would follow suit almost immediately." The goal, in Harnad's electronic redaction (and typeface):

> The literature will all be interconnected by citation, author, and keyword/subject links, allowing for unheard-of power and ease of access and navigability. Successive drafts of pre-refereeing preprints will be linked to the official refereed draft, as well as to any subsequent corrections, revisions, updates, comments, responses, and underlying empirical databases, all enhancing the self-correctiveness and interactiveness of scholarly and scientific research and communication in remarkable new ways.

From the disputations, three considerable questions emerged: the editorial process, including peer review; costs; and the problem of the definitive version. These overlap. In particular, costs depend on what of the editorial processes is to be preserved or replaced. Ginsparg's model in its purest form entails no refereeing, no other editing, judgement of the quality of individual submissions left to the responses of others "in the form of these intellectual overlays," and everything electronic. He points out, though, that "there has always been a barrier to entry. The intent has been to restrict participation to the known community, namely, people at known institutions and the people they know and trust. It's never been an open, usenet-type forum." On his model, direct costs of publishing—as against the research and writing, which of course remain—reduce to pennies per paper. Access to the Internet, for this and all other purposes, costs private individuals the fees charged for high-speed transmission, but for scholars these are almost always borne by their institutions. "If the Net is giving a journal a free ride," Harnad observed, "it gives—as I delight in showing audiences in (numerical) figures—an incomparably bigger free ride to porno-graphics, flaming, and Trivial Pursuit."

Harnad addressed the costs of open electronic publishing in two ways. His dispersed model is cheap, too—though it requires rules, an agreed protocol, so that all those globally dispersed individual archives remain accessible. But he has also held that at a level more formalized than those local archives some editing is necessary, especially some form of refereeing. (Ginsparg sniffs at this. Yet he has implicitly recognized the need in his observation that some types of journals may have audiences that outnumber potential contributors by orders of magnitude.)

Harnad wrote, "There are two distinct questions here: Publishing of the refereed literature, and 'publishing' of the preprint literature. My subversive proposal was intended to make the latter break down the doors for the former." He estimated that a journal entirely electronic will save at least three-quarters of the cost of a comparable ink-on-paper publication. The remaining twenty-five per cent he believed to be a maximum figure for the essential editorial functions. A number of journal editors have objected vigorously to Harnad's seventy-five/twenty-five split. Their estimates regularly put printing and paper at about fifteen per cent of the total, and the "first copy," or prepress, costs as high as eighty-five per cent. (Ginsparg, too, scoffs at Harnad's numbers as imaginary: "That percentage of cost due to ink-on-paper is not the case for any known publisher.") In short, to suppose that electronic publishing of refereed journals would be much cheaper is plausible, but in fact no firm estimates exist as yet.

Considering electronic publishing in either mode, archives or formal journals, the issue that alarms established editors is peer review. This is most extreme for editors of biomedical publications. Simply, brutally, they don't believe that the mass of physicians reading their journals are capable of evaluating raw papers not filtered for quality. Craig Bingham, at *The Medical Journal of Australia,* is in fact a strong advocate of a structured form of open peer review on the Internet; yet in 1999 he wrote, "Publishing any article before it has been peer reviewed involves the risk that it is rubbish. Publishing without peer review requires all readers to take on the jobs of selecting and criticizing the literature that are currently performed for them by reviewers." George Lundberg was for seventeen years editor of *JAMA.* In 1999, in a dispute rooted in national politics, the board of trustees of the American Medical Association fired him. He became the founding editor of *Medscape,* an electronically published medical journal. At one point in the on-line discussion, from his editor's computer at *Medscape,* Lundberg intervened:

> but there is another fly in Stevan's ointment large numbers of articles submitted to biomedical journals NEVER appear in print after being unfavorably reviewed they do not deserve publication editors and

> reviewers REALLY DO protect both readers and author in this process what is then done with those "pre-print" versions that are available to all?

Yet received opinion scants Lundberg's major premise. Richard Smith, editor of *BMJ,* said it simply during our conversation in London in August of 2002: "There are so many journals that almost any article will find a publisher."

The problem arises of how to pay for the irreducible editorial functions. Archives in Ginsparg's mode, to be sure, have essentially no editorial functions, or, rather, as we have seen, the processes of evaluation are taken over by the audiences. Costs are clerical or for software development, with an occasional item for travel to one of the innumerable conferences on electronic publishing; they are covered by grants or institutional overheads and are as invisible to the user as is the cost of using the Internet. For peer-reviewed, open-access, purely electronic journals, Harnad and Odlyzko said from early on that the residual costs must be shifted from readers to the authors, a fee paid for each paper and in most cases borne by their institutions and so eventually out of the research grants. After all, many print journals, even those that are not published for profit, already impose what are called page charges (with the anomalous consequence, under United States law, that such articles must be labelled "advertisements"). The savings to institutions from cancelled print subscriptions, they assert, will cover such costs many times over. This seems to have become the consensus. The process is indeed well begun.

More generally, though, the fluid and cumulative electronic process as envisaged by Ginsparg, Odlyzko, Harnad, and scores of others by now, building towards that global knowledge network—this itself frightens many editors. Anyone who has tried to follow an extended branching discussion of serious issues of any sort on an Internet list server will be aware of the potential for chaos. (Try SCIFRAUD!) Several such sites have carried the disputations about electronic publishing, and from these several anthologies of hundreds of pages of e-mail messages have been compiled and, indeed, edited with remarkable care. They are difficult to use. Lundberg again:

> in the Harnad approach all articles are works in progress perhaps never completed since science itself is a work in progress there is some appeal to the concept that the literature of science should also be thought of as a work in progress but i believe that for most human beings a more orderly trackable and ??valid?? approach would include "end points" for some recorded work and beginning points for others rather that [*sic*] a continuing sequence of changing and incomplete communications

One e-correspondent asked querulously, "How is the user to know which is the copy of record? Which should be cited?" Certainly, as many scientists and editors conceived of the process it moved asymptotically towards definitive versions. Here was the complementary functional relationship between screen and print: the final form, the one to be cited, would be published on paper. Harnad, as he favors a degree of editing at some stage, replied to the querulous correspondent, "The copy of record for the refereed postprint is the draft published by the refereed journal in which it appears, just as it has always been!" Although the matter is not yet settled, increasingly this looks like a transitional solution. One respondent wrote that the e-environment "includes versioning, allowing updates, creating links to subsequent and related works, etc. Fixing versions is an artefact of print"—embracing Lundberg's "work in progress."

A miscellany of other questions that once seemed critical we can recognize as transitional also. Consider the dilemma of the forlorn assistant professor seeking tenure, needing a record of serious publication. Although a young historian can't make it without a book peer reviewed and published in perhaps a thousand copies by a university press, at least among scientists in more and more fields every member of the tenure committee is already publishing electronically.

To some, at first glance, copyright was menacing. Yet recall the fundamental difference: copyright protects the earnings of writers whose writing is itself a source of income, whereas scientists don't expect to get paid for their research reports. They want publication to which readers have maximum access. In the sciences, what copyright supposedly protects is the profits of journal publishers, even while it restricts access. It might seem to offer a residual value if it somehow discouraged misuse of published material; a charge of copyright infringement might be a useful threat. But plagiarism is still fraud whether the words and ideas stolen are formally copyrighted or not.

Libraries will have to shrink, Odlyzko wrote in 1994.

> If the review journals evolve the way I project, they will provide directly to scholars all the services that libraries used to. With immediate electronic access to all the information in a field, with navigating tools, reviews, and other aids, a few dozen librarians and scholars at review journals might be able to substitute for a thousand reference librarians.

Librarians have indeed found themselves threatened. For decades, they have faced a growing set of problems: space, costs, staff, and beyond that a change of their functions. Books will continue to be printed, bought, and shelved. Marshall McLuhan was wildly mistaken when he asserted, forty years ago, that print is linear and restricting, electronic media spacious and flexible. No: books are easier to use because besides being portable they offer many dimensions of access, which is to say, you can move around in them, skimming, browsing, dipping in and out. Over breakfast you and I may both be rustling the sheets of today's *New York Times,* but in effect we're reading two different newspapers. The subset I pay attention to will omit the sports except during Wimbledon and include little of the business section save the latest news of corporate fraud—and you?

Suppose printed journals die out. That will kill the libraries' subscription budget—at least, if electronic journals are not pay-to-view. Yet that may not save space; university libraries may need roomsful of computers for undergraduates, and though by now almost every student has a laptop, they need someplace to set them down. Functions of staff are changing. David Goodman, biology librarian at Princeton University, stated this clearly in the ongoing e-dialogue several years ago:

> The distinctive characteristic of librarians at one time was our ability to organize paper documents (both physically and intellectually). We pretty well understand how to do this, though it is true that we have never found a way of making it particularly easy for the inexperienced user. This role will continue to a limited extent; for one thing, it will be many years before all documents are issued in electronic format, and all older documents converted. For another, I suggest that it will remain advisable to maintain microfilm and paper copies of documents in limited numbers as an ultimate backup. Many of those in the e-pub area do not recognize the need for this, but I think it will be both prudent and inexpensive.

> But now the distinctive ability of librarians is their ability to help users find documents. This is even more true in an electronic environment, and I think the users know this. My experience in working with a reasonably sophisticated and intelligent community of patrons is that an expert can find things both more comprehensively and faster than the average web-savvy user.

In any case, libraries and other traditional scholarly archives are being outflanked. Examples spring to mind. At Tufts University, the entire corpus of classical Greek texts, edited, annotated, accessible, enjoyable, there to be down-loaded and printed out, has been put on a Web site called Perseus Digital Library. Perseus is adding the Latin classics now, and texts and secondary sources for English Renaissance literature. Mirror sites—duplicates—have been established at Oxford, Berlin, and Chicago. At Oregon State University, Linus Pauling's papers and correspondence—he was an alumnus—have been scanned in, equipped with hypertext cross-references, and on his birthday in February of 2003 were opened to the wide world on a Web site. At the National Library of Medicine, in Bethesda, Joshua Lederberg's papers published and unpublished, correspondence, laboratory notebooks, and all are going onto a Web site, thoroughly cross-linked. Lederberg himself, sitting at his desk at Rockefeller University and wanting something from his own older files, gets it off that Web site and can print it out and hand it to you. In Italy, a team of scholars has put everything that survives of Galileo's—books, manuscripts, correspondence, and all—on a Web site, scanned in in facsimile, transcribed where handwritten, and garnished with a translation from the Latin or Italian into English. At the Library of Congress, which owns one of the three surviving perfect copies of the Gutenberg Bible, the entire volume has been scanned in and issued as a compact-disc set; the user can highlight any of the Latin and see an

English translation. At the Max-Planck-Institut für Wissenschaftsgeschichte, in Berlin, Jürgen Renn, one of the three division chiefs, told me several years ago that he will put the totality of source materials for the history of science onto a Web site, scanned, indexed, cross-linked, and so on. (Renn does have a copyright problem.) For historians and other scholars the nature of archival research is being forever transformed: less dust, less providential discovery, less excuse for putting off analysis and writing.

In the fall of 2003, a tricky issue popped up to perturb open archiving, involving Ginsparg's arXiv—a question of libel. Martin Rees is a cosmologist at the University of Cambridge, and is Britain's Astronomer Royal; he has been a trustee of top scientific and educational institutions; he is Master of Trinity College; and besides nearly five hundred scientific papers he is author of five highly regarded, highly readable books for the general audience: a scientist at the peak of power and accomplishment. He had posted a paper to arXiv, while awaiting peer review, announcing certain results in gamma-ray astronomy. Alvaro de Rújula is a Spanish theoretical physicist working at CERN, in Switzerland. He, too, is expert in gamma-ray astronomy, and has a close colleague, Arnon Dar, at the Technion Space Research Institute, in Israel. On October 27, Rújula posted an article—well written, evidently long pondered—attacking Rees for failing to cite work by others, including Dar. On November 2, after the controversy had blown up, Rújula posted a revised version. His title asked if this is "a case of unethical behavior." The abstract, version 2, read, in full:

> One might have hoped that the immediacy and completeness of scientific information provided through the internet would have made the wrongful appropriation of someone else's ideas—now

> so easy to detect and document—a sin of the past. Not so. Here I present evidence, which the reader should judge, of such an apparent misconduct by Sir Martin Rees, the British Astronomer Royal, and others. Unethical behaviour is not unknown in science. My sole intention is to call attention to the problem by way of example, in an attempt to contribute to a more ethical atmosphere, which would in my opinion be beneficial to the field.

His text read, in part:

> [T]his note is less concerned with scientific issues than with questions of academic integrity: the proper procedure by which novel ideas may be adopted from the works of others, the appropriate attribution of priorities, and the example to be set by those in high office.

Weeks earlier, indeed, Dar had asked Rees to fix the citations. Rees told a reporter for *Nature,* Jim Giles, that he had intended to do so as soon as the paper had been reviewed. But on October 29 he went ahead with the changes.

As we have seen, arguments over credit and authorship can be the nastiest in science—and the fact that this one was resolved briskly and publicly could encourage others to complain. Worse, though, Rújula's language was harsh, and more so in the original version. The law of libel varies greatly from country to country. In the United States, it is hard to libel a public figure, for the plaintiff must prove actual malice—though the mere hint of such a suit, with its legal costs, even if unwinnable is intimidating. In Britain, the law of libel is notoriously favorable to the plaintiff. Postings to preprint archives are publication. The question, then: Are they subject to the local law of libel wherever they can be opened? The point has not been resolved. But if so, would not only Rújula but Ginsparg and Cornell be potentially liable?

Reflecting on the matter, Ginsparg pointed out that here as in many imaginable legal issues involving the Internet "there is no global court to adjudicate these things. The journals are not equipped to do it, and are too slow, as are the institutions at which people work." The very fact that a scientist is prominent would make a problem of plagiarism difficult to pursue in the ordinary way. "What should an author like De Rújula do in a case like this, if every other avenue has failed?"

Rújula's article is also a remarkable testimony to the power the archives now exercise. "What I shall discuss here," he wrote, defining his complaint, is "the ways by which some scientists rewrite history by de-emphasizing the contributions of others relative to their own, or by imposing their own modes of thought upon the community." More particularly, he asked whether the main concepts of the developing "new paradigm" in gamma-ray astronomy "will be attributed to their creators." He then went on:

> Today, because scientific articles are virtually instantaneously 'posted on the web' in freely accessible and inviolate electronic Archives, one might naively suppose an affirmative answer. But science is a profoundly human endeavour and what may appear obvious is not always so. Whether or not a note such as this should be posted in the Archives is a debatable question, alternatives to which I have discussed with many colleagues. But a "higher court" of science does not exist. Moreover, the Archives now have a much greater impact than journals: it is mostly the

> original version of posted papers that people read, without checking the journal versions for changes, even when the journal version is the one they quote. In my opinion, what is freely posted in the Archives ought to be discussed in the Archives, since it is the sole responsibility of its authors.

*

Electronic publishing, whether by formal journals or in one or another variation of Ginsparg's anarcho-democratic self-governing mode, can overcome the two worst problems that plague ink-on-paper journals. The most obvious advantage is speed. To be sure, even in print the hot-news journals have given the most significant submissions special treatment. Of course they do: a necessary exercise of editorial judgement. In the instance that must be the record holder, when David Baltimore and Howard Temin, in early and mid-June of 1970, sent *Nature* their reports of their independent, simultaneous discovery of what we now call reverse transcriptase, they were in print in a matter of days, the issue with the cover date 27 June 1970. This was so quick that I have wondered whether they were peer reviewed at all: John Maddox, editor at the time, when questioned two decades later was uncharacteristically vague. Normally the lag time from submission to print publication has been months, even in the best-run journals, and if revisions are called for can stretch to a year or more. Before the coming of electronic submissions and editing, much of this lag could be blamed on less-than-instantaneous communication between editors, staff, referees, and authors. Today the lag is due in part to the mechanics of printing and mailing—but also to the backlog of other papers ahead in the queue.

The second problem of print, more important to maintenance of the fabric of science, is the pressure, already described, from editors on authors to condense, to simplify, and to excerpt the data itself. Short and punchy: even where punchy—strong unqualified assertion—is not explicitly demanded by the editor, tight space works towards a similar result. Electronic publication ends such stringencies. During the transition, printed papers are coming to carry the note that a more extended discussion and a full presentation of the data are available on line. George Lundberg, when he was editor of *JAMA,* long ago began requiring that authors agree that upon any reader's request they will supply the raw data. In October of 1995, under the headline "*Nature* on the Internet (at last)!," Maddox asked in the lead editorial that readers charitably regard what they would find on visiting the new Web site "as part of a learning process." He wrote, in part:

> It is the standard complaint of the authors of research that journals never give them enough room to say what they have to say. *Nature* is deemed particularly niggardly in this respect, although the facility for accompanying a research article with 'supplementary information' has been available for some years. Hitherto, copies of such material have been mailed free of charge to those who have asked for it. From now on, it will also be made available without restriction on www.nature.com.

In short, as I noted previously, electronic publication can eliminate forever that problematic parenthesis "(data not shown)." (We shall still have "data not read.")

As early as April of 1994, one publisher began offering papers in two of its physics journals electronically as soon as they were accepted, thus getting them to subscribers to this new service fully five months before they appeared in print. By the end of 1995, more than one hundred peer-reviewed journals in the sciences, technology, or medicine were on line. Publishers were compelled to keep up with rivals yet were desperate to maintain circulation, satisfy advertisers—and retain some control of their content. A year later, the number had increased by a factor of ten.

Coming awake to the power of the Web, yet still holding back, a number of the high-impact journals have taken the next step: pushing for general-media coverage, they have at least come so far as to place articles of exceptional or urgent interest on line days before the print version is out. *Science* and *Nature* do this. Thus in mid-August of 2002 scientists at Leipzig and Oxford announced that they had identified a gene, tagged *FOXP2*, that is distributed among mammals in essentially identical form—except that compared to mice chimpanzees have two mutations in the sequence while humans have one more, which chimps haven't got. A defect at another point in the human sequence of the gene was known since the previous summer to block acquisition of normal speech and even the comprehension of grammar. The third sentence of the report: "*FOXP2* is the first gene relevant to the human ability to develop language." If correct, the discovery of a talking gene is momentous. The paper was on the *Nature* Web site days before the printed journal was out.

Again, on 9 October 2003 *The New England Journal of Medicine* posted to their Web site a report for early release, "A randomized trial of Letrozole in postmenopausal women after Tamoxifen therapy for breast cancer," and accompanied it with a brace of editorial commentaries. The fact had been known for some time that women who had been treated for breast cancer, and had then been put on the drug tamoxifen to make recurrence less likely, sometimes after five years became susceptible again. The drug letrozole was undergoing a long-term trial, matched against a placebo, as a replacement: the new therapy had turned out so successful that the trial had just been terminated early to make the drug available at once. The editors judged the paper urgently important.

In 2000, Harnad wrote, "It is a foregone conclusion that all refereed journals will soon be available online; most of them already are. This means that anyone can access them from any networked desk-top." The move has taken forms along a spectrum. Many high-impact journals—*Nature, Science, JAMA, BMJ*—began by offering stripped-down teasers on the Web, tables of contents, abstracts, a few reports. Other journals did the obvious from the start, putting up straight electronic versions—clones. Over the years, editors who had held out found themselves forced to move towards full, simultaneous electronic publication, with what you see on-screen adapted more or less effectively to the radical differences in format. Thus by the turn of the millennium, when Harnad made his prediction, several of the big publishers of traditional journals had gone to parallel publishing on the Web, Reed Elsevier with more than twelve hundred journals, Springer-Verlag with three hundred and sixty, Academic Press with nearly two hundred. To hang on to that so-lucrative market, most have imposed restrictions on access with subscription fees. Reed Elsevier has always been notoriously rapacious. The conspicuous, high-impact journals have adopted various practices. Thus today, like *The Economist, Nature* lets subscribers browse the latest issue and search the archive; the *BMJ* does the same even if you don't subscribe. An increasing number of journals are published electronically only.

In December of 1998, when Paul Ginsparg spoke to the biologists meeting at Cold Spring Harbor to organize a depository for data on gene expression, among those listening were Patrick Brown, a geneticist at Stanford, and Richard Young, at the Whitehead Institute for

Biomedical Research, at MIT. Ginsparg urged them to think more expansively. The meeting proved to be a catalyst. Over the next months, discussion spread, developing proposals for a preprint server for molecular biology and biomedical research, more or less on the lines of Ginsparg's physics archive.

The idea attracted Harold Varmus, director of the National Institutes of Health. In the first week of March of 1999, Varmus announced at a congressional hearing that he was exploring how to set up an electronic repository for biomedical papers. Others, meanwhile, were converging on the idea, notably biologists in England and on the Continent centered on the European Molecular Biology Laboratory, at Heidelberg. After circulating an earlier sketch, the first week in May Varmus went on line with "E-BIOMED: A Proposal for Electronic Publications in the Biomedical Sciences." He had consulted with many, including Brown at Stanford, and David Lipman, a colleague at the NIH, director of the National Center for Biotechnology Information. Varmus had written the proposal himself—and it bears the imprint of a clear, original intelligence, politically astute. He began with a disclaimer: "It is important to emphasize at the outset that in no sense would the NIH operate as the owner or rule-maker for this enterprise." (Members of Congress, please note.) "The essential feature of the plan is simplified, instantaneous cost-free access by potential readers to E-biomed's entire content in a manner that permits each reader to pursue his or her interests as productively as possible." He proposed publishing on the Net in two tracks. Scientific reports could be submitted to editorial boards, which would function much as journals traditionally have done. This track would include refereeing, editing, and selection of what was to appear or not—while reaping cost savings as Harnad and Odlyzko have predicated. Rejected papers could be submitted to other journals within E-biomed. But the novelty was the second track. "Authors would also have the option of entering scientific reports directly into the E-biomed repository without soliciting endorsement by one of its editorial boards."

In effect, to pull the chariot Varmus yoked Harnad's ox with Ginsparg's horse. But he wanted safeguards along the second track. "Before publication in the database, each report would need to be approved by two individuals with appropriate credentials. These credentials, to be established by the E-biomed Governing Board, should be broad enough to include several thousands of scientists, but stringent enough to provide protection of the database from extraneous or outrageous material." Criteria, he wrote, "must be sufficiently firm to guard against gross abuse of the E-biomed repository, but sufficiently flexible to permit rapid posting of virtually any legitimate work."

Varmus endorsed the electronic vision, pushing it farther.

> With the dramatic expansion of space, it will be possible to present much larger data sets (including detailed photographs and movies), provide more extensive analysis, and describe methods in the precise detail necessary to recapitulate experiments. Moreover, electronic formats allow layered viewing at increasingly greater levels of detail, so that readers can first get a concise message and then pursue information in proportion to need and interest.

With the enthusiasm of a convert, he wrote, "E-biomed is designed to evolve in ways that might affect the way we practice science." These could include a more open reviewing process, career rewards for effective commentaries on the work of others, posting of synopses or even the full texts of talks at meetings and symposia, above all the means to amplify and amend reports. "One further, less tangible benefit might also occur as a natural outcome of E-biomed: a heightened sense of community among biomedical scientists."

Varmus has unrivalled standing in the world biomedical community, and as director of the NIH held a position of the greatest influence in the science and in the politics of science. The proposal aroused opposition of two sorts. Peer review was the first obstacle. The unrefereed, Ginsparg-style track had to be abandoned—a public embarrassment. Secondly, publishers, threatened by the prospect of electronic competition from a government agency, lobbied to cut off funding. Under these pressures, the original proposal was drastically modified. Plans to publish original articles were abandoned; under a new name, PubMed Central, it became a digital archive of existing journals in the life sciences, in effect the electronic equivalent of a conventional library's stacks, but able to be cross-referenced and searchable from any personal computer with a few keystrokes. PubMed is run by the National Library of Medicine. In this form, it has signed up more than one hundred journals and is becoming a useful reference tool. But it is not electronic publishing.

In London in 1999, Vitek Tracz, an entrepreneur of biomedical publishing, began work on an electronic publishing venture conceived differently, and in the first place commercially. He calls it BioMed Central. (He settled on the name before PubMed Central.) His Web site (http://www.biomedcentral.com) promises "full peer review / immediate publication / open access." We talked late in August of 2002, at his office in Soho, north of Oxford Street in London, and then over lunch. Tracz (rhymes with crotch; the "cz" is as in Czech Republic) has had a checkered career. "I was born in Poland during the war but grew up in the Soviet Union—without leaving my home village," because Stalin appropriated nearly half of Poland, shifting the Polish border with the Soviet Union west. He spent four years in Israel, and was making documentary films, but "I got into medical publishing in 1979." His umbrella organization, Current Science Group, runs a stable of print journals, notably *The Scientist* and a series of monthlies, sold by subscription, on medical specialties from *Current Allergy and Asthma Reports* through, for instance, *Current Opinion in Molecular Therapeutics,* down to *Current Urology Reports* and *Current Women's Health Reports.* Tracz, it must be said, is a hard-driving promoter and salesman of ideas, bullet-headed, tough and assured, but unquestionably original and indeed brilliant. He has won the respect of established competitors. That same afternoon was when I saw Richard Smith, of *BMJ.* Smith, a tough man himself, spoke admiringly of Tracz.

BioMed Central published its first article on line in 2000. In our conversation, Tracz said they had just published their thousandth, and he expected to reach the second thousandth in a matter of months. The principle, Tracz said, is absolute open access. "Every paper published is available to anyone for any use so long as its integrity is not destroyed." The venture is divided into a series of specialized journals, each edited separately, mostly in a series *BMC Anesthesiology* to *BMC Surgery.* As anyone can look at any of them, the point of the series titles is to guide readers to what they want. Submissions are electronic, of course. After an initial cull, some editors post papers before peer review, others go through the process conventionally.

BioMed Central has proliferated and subdivided. By the end of 2003 it was putting out some one hundred and thirty electronic journals—but this is a moving target, blink and the number has increased. The latest addition was the *Journal of Circadian Rhythms.* In any event, Tracz says, the individual article is the important thing, while the notion of a journal in a world of hypertext links is becoming little more than a convenient way to group articles for initial access by related content. The policy is still that research articles are immediately and permanently available without charge. (However, a number of other Current Science Group journals—some thirty-one—require a subscription to open other kinds of content, such as reviews; most of these are the on-line versions of the print journals in the *Current Reports* series.) By mid-2003, BioMed Central claimed more than two hundred and twenty thousand registered viewers, who clicked into individual reports an average of a thousand times during their first four months on line; some articles in BioMed Central's *Journal of Biology* have reportedly scored some twenty thousand hits.

The problem, therefore, is where to make the money. "Where is the value added," Tracz asked, "and how can we charge for it?" We all know that print journals are immensely profitable; they make their money through subscriptions. "But their day is over." The only thing slowing the transition is the scientists. "It's the scientists who are extremely conservative." The answer is just what Odlyzko and Harnad foresaw, shift of charges to authors. Though BioMed Central content is free, authors pay a five-hundred-dollar processing fee. As expected, the fees are being picked up by institutions. Two examples. In Britain in June of 2003, the Joint Information Systems Committee, a government-sponsored advisory organization whose members include all British universities, struck a deal to pay the equivalent of one hundred and thirty-eight thousand dollars to cover fifteen months of BioMed Central fees for scientists at British universities—some eighty thousand. The committee had calculated that British universities spend the equivalent of one hundred and twenty-eight million dollars a year on journal subscriptions. In the fall of 2003, the Ohio

Library and Information Network, a consortium, bought academic authors at eighty-four Ohio institutions, starting in January of 2004, the right to publish in BioMed Central journals, paying no individual fees.

Author fees, then, are one way to recoup the costs of the editorial processes. Tracz has invented others. Audience research at Elsevier, he told me, has uncovered the eyebrow-raising fact that the typical article in their journals is read thoroughly by about five people. "What everybody reads is the abstracts; that's where most of the learning takes place." So perhaps one should charge only for abstracts?

More promising is a scheme Tracz calls post-publication selection—which adds value by having experts rate the quality of research reports after they have been published. He has devised a delightfully ingenious way to do this, and has packaged it and put it on the Web as Faculty of 1000 (http://www.facultyof1000.com).

The structure first, then how it functions. The structure is hierarchical. It divides all of biology into faculties, meaning major subject areas, sixteen at the end of 2003, biochemistry, cell biology, developmental biology, genomics & genetics, molecular medicine, and so on. Each faculty is run by two, three, or four scientists designated heads of faculty. Each faculty is subdivided into anywhere from three to twelve sections, seventy-three in all last time I looked. Each section, in turn, has two or three section heads and anywhere from ten to fifty other faculty members. The faculty immunology, for example, has twelve sections, from allergy & hypersensitivity through genetics of the immune system to leukocyte signalling. The subtopic gene therapy, for example, is one of four in the section medical genetics, within the faculty of evolutionary biology. All told, about fourteen hundred scientists are involved; Tracz has signed up some top people and many younger scientists of promise.

Despite the terminology, they don't teach. What they do is nominate research reports to be posted on the site within their sections. Papers can come from any journal, however well or little known. They can come from last week's published work, or from older research. (I would nominate, for example, T. H. Morgan's polemical review article "The theory of the gene," originally published in 1917.) And they give each paper a rating from one to ten for quality and importance. (I'd rate Morgan's paper a ten: it is powerfully written and immediately relevant to debates today.) Nominations are vetted by section heads. High-scoring papers are flagged for attention: one can look only at the top ten in the section of interest, or only at the top ten in the whole system. Where copyright interferes, hypertext links take the reader to other services that may allow the paper to be opened. Cross-referencing is ample, browsing and searching methods flexible with many modes and dimensions of access. A personal subscription costs fifty-two dollars a year, a dollar a week. A year's personal subscription to *Nature* will presently cost you one hundred and fifty-nine dollars.

Harold Varmus is resilient, resourceful, and not easily discouraged: indeed, that's a textbook example of ironical understatement. At the end of 2000, he and allies organized a coalition of research scientists under the rubric Public Library of Science, or PLoS. The first act was to circulate an open letter calling on scientific publishers to make their primary research articles available, free of charge, depositing them in electronic archives like PubMed Central. More than thirty thousand scientists, from some one hundred and eighty countries, signed the letter. Some publishers coöperated to some extent; but, Varmus wrote, "in general the publishers' responses fell short of the reasonable policies we advocated." So he and allies concluded that they must publish a series of PLoS journals, open access, online and in print. They secured a grant of nine million dollars from a private foundation. In October of 2003, they launched the first issue of their first journal, *PLoS Biology.*

Vitek Tracz wrote them a nice letter of congratulation.

Over the past decade, multiple conferences on open-access publishing have taken place, at UNESCO in Paris, in Budapest, at the NIH in Bethesda, at the Howard Hughes Medical Institute, outside of Washington, in April of 2003, and the latest in Berlin in October of 2003. What began as proselytizing has modulated to the setting and evolution of technical standards—even as the support for the movement has grown. Thus, the meeting at the Howard Hughes headquarters was attended by twenty-four people from the United States and Europe. The luminaries were there: scientists; publishers including the BMJ Publishing Group; librarians; representatives of scientific societies and of rich foundations including Wellcome and Rockefeller; and of course BioMed Central and the Public Library of Science. Among much else, they issued a definition.

> An Open Access Publication is one that meets the following two conditions:
>
> 1. The author(s) and copyright holder(s) grant(s) to all users a free, irrevocable, worldwide, perpetual right of access to, and a license to copy, use, distribute, transmit and display the work publicly and to make and distribute derivative works, in any digital medium for any responsible purpose, subject to proper attribution of authorship, as well as the right to make small numbers of printed copies for their personal use.
>
> 2. A complete version of the work and all supplemental materials, including a copy of the permission as stated above, in a suitable standard electronic format is deposited immediately upon initial

publication in at least one online repository that is supported by an academic institution, scholarly society, government agency, or other well-established organization that seeks to enable open access, unrestricted distribution, interoperability, and long-term archiving (for the biomedical sciences, PubMed Central is such a repository).

Similarly, the Berlin meeting was called by the presidents and chairmen of seven German national scientific organizations, including the Max-Planck Gesellschaft, the Science Council, and the powerful Deutsche Forschungsgemeinschaft, or German Research Foundation. Participants came from all over Europe. They labored and brought forth, in German with an English translation, a Berlin Declaration on Open Access to Knowledge in the Sciences and Humanities. As such documents go, the declaration is reasonably terse and clear (though the English translation is mysterious in places). The aim, they said, was to promote the Internet as a functional instrument for a global base for scientific knowledge and intellectual interchange. The heart of the declaration was a pair of standards for open-access publication, set out in lawyerly language that if anything emphasizes the breadth of the claim asserted. As was obviously necessary, it closely resembles the Bethesda declaration. First, authors and other copyright holders must "grant to all users a free, irrevocable, worldwide right of access to, and a license to copy, use, distribute, transmit and display the work publicly and to make and distribute derivative works, in any digital medium for any responsible purpose, subject to proper attribution of authorship." Secondly, a complete version of the work and all supplemental materials, "in an appropriate standard electronic format," must be "deposited (and thus published) in at least one online repository . . . that is supported and maintained by an academic institution, scholarly society, government agency, or other well established organization that seeks to enable open access, unrestricted distribution, inter-operability, and long-term archiving."

Print-journal subscriptions are eroding. Even high-impact journals are losing. One that at its high point sent out seventeen thousand copies an issue now has ten thousand subscribers—but their on-line readers download hundreds of thousands of articles a month. Libraries are cutting back ruthlessly—and, one imagines, with a certain vengeful glee. One example says all. In October of 2003, the Cornell University Library announced that it is cancelling subscriptions to hundreds of Elsevier journals. Their explanation affords a glimpse of the turmoil in print publishing. The announcement, in part:

> We can no longer subscribe to so many Elsevier journals (including duplicates) that we no longer need. We must now free up some of the money spent on Elsevier journals to pay for journals published by other publishers that are more needed by our users. We have explained this to Elsevier in lengthy discussions, both through our research library consortium and then independently. We have tried in these discussions to broker an arrangement that would allow us to cancel some Elsevier titles without such a large price increase to the titles remaining—but Elsevier has been unwilling to accept any of our proposals. We are therefore planning to cancel several hundred Elsevier journals for 2004. The decisions on cancellations will be made on the basis of faculty input, as well as several years of statistical information on individual journal use.

*

Having browsed through the decade and more of discussions, proposals, and varied electronic publishing ventures, having interviewed a number of the chief advocates, aside from the occasional glancing reference to the detection and so the discouragement of plagiarism I have encountered no awareness of the effects of such publishing on the nature or incidence of fraud. Yet the effects will be great. Once again, but this time favorably, they are structural, built in. Plagiarism, that most pathetic of frauds, will not automatically be detectable with each submitted paper—or not yet. We saw in Chapter 7 that given the right search algorithms today's high-speed computers could do it easily. After all, Google claims today that to satisfy your most trivial inquiry it matches your keywords searching more than three billion Web pages. Be that as it may, once suspicion arises, powerful programs for comparing strings of characters are available; in a trivial example, any college teacher can call up, from the desktop, software that will match suspect passages in a student essay to all the likely sources.

How satisfying it is to tell the hovering shade of Charles Babbage how computers will disable cooking and trimming! The less sophisticated forms of inventing or adjusting outcomes become far more difficult when the published report must make all the experimental methods reproducibly clear and offer to make all the data available. Harold Varmus did not have fraud in mind when he said in his proposal for E-biomed that Internet publishing encourages "much larger data sets," while "electronic formats allow layered viewing at increasingly greater levels of detail." But Thereza Imanishi-Kari when Weaver et al. was published apparently had the barest scraps of data.

*

Details vary and are evolving. The shape of the future has been clear for ten years or more. At that conference in Chicago on peer review in November of 1993, addressing the auditorium full of biomedical-journal editors, here is what I said in peroration.

The modes of electronic publication appear to be converging. Software running on geographically dispersed computers will be able to link together an entire research corpus. The time is soon coming when the act of publishing a report electronically will be but a preliminary step. You the reader will be able to scan not just the compressed references to other articles that we now see in the notes but, if you wish, hyperlinks will bring you the abstracts of those articles or the texts themselves. You will be able to demand the referees' reports, perhaps the raw data. Publication will invite open criticism, suggestions, rebuttals; the article so published will be able to go through revisions on the screen, with comments and old versions retained for reference. When anyone calls up the article months or years later it will bring with it indexes of all these responses and changes that have accrued, available to anyone who wants to look. Corrections and retractions will similarly be forever linked to the article itself. The capacity to compare texts across journals, scientific fields, and years will be a powerful restraint on plagiarism, at least of the simplest kind. Every time the anchor is pulled up, all the seaweed will come with it. Beyond that stage, we must imagine that remarkable interactive options will develop for readers of scientific articles published electronically.

I go into this detail about the potential of electronic publication because we must recognize how revolutionary it really is. We are not talking of the substitution of one medium for another, of the replacement of the printed page by the screen, with everything else—including editing, refereeing, revision before publication, and readers' correspondence—to go on as before.

We cannot dismiss the problem of quality control in electronic publication. Certain journals, at least, must adapt the speed and potential openness of the network in some fashion to preserve—to enhance—the values that can be added by skilled editors and thoughtful expert commentators. These values are what make some journals authoritative. They will continue to be desirable as one element of the new dialectic of publication.

At the same time, though, journals and their editors are powerfully placed in the social system of medicine and the sciences. Both in teaching and in practice, this social system is steeply hierarchical and in many respects authoritarian. At many levels of the system, then, the transformation brought in by electronic publishing will seem threatening, to be resisted. Yet profound change is inevitable. Research processes themselves evolve with the emergence of new questions and technologies, as well as in response to the transformations I have sketched. A new generation of journal editors will arise who have grown up with electronic editing and publishing. In ten years' time, although procedures will be followed that some journals will still label refereeing or journal peer review, these procedures will be startlingly different from those put into place in the years after the second world war, which, despite their brief history, seem so monolithic and unchangeable today. To predict the details of the new procedures is no doubt foolhardy. Yet I venture to say that the transformation will open up the processes by which scientists judge each other's work, making the system less anonymous, capricious, rigid, and subject to abuse, and more thorough, responsible, and accountable. It will oblige the readers of journals to take a more active part in the intellectual assessment of published work. Eventually—but sooner than you can easily imagine—we will see an evolution towards a form of publication that will be a continuing open dialogue and collaboration among contributing scientists, editors, expert commentators, and readers.

What will the editors and readers of those journals, looking back, be likely to think of the system of publication that they have replaced?

9

LABORATORY TO LAW

THE PROBLEMS OF INSTITUTIONS WHEN MISCONDUCT IS CHARGED

In this age of science, science should expect to find a warm welcome, perhaps a permanent home, in our courtrooms. The reason is a simple one. The legal disputes before us increasingly involve the principles and tools of science. Proper resolution of those disputes matters not just to the litigants, but also to the general public—those who live in our technologically complex society and whom the law must serve.

—*Stephen Breyer, Associate Justice of the Supreme Court of the United States*

At first glance, many instances of scientific fraud seem the result of failure of supervision: the big laboratory, the remote boss. Vijay Soman at Yale was working in the laboratory of Philip Felig, but Felig's office, as vice chairman of the Department of Medicine, was two blocks away. Darsee at Harvard was working in one of two laboratories in the large-scale science enterprise run by Eugene Braunwald, who was one of the committee directing the Department of Medicine. Stephen Breuning was first working in Coldwater, Michigan, at a residential facility for the mentally retarded, and after two years moved to the University of Pittsburgh, while throughout that time his senior collaborator, Robert Sprague, was at the University of

Illinois in Urbana-Champaign. Thereza Imanishi-Kari, at the Massachusetts Institute of Technology, ran a lab in a different building from David Baltimore's, and it appears that he never scrutinized her data for Weaver et al. before the paper was published. Jan Hendrik Schön became a permanent research scientist at Bell Labs in December of 2000, but had to return to Germany to await a new visa; staying in touch with his collaborators in New Jersey, for five months he performed his experiments at the University of Koblenz—and even after returning to Bell Labs he went back to Koblenz from time to time, ostensibly to use instruments there. After Schön's fall, *The New York Times* reported that "some outside scientists, including former Bell researchers, said they thought such a scandal would not have occurred at a Bell Labs of an earlier era, because scientists scrutinized each other's work more closely."

Failure of supervision, however, must be understood as but one aspect of failure of mentorship. Recall Robert Merton's analysis of the norms of science. They are "a distinctive pattern of institutional control" of the behavior of scientists, and "it is to the interest of scientists to conform on pain of sanctions and, insofar as the norm has been internalized, on pain of psychological conflict." In response to the increasing awareness of misconduct, the National Institutes of Health since 1989 have required that research institutions receiving grants hold regular courses in ethics, which doctoral students are required to attend. At universities across the country, scientists have developed such courses, and some make inventive use of case studies, small-group discussions, and role-playing. In response to the Schön scandal, Pierre Hohenberg, associate provost at Yale, suggested to the American Physical Society that physicists should be required to have training in research ethics.

To be sure, instruction in ethics can prepare otherwise naïve young research scientists to recognize some of the dilemmas and plausible temptations that may take them by surprise. All to the good; yet it is not obvious that norms can be inculcated by precept. What works to internalize norms, to set the social conditions for psychological discomfort, is not a class once a week for a semester, but the day-to-day practice of a well-run laboratory—the example in action of a mentor. "In those days, when you came to a new lab, the first thing you did was calibrate the weights," Alfred Nisonoff said, and the remark contains not nostalgia but something much larger however laconically stated, a stern, plain realism about the way good science proceeds.

That wooden chair by Howard Temin's desk at the University of Wisconsin was uncushioned, numbingly hard: I sat on it for hours at a time in two days of interviews. Temin's laboratory was small, his scientific imagination profound, his personality forbidding, and each Friday the workers in his lab came one by one to sit on that chair and show what the week had achieved. This was not simply tight supervision but style. Temin insisted as a matter of his personal ideology of American science that other styles were also valuable; he defended Robert Gallo, and Baltimore talked to him repeatedly during that affair.

Style in the doing is the essence of mentoring. Temin was himself one of a younger generation of a great lineage in science. This line goes back to the early nineteen-hundreds and Christian Bohr, a Danish physiologist of blood, father of the early quantum physicist Niels Bohr. In the nineteen-twenties and early thirties, Niels Bohr had a circle of younger quantum physicists, and his style of doing science became enduringly influential. One in that circle in the early thirties was Max Delbrück, from Berlin, who got to the California Institute of Technology in 1938 and switched to what became molecular biology. Delbrück's understated, skeptical, relentless style owed much to Bohr and shaped a generation of younger scientists in *his* circle, famously James Watson, among others Gunther Stent, Matthew Meselson, Renato Dulbecco. Fraud would have been unthinkable in the Delbrück group. "We worked to win Max's approval," Meselson said to me—a definition of charisma. Temin was at Caltech, became a friend of Meselson's, later worked as a postdoc in Dulbecco's laboratory. For a young scientist on the threshold of a career, waiting to be initiated into the community as an adult in the tribe, such influences drive the norms deep.

That said, of course the whole point is that in practice mentoring is not perceived in dramatic terms. Walter Gilbert had taken two years away from Harvard to found and run Biogen, a biotechnology company. He got rich, returned to Harvard, and started collecting classical Greek vases. He told me that in the period he was away graduate students and postdocs had stopped keeping notebooks. So he started them up again, and prepared a brief instruction sheet explaining how and why to do so. The first reason, he said, was to guard against self-delusion. One remembers successes. At the bench, failures are more common than successes, but they are naturally forgotten. A well-kept notebook will remind you of the things that went wrong. At Stanford University some years ago I was talking to a young graduate student about keeping notebooks, and told her Gilbert's story. "Oh yes," she said, "I've started keeping a notebook. My boyfriend worked last year in Dr Gilbert's lab."

With the growth of the sciences the lineages have been swamped out, mentoring has been attenuated. Baltimore was never part of a great lineage—although, in justice be it said, postdocs and junior faculty have held him in regard. Summerlin of the painted mice, itinerant and a cheat from the start, Darsee, Slutsky, countless others have not been brought into the community by strong mentors. Imanishi-Kari actively fought off the possible influence of Olli Mäkelä in Helsinki and Klaus Rajewsky in Cologne.

*

When some member of a laboratory approaches a department head, an associate dean, or some other senior figure with a tale of misconduct, certain initial reactions are typical and guarantee that problems will follow. They are variously the product of alarm, disbelief, and sometimes disgust. On the one hand, the whistle-blower is likely to be a junior scientist, even a technician, often angry and frightened, almost surely disgruntled, perhaps someone who could be brushed off as disturbed. Further, the laboratory where the complaint arose is likely to be in trouble, ill-run, rife with conflict. That was certainly true of Imanishi-Kari's lab, as everybody in the MIT Center for Cancer Research was acutely aware. Morale was terrible, she was a screamer, and productivity was far below the MIT standard. Margot O'Toole was certainly disgruntled, after six months of increasingly heated conflict that not many outside the lab knew of, and those to whom she brought her discovery of the problem with Weaver et al. soon realized that she was clear-minded, outspoken, and stubborn. On the other hand, those to whom the complaint is made naturally feel a painful mix of responses, alarm, disbelief at least at first, a wish to suppress scandal, loyalty to longtime colleagues and friends. Often the danger is worse. Grants and contracts may be at stake, scientists' reputations and careers threatened. The motives to downplay the charges, cover up the misconduct, are strong—inevitable, almost universal. The most grandiloquent instance in modern history is Watergate, during which Nixon and his co-conspirators in cover-up made such elegant contributions to our language as "third-rate burglary" and "I am not a crook!" and "limited partial hang-out" and "twisting slowly, slowly in the wind."

Such an assortment of motives surely led to Henry Wortis's and Herman Eisen's attempts to suppress O'Toole's charges; they nearly succeeded. Some such mix surely was at the root of Baltimore's implacable defense of Weaver et al.—all that and arrogant pride, for which he was known. We have seen that a quiet announcement that the paper was to be re-examined and possibly withdrawn would have quenched the flames overnight. Every good public-relations person knows that the only way out of such a problem is to acknowledge it fully and candidly, end the talk and hope to move on. Yet even so intelligent a man as Baltimore refused for months to be interviewed. When I arranged to be invited to a dinner at which he was to speak, and accosted him directly, he said he couldn't talk because "in another few months the story might be different." Baltimore was president of Rockefeller University by then: only when David Rockefeller turned to his personal public-relations man, Howard Burson of Burson-Marsteller, did I get to turn on the tape recorder.

Many research universities now have senior administrators assigned to deal with such problems, and can do so more or less effectively, although typically after the institution has been burned at least once. One of the earliest and most experienced is C. K. (Tina) Gunsalus, at the University of Illinois, who as associate vice chancellor for research until 2002 has handled scores of allegations of misconduct, in the sciences and humanities both. Sponsored by the American Association for the Advancement of Science and a spectrum of associations of colleges, universities, and medical schools, she has conducted practicums and seminars across the country. They use a small panel of experts besides Gunsalus herself. They attract large audiences of anxious administrators and faculty. I have attended two of them. They are brisk and brutally frank.

Gunsalus's university is large. She was catching twenty to forty allegations of misconduct a year, sciences and humanities together. Of these, one or two a year had to be taken as far as a formal investigation. "But the numbers are greatly variable," Gunsalus said at a practicum in San Francisco, attended by just shy of one hundred participants. The complaint sometimes comes anonymously: What then? On this issue as elsewhere, "There are no cut-and-dried answers—just hard questions and balances to strike." A lawyer on the panel said, "In 'external' law, anonymous allegations must, often, be taken seriously." Gunsalus's comment: Correct, "But it is also important not to embark on what could be perceived as a witch hunt on the basis of purely anonymous and unverified statements."

The initial response will be crucial, and is in many ways the most interesting because it is the least structured part of the process. Gunsalus gave rules for that first encounter. Most important, "You cannot promise absolute confidentiality." Therefore the advice to the person bringing the complaint must be to think long and hard before pressing a charge. "Anonymity is difficult to protect on a practical level, and becomes increasingly troubling from the perspective of fairness as the process proceeds." Secondly, "You cannot become the friend, counselor, confidant of the person."

Then begins the most difficult part, because the moment has not yet arrived, and may not, when a formal inquiry must be started. Cases are bound to be messy. Accusations can range from sloppiness to scientific misconduct in any of its forms to financial or sexual misconduct to exploitation of students. Several of these may need to be disentangled in a single case.

(At this, Donald Buzzelli, who was then head of investigations in the Office of Inspector General of the National Science Foundation, intervened from the audience. "If you turn up things like financial irregularities or misuse of human subjects or animals, you run into overlapping jurisdictions, and what you've found may have to be taken *at once* to the appropriate agency.")

Much at this preliminary stage will seem counter-intuitive. The first interview may be difficult, Gunsalus warned. Calm can be deceptive. First impressions may be biassed. "We don't like whistle-blowers in general, we don't like tattletales." One must keep in mind that respectable, established, senior scientists have been known to commit fraud, while among the variety of accusers one encounters, "even a flake may turn out to be right."

Evaluating the complaint exposes a nest of problems, of two general classes. The first is to determine the nature and seriousness of the claimed misconduct. Often, initial indications are not clear, not sharply drawn, vague, and not pressed home. When sorted out, the complaint can range from insubstantial to positive evidence of serious fraud. Second is the social setting, the micro-anthropology of the laboratory: personal relations are likely to complicate judgement and compound stress. "What if the person who expresses the concern is so fearful of retaliation that he or she demands anonymity, or withdraws the concern? What if the person who expresses the concern appears vindictive, overwrought, or unstable? What if the respondent"—the accused—"is known to have a volatile or vindictive personality?" Motives of the several players are likely to be complex and partially opaque.

These two classes of problems, seriousness and setting, interact in the judgement of what action to take. To begin with, if a charge is serious and unmistakably one of fabrication, falsification, or plagiarism, then the institution is obligated to begin an inquiry, the first stage of the formal process, at once. Short of that, Gunsalus said in San Francisco, "The more serious the allegation, the more volatile the situation, the greater the imbalance of power, for example pitting the word of a student against a lab chief, and the more there have been previous problems at that lab, even rumors—the stink of misconduct—the more likely I am to move *promptly* to an inquiry." Even before an inquiry and throughout the process one must protect the whistle-blower. Yet one must always, or almost always, soon let the accused know.

Carpe data! "It's a delicate problem; but you should seize data promptly. One touchstone of the need to go to a formal process is the malleability, changeability of the data!" When a complaint was to be taken seriously, Gunsalus first went directly and in person, not by telephone, to the accused's superior, lab chief, department head, dean. The problem broached, that person was to telephone the suspect to say that they were coming over straightaway. "I would never do it alone!" Once there, they explained the problem and asked for the data on the spot—notebooks, slides, photographs, everything. (One time, she said, the complaint involved transgenic plants, and only one facility was allowed to hold them—with forty-eight keys.) Detailed written records are essential at this point. The loose-leaf workbook distributed at the practicum provided model forms. Many scientists are not aware that in federally financed research the data belong not to the individual but to the institution. Learning this can provoke outrage. None the less, "I usually don't have to explain more than twice that securing the data immediately is the only way to be sure that if you do *clear* the scientist it is unequivocal, the suspicion can never arise that data were tampered with."

The next phase, if the complaint warrants it, is an inquiry—"an expeditious and confidential review of allegations or concerns." Its purpose is to distinguish between allegations of scientific misconduct that have no merit and those that require closer examination.

The inquiry is roughly like a grand jury, where the "sole purpose is to determine whether there is sufficient cause to proceed." If yes, the phase after the inquiry is the formal investigation.

When an inquiry begins, the first step is to appoint a faculty committee. Gunsalus's practice was to recruit two members, one within the department, the other from another division of the university—plus herself. From the audience, somebody said that panels at his place might number as many as seven; but that, others said, is unwieldy and rare. (Recall the Darsee case, where the dean of Harvard Medical School appointed three from other departments, another from arts and sciences, one from the law school, three from institutions other than Harvard, and as chairman the dean of Johns Hopkins University School of Medicine.) Elsewhere, the practice is to draw two or three from a standing group. Gunsalus, though, said her approach was to ask new people each time, for this educates the faculty. To which Paul Friedman, on the panel, the dean for academic affairs at the University of California, San Diego, said with a shake of his head, "Education of the faculty takes a while."

The steps seem endless, from selecting the panel and charging it, through notifying the respondent and any collaborators, choosing witnesses and securing other evidence, up to generating a final report and deciding whether to go on to an investigation. An inquiry may take a hundred and twenty man hours, but often more; an investigation can devour several thousand hours. Many dangers lurk. Gunsalus warned of one: "The role of the accuser is to provide information—that only. They are not the prosecutor. A witness like any other. For example, the accuser is absolutely not to be present all through." A tough problem arises if the respondent refuses to coöperate. Ideally, Gunsalus had written, the institution will have policies in place that make clear that employees must coöperate. At this point, Andrew Schaffer, another of the expert panel, vice president and general counsel at New York University, intervened: "The presumption of innocence is imported from the criminal-law model. An inquiry, or an investigation, is equivalent to a civil-law proceeding. So one can—and sometimes should—make clear to the respondent that refusal to coöperate can weigh against the defense."

The charge to the inquiry panel is crucial and must strike a balance. "I have never written a charge I was totally happy with," Gunsalus said. "One must word the charge specifically enough to allow response, to allow a defense, yet broadly enough to allow new problems, if they turn up, to be pursued." To which a lawyerly colleague added an example, a charge of fabrication he knew of that had recently been levelled: "The inquiry found evidence of fabrication two years earlier. Even if this was not in the charge, it is relevant to the later problem. It is evidence of a pattern of behavior."

Whichever way the facts point, the panel writes a report. To allow all parties to review the draft "is the best possible approach for achieving maximum accuracy and fairness." If the inquiry found no clear misconduct, "To reach closure, the accuser must feel that the complaint has been heard out, fully aired," Gunsalus said. As for the innocent accused, "The most important fact about restoring reputations is that basically you can't." Therefore, the accused must take strong preëmptive action, she said. "It's counter-intuitive—but the innocent respondent cannot simply express outrage, 'Why are you accusing me?!' He or she must demand a full investigation. The institution owns the process, and if the investigation is thorough, then at the end they can tell everyone that the matter has definitively been cleared up; they can stand behind the respondent."

Is there sufficient cause to proceed? That is the question before the inquiry. "At Illinois, the moment we have reasonable evidence on even *one* charge, the inquiry is over. But panel members always agonize."

The discussion at the practicum made evident that as the supposed typical case progresses the emphasis shifts. Note the change in the sort of language used. Gunsalus again: "You can't stop the investigation when the respondent agrees early to one or two charges." More darkly, "You can't plea-bargain." Again, "There is no statute of limitations on scientific misconduct." The initial stage and then the early steps in the inquiry were digging for facts, but at a certain point the balance swung from fact seeking, whether indeed something had gone wrong with the scientific process, towards the judgement of the guilt or not of an individual, a reversion to the adversarial mode of Anglo-American proceedings.

*

One more case study. By now, most of its features will be familiar, but it generated a thoughtful legal commentary. In 1982, the National Institutes of Health and the University of Mississippi received anonymous letters charging that data was probably fabricated in three published papers. Author or co-author of these was Robert McCaa, a professor of physiology and biophysics at the university's medical center. His field was hypertension and blood chemistry. He had been at the university since the nineteen-sixties and had been receiving grants since 1977 from the National Heart, Lung, and Blood Institute. The university began a formal inquiry. This took six months. The case then moved up the administrative ladder. The university's findings prompted the institute to start an investigation, appointing a panel of scientists. Its establishment took three months. The panel's work took nineteen months. Their report in turn led the university to fire McCaa, and the National Institutes of Health to recommend—after another eight months—that McCaa be barred for three years from receiving further grants and that the Department of Health and Human Services Grant Appeals Board (not the same as the Departmental Appeals Board) hold a hearing. After voluminous submissions of documents and a seven-day quasi-judicial proceeding with witnesses, the hearing found against McCaa. Preparation for the hearing and then writing the decision took sixteen months. After still another four months, in April of 1987 the deputy assistant secretary for procurement assistance and logistics—designated as the debarring official—uttered the final decision. Commendably prompt, Susan Lauscher, a lawyer with a Washington firm who earlier, as chief attorney for the Grant Appeals Board, had worked on the McCaa case, published in the September 1987 issue of *Grants Magazine* a flat, damning analysis of the case and its handling, with recommendations for dealing with future charges of misconduct.

McCaa's part in the first doubtful paper, Lauscher said, had been limited: he "claimed to have done some of the chemical tests." He misreported the experimental procedures "based on what he recalled or surmised, without bothering to check." He misreported the data, and may not have had a complete set. "In one instance, Dr. McCaa invented data based on his extensive knowledge and understanding of the field." As lead author of the second paper, he misreported the number of test subjects in one section, failed to correct this when it was pointed out, produced a graph that misrepresented the procedures, the data, and their statistical significance, and "intentionally continued to publish the inaccurate data in subsequent papers." For the third paper, a review of work in his field, he invented an observation "made in the course of an experiment when in fact there had been no experiment."

Analyzing the McCaa case as a lawyer who had been close to it, Lauscher wrote:

> After making findings as to what really happened, the second most difficult task was discovering the standards of conduct that Dr. McCaa's actions should be measured against. From a legal

> standpoint, there was a conspicuous absence of written guidance as to standard scientific practices, and much of the testimony by witnesses was contradictory, for example, on issues of the significance of being the first author on a paper or how much and what documentation a scientist is expected to retain after a paper has been published.

This was in the mid–nineteen-eighties, yet some of these issues remain unclear to this day. Lauscher went on:

> The principles finally used were gleaned from the testimony of witnesses who talked about scientific practices, all of whom appeared for the NIH, and implications from a policy statement and the wording of a grant award. From this rather meager evidence HHS [the department] measured Dr. McCaa's conduct by four standards that it found should be exercised by an ordinary research scientist:
>
> 1. Reporting experimental procedures accurately;
> 2. reporting experimental results accurately;
> 3. reporting experimental results only when there is supporting empirical data, when those results are not represented as hypothetical or preliminary; and
> 4. while personal involvement in reported experiments may not be required, there needs to be a degree of personal knowledge of data and procedures that will ensure accurate reporting.

So much for the ordinary research scientist. Beyond that, "HHS found that the principal investigator must accept more than just the responsibility for his own work, but must also oversee or otherwise ensure the accurate reporting of any work actually carried out by others with the support of federal grant funds."

Lauscher was scathing about the delays.

> Notwithstanding the important goals of doing a thorough investigation and not accusing someone falsely of professional misconduct, four years is too long a period of time for someone's professional and personal life, as well as a sponsoring institution's stance toward the researcher, to be in limbo. In addition, from a legal standpoint, protracted proceedings might be viewed as a denial of due process.

Look back to some of the cases presented in earlier chapters. Obviously, the actions that Henry Wortis and Herman Eisen took, separately, in response to Margot O'Toole's complaint do not remotely qualify as inquiries. At the other extreme, although Francis Collins acted swiftly and with entire openness when fraud by the student Amitov Hajra was uncovered, after all the consequences for Hajra were catastrophic, and some have held, privately, that Collins was precipitate: while nobody questions the outcome, in other such cases a proper inquiry would provide a necessary assurance of fairness. Something close to the ideal balance was struck, for example, in the affair of Hendrik Schön at Bell Labs: prompt initial response, a thorough brisk investigation by a strong panel, a report, and openness throughout. But Bell Labs is not a university, and the research in question was not federally funded.

*

What can the law do for science?

A first step would be to address the requirement—which the Departmental Appeals Board raised in the Imanishi-Kari appeal—that to prove scientific fraud one must show motive, one must prove intent. This let in the sloppiness defense: "She didn't mean to do it." In criminal law, as we noted, the definition of fraud includes *mens rea,* the element of intent. When fraud is at issue in civil law the rules can be different. "One of the principal weapons the IRS has for use against taxpayers who would undermine the honor system of tax reporting," writes the scholiast, is "the heavy civil fraud penalty," and the courts have held that when it comes to taxes, sloppiness itself—if bad enough, egregious sloppiness, failure to keep or to produce records—can suffice as proof of fraudulent intent. Although the level of sloppiness is judged case by case, yet at a stroke the subjective and difficult requirement to demonstrate intent is transformed into something of an objective test. To apply such a principle to scientific misconduct would amount to saying that you may keep records as you like, heap your desk as high as you like, but if your work is challenged and you cannot produce respectable notebooks or the equivalent, then the burden of proof shifts to you. This would go some way towards restoring the honor system in science.

We must go deeper. The American legal system has great trouble dealing with science. Yet the need has grown urgent. Cases both criminal and civil that draw upon the sciences, but especially some kinds of civil suits, have steadily become larger and more complex. Expert scientific testimony has come to play a greater, often a decisive role in legal decisions. At the same time, the novelty and technical difficulty of scientific issues has been increasing. Writing in *Science* in 1998, Stephen Breyer, an associate justice of the Supreme Court of the United States—a subtle and amusing man with wide-ranging interests—said, in part:

> The practice of science depends on sound law—law that at a minimum supports science by offering the scientist breathing

> space, within which he or she may search freely for the truth on which all knowledge depends. It is equally true that the law itself increasingly requires access to sound science. This need arises because society is becoming more dependent for its well-being on scientifically complex technology, so, to an increasing degree, this technology underlies legal issues of importance to all of us. We see this conclusion illustrated throughout the legal system.

That's very well, as far as it goes. Yet Breyer's formulation here makes the relationship between science and law seem reciprocal. Though I am an amateur of the law—amateur in the word's old, best sense, a lover of law and legal reasoning but not a professional—it seems to me that in certain respects the law and the sciences are incommensurate. In aims and methods they often don't go together. Perhaps the deepest reason is that the law deals with the application of principles to particulars while the sciences strive for generalities—in the extreme, the physicists' search for "the theory of everything." The distinction between the sciences and their applications goes back to Aristotle. Anyway, the lack of fit between science and law begins with the way law employs science in the courtroom. It becomes crucial in dealing with scientific fraud. Certain trends in American jurisprudence suggest ways to improve that dealing radically.

The law looks to the sciences in the first place for evidence. Experts must testify, that is inescapable. The Anglo-American adversarial mode exacerbates the problems. Prosecution and defense, or plaintiff and respondent—each side brings its chosen, coached experts to the hearings, to the witness box. The opposing experts do battle. Juries are baffled, judges bemused. Certain reforms have been made, others proposed. Although they vary, all have the effect of enlarging the function and power of the judge, when scientific matters arise, in relation to the attorneys for the two sides.

Beginning in 1993, the Supreme Court has ruled three times on the powers and obligations of trial judges in admitting expert testimony, and thus on the nature of sound science. Before then, in federal courts the dominant standard for the admissibility of expert testimony had been set in 1923—a remote age—in a case called *Frye v. United States. Frye* allowed only one test, that the basis of an expert's testimony must "have gained general acceptance in the particular field in which it belongs." To be sure, judges had the power to exclude an expert's testimony, yet the power had fallen into desuetude.

On 28 June 1993 the Supreme Court decided *Daubert v. Merrell Dow Pharmaceuticals, Inc.* William Daubert and his wife had sued, claiming that a drug marketed by the defendant and prescribed to suppress nausea in pregnancy had caused their son to be born with crippling birth defects. The court's decision boldly transformed the role of trial judges. They were henceforth required to determine "the subjects and theories about which an expert may testify." The judge now became the gatekeeper, obliged to decide whether proffered expert testimony is admissible, and this means "not only relevant but reliable." *Daubert* thus frog-marched evaluation of the quality of science into the courtroom. Reliability, the court said, means that "in order to qualify as 'scientific knowledge,' an inference or assertion must be derived by the scientific method"; scientific evidence or testimony must be grounded "in the methods and procedures of science." This restriction is necessary because unlike an ordinary witness "an expert is permitted wide latitude to offer opinions, including those that are not based on firsthand knowledge or observation." The decision was emphatic: "In a case involving scientific evidence, *evidentiary reliability* will be based upon *scientific validity.*"

The decision pressed on, to show how conformance to scientific method may be ascertained. "Many factors will bear on the inquiry, and we do not presume to set out a definitive checklist or test," *Daubert* said; the two later decisions reinforced this point. But *Daubert* then offered four considerations. First in importance, "Ordinarily, a key question to be answered in determining whether a theory or technique is scientific knowledge . . . will be whether it can be (and has been) tested." The justices had been reading philosophers of science. Among others, they quoted the *locus classicus,* Karl Popper: "[T]he criterion of the scientific status of a theory is its falsifiability, or refutability. . . ." Encountering the criterion of falsifiability for the first time, many people find it paradoxical. What it means is that in order to have a claim to be right, a scientific theory or technique must imply tests by which it could be proved wrong. Breyer, commenting on *Daubert* elsewhere, quoted the physicist Wolfgang Pauli, who when asked whether a particular paper was wrong replied, "Certainly not. That paper isn't even good enough to be wrong!" *Daubert* imposes on trial judges the obligation to rule out junk science.

A second consideration, though in some instances it might not apply, "is whether the theory or technique has been subjected to peer review and publication." A third applies more narrowly, to particular scientific techniques, about which the court "should consider the known or potential rate of error," in other words, the likelihood of the technique producing false positives or false negatives. Finally, although "general acceptance" by scientists within the particular specialty, the sole standard of *Frye,* was demoted, it could still be a consideration. Justice Harry Blackmun wrote the opinion; he was joined by six others, while Chief Justice William Rehnquist and Justice John Paul Stevens dissented in part. The language of *Daubert* allows flexibility. But it was backed by the weight of the Supreme Court.

"We are confident that federal judges possess the capacity to undertake this review." (One imagines Breyer's ironically cocked eyebrow.)

Daubert engendered consternation among judges, chagrin among trial lawyers facing the prospect that their chosen experts might be excluded, and commentary in torrents by legal scholars. Joe Cecil is a jurist at the Federal Judicial Center, in Washington, D.C., who is a leading authority on expert testimony in court. The center provides research and planning assistance to the Judicial Conference of the United States, the body that supervises the working of the federal court system, and actively develops proposals for reforms. Cecil summed up the initial reactions to *Daubert*:

> Immediately after the *Daubert* decision the winner was unclear. The plaintiffs declared victory because the ruling by the court excluding the evidence under the *Frye* standard was reversed. The next day the *Wall Street Journal* echoed this view with the headline, "Justices Rule Against Business in Evidence Case." The defendants declared victory because the Supreme Court placed judges, not juries, in charge of assessing the validity of the evidence. *The New York Times* echoed this view with the headline, "Justices Put Judges in Charge of Deciding Reliability of Expert Evidence." Of course, it is now clear that the decision works to the detriment of the party with the burden of proof, usually the plaintiff.

Four years later, on 15 December 1997, the Supreme Court clarified and strengthened *Daubert*. The case is *General Electric Company v. Joiner*. At question was what standard an appellate court is to apply to trial judges' decisions to admit or exclude expert scientific testimony. Robert Joiner had sued, claiming that chemicals he was exposed to had aggravated his lung cancer. The trial judge, following *Daubert*, had thrown Joiner's experts' testimony out as insufficiently based: in effect, the reliable science was not relevant and the relevant science was not reliable. The experts' opinions were "unsupported speculation." The court of appeals reversed the decision, stating that the judge was required to have interpreted *Daubert* narrowly, admitting expert testimony whenever possible; therefore "we apply a particularly stringent standard of review to the trial judge's exclusion of expert testimony." Not right, said the Supreme Court, reversing the reversal: the standard for exclusion is no more stringent than that for admission and the appeals court should have deferred to the trial judge. Furthermore, in what may be the passage from *Joiner* most quoted in legal briefs challenging expert testimony, the opinion stated that in science "conclusions and methodology are not entirely distinct from one another," and went on:

> Trained experts commonly extrapolate from existing data. But nothing in either *Daubert* or the Federal Rules of Evidence requires a district court to admit opinion evidence that is connected to existing data only by the *ipse dixit* [mere say-so] of the expert. A court may conclude that there is simply too great an analytical gap between the data and the opinion proffered.

The court's sophistication about how science works had improved in that four years. Chief Justice Rehnquist wrote the opinion; all nine justices agreed in the finding.

For laymen, the most interesting aspects of *Joiner* came in Justice Breyer's concurring opinion. *Daubert,* he began, "will sometimes ask judges to make subtle and sophisticated determinations about scientific methodology and its relation to the conclusions an expert witness seeks to offer—particularly when a case arises in an area where the science itself is tentative or uncertain." Yet "judges are not scientists and do not have the scientific training that can facilitate the making of such decisions." None the less, "Of course, neither the difficulty of the task nor any comparative lack of expertise can excuse the judge from exercising the 'gatekeeper' duties." So judges need help. Breyer noted that the rules of evidence in civil trials already gave them the power to appoint their own independent experts, to order pretrial hearings where potential experts are subject to examination by the court, and to appoint special masters and specially trained law clerks.

The scientific community responded to Breyer's lead. At the practical level, the American Association for the Advancement of Science has established a pilot program to help federal judges find suitable experts. More generally, the National Academy of Sciences has set up a Science, Technology, and Law program to consider issues on how the law affects the conduct of science, and how the law uses scientific information.

The third case in what has been called the Supreme Court's *Daubert* trilogy is *Kumho Tire Company v. Carmichael,* decided less than a year after *Joiner* and again reversing a reversal. Never mind the details, the principle at issue was whether *Daubert* applies only to scientific expertise. Justice Breyer wrote this opinion, but the court was unanimous that "*Daubert's* general holding—setting forth the trial judge's general 'gatekeeping' obligation—applies not only to testimony based on 'scientific' knowledge, but also to testimony based on 'technical' and 'other specialized' knowledge." Furthermore, Breyer wrote, the object of the gatekeeping requirement "is to make certain that an expert, whether basing testimony upon professional studies or personal experience, employs in the courtroom the same level of intellectual rigor that characterizes the practice of an expert in the relevant field."

Daubert may be one of the more revolutionary rulings in a century of American law. It is part of a growing movement among jurists to reshape the use of expert testimony in complex technical cases.

In June of 1997, an extraordinary experiment in judicial control of scientific evidence took place in Birmingham, Alabama. By that time, in federal district courts across the country tens of thousands of lawsuits had been filed claiming that silicone-gel breast implants cause a variety of systemic connective-tissue diseases, seriously damaging the women's health. Damages were potentially huge, legal strategies developed by plaintiffs' and defendants' attorneys were sophisticated, and the quality of the science was of course contested. That June, Sam Pointer, Jr, chief judge of the state's northern district, presided over a week-long hearing to evaluate the scientific evidence and testimony. Some twenty-seven thousand cases had been brought together before Pointer for this evaluation. He had appointed a national science panel of four recognized scientific specialists, in immunology, epidemiology, toxicology, and rheumatology. (Three of the four were women.) Plaintiffs and defense put up their expert witnesses. Members of the panel questioned them, and elicited responses that were full and wide-ranging. At this stage, the lawyers for the parties to the litigation sat mum—no objections permitted, and no limits on the fullness of the witnesses' answers. Only when the panel was done with a witness were the lawyers allowed to examine the witness in the conventional manner. According to observers, it quickly became clear that the questioning by the panel members was far more productive and effective.

The hearing in Birmingham was the small but visible part of a larger experiment in selecting and using specialized and independent experts to advise the court which had begun in 1996 and lasted into

1999. The experiment grew from earlier experiences of Pointer and three other federal judges, one from Oregon and two from New York, with breast-implant litigation. It was sparked by a request to Pointer from a committee guiding the multiple plaintiffs that he consider appointing such a panel of experts. At the end of May of 1996, he agreed conditionally. In August, the experiment was authorized and funded by the Judicial Conference. Towards the end of the year, the Federal Judicial Center was asked to evaluate it.

The evaluation, *Neutral Science Panels,* appeared at the end of 1999, the length of a small book. It began, "The use of such panels of appointed experts represents a marked departure from the traditional means of presenting and considering expert testimony." The authors promised "sufficient detail to permit others to understand the procedures that were used, the benefits that resulted, and the problems that arose."

The first problem was selection of the panelists. This had a tortuous history, which makes abundantly clear the difficulty of finding experts of quality and genuine independence. After consultations among Pointer and the three other federal judges, a six-person selection panel of scientists was put together to nominate the four-person expert panel. The need was to satisfy the parties to the litigation that the experts chosen were indeed independent. This took the better part of a year. To insulate the panelists from influence by the opposing attorneys—and from himself—Pointer then named John Kobayashi, an attorney in Denver, as special counsel to protect their interests. The plan was that all communication to them, and even among them, was to go through Kobayashi, but this proved impractical.

The four experts' principal task, well beyond that public hearing, was to scrutinize the scientific literature in their several specialties and prepare a report evaluating the reliability—that is, in *Daubert's* term, the validity—of the sciences. This they produced in December of 1998, four chapters, each by a panel member, the whole tied up with an abstract (in the fashionable jargon, an "executive summary"), an introduction, and a conclusion. The central question was whether the science demonstrated that silicone gel caused the disorders claimed. The panelists reached consensus: No. The Federal Judicial Center's report excerpted details. They are of highest interest in themselves and cumulatively demolished the plaintiffs' position.

> Specifically, the toxicologist concluded that "[t]he preponderance of data from [animal] studies indicate that silicone implants do not alter incidence or severity of autoimmune disease." "Considering the broad range of testing systems that have been used in the study of silicone effects, the toxicologic and immunologic responses are few in number and questionable in significance." The immunologist found that many of the studies available for analysis were methodologically inadequate with ill-defined or inappropriate comparison subjects or unorthodox data analyses. Even with these limitations, among others, she concluded from the existing studies that women with silicone breast implants do not display a silicone-induced systemic abnormality in the types or functions of cells of the immune system. The epidemiologist found "no association between breast implants and any of the individual connective tissue diseases, all definite connective diseases combined, or the other autoimmune/rheumatic conditions." Finally, the rheumatologist found problems analyzing many of the studies. For example, "the same complaint appeared in more than one disease category; self-report was not verified; timing of the complaint in relation to

> the implant was not known; indication for the implant was ignored; and in individual studies, the number of affected women was small. Furthermore, many of the rheumatologic complaints reported are common in the general population and . . . in physicians' offices. No distinctive features relating to silicone breast implants could be identified."

Parenthetically one wonders whether in dragging such stuff into the courtroom the attorneys were not themselves perilously close to a kind of scientific fraud. Anyway, *Daubert*'s power was demonstrated. *The Wall Street Journal*'s editorial page exulted. The plaintiffs' attorneys, aghast, searched to disqualify the experts and to upset the report. The panelists faced a week of grilling by the parties, a stage called discovery depositions. They were puzzled about the need for this stage, several of them outraged by the intrusiveness of the questions. One, the center's report said drily, "perhaps failing to acknowledge the needs of the legal system, asserted that 'science was not served by this adversarial proceeding'." The report went on, "The panelists' reactions represent one measure of the gulf separating the scientific and legal spheres." Pointer kept a tight hold on the proceedings and ruled against withdrawing the report. Then came the culminating stage, trial depositions. The four experts were examined for eight days, each one leading off with a presentation, in effect a lecture, led through it by Kobayashi. The proceedings were transcribed by a court reporter and videotaped.

The end product, then, was the expert panel's report and a set of videotapes of the trial depositions. As those thousands of individual lawsuits were sent back to their original courts, this package of scientific expert testimony accompanied them, for use before the juries. The stakes were high, the quality of the product outstanding. Three years was perhaps, after all, not an excessive time. But the final problem was the experiment's cost. Undoubtedly, nationwide across all the individual trials the experiment provided incomparably better expertise and considerable savings of time and expense. Yet the total cost of the experiment itself was two million dollars (Kobayashi's fee accounting for half). The experiment has not been repeated.

The adversarial trial: in the United States and Britain, by legal practice, news stories, the countless courtroom dramas, real and fictional, that we read about and watch, we are so deeply indoctrinated into the adversarial model of legal proceedings, whether criminal or civil, that we are unable to imagine another way of resolving serious accusations of bad conduct. We have not even got words in common currency to express alternatives. Yet for three decades or longer, eminent American and British jurists have been raising grave criticisms of this model and suggesting remedies. In 1974, in an address before the Association of the Bar of the City of New York, Marvin Frankel, a federal judge, began with a deceptively simple observation. "My theme, to be elaborated at some length, is that our adversary system rates truth too low among the values that institutions of justice are meant to serve." He went on:

> Because the parties and counsel control the gathering and presentation of evidence, we have made no fixed, routine, expected place for the judge's contributions. We should begin, as a concerted professional task, to question the premise that adversariness is ultimately and invariably good. For most of us trained in American law, the superiority of the adversary process over any other is too plain to doubt or examine.
>
> Our commitment to the adversary or "accusatorial" mode is buttressed by a corollary certainty that other, alien systems are inferior. We contrast our form of criminal procedure with the

> "inquisitorial" system, conjuring up visions of torture, secrecy, and dictatorial government. Confident of our superiority, we do not bother to find out how others work.

The inquisitorial system prevails on the European continent, in Japan, and in most other non-English-speaking countries, in forms that vary somewhat. It is sometimes called "truth-seeking," but if the usual term seems sinister the alternative sounds hopelessly naïve. The central feature, the one that most markedly sets the inquisitorial mode off from American and British practice, is that the judge, and not the lawyers for the two sides, is the one who gathers the facts. The judge calls the witnesses and questions them, the judge decides whether expert witnesses are necessary and appoints them, the judge determines the course of the trial. And that's about all that most Anglo-American lawyers know. Frankel went on, "It is not common knowledge among us that purely inquisitorial systems exist scarcely anywhere; that elements of our adversary approach exist probably everywhere; and that the evolving procedures of criminal justice, in Europe and elsewhere, are better described as 'mixed' than as strictly accusatorial or strictly inquisitorial."

Since Frankel's address, legal scholars have examined the inquisitorial system at length to see what we might learn and adopt or adapt. Frankel was of course correct about commitment to the accusatorial mode: controversy has been hot, sometimes nasty. My interest—again I stress, as an amateur of the law—is narrower. Judge Pointer wrote at one point of a "cultural chasm" between scientific and legal approaches, and although he was referring specifically to the relative decorum of scientific controversy compared to the hurly-burly of courtroom cross-examination which had shocked his experts, we can generalize. When it comes to fraud, the incommensurability of law and the sciences is rooted in the organization of the legal system to determine the guilt or not guilt of individuals, contrasted to what the sciences most require, protection of the scientific process and of the integrity of the scientific record, of which the punishment of individuals who have violated the standards is a small though necessary part. The *Daubert* trilogy and the experiment of the panel of experts in the breast-implant litigation were concerned with the proper use of scientific evidence that the law needs. They don't address the needs of the sciences. What they do indicate is that reforms in the adversarial mode have been making the role of the judge more active, and correspondingly lessening the reliance on the lawyers for the adversaries to control and direct the proceeding.

The most full and persuasive explication of the inquisitorial mode in the American legal literature is an essay by John Langbein, a professor of comparative law at the University of Chicago Law School. "The German advantage in civil procedure" appeared in *The University of Chicago Law Review* in 1985. Being most familiar with West German (as it then was) civil procedure, Langbein used that as his example. The contrast he drew was with the American "lawyer-dominated system . . . a system that leaves to partisans the work of gathering and producing the factual material upon which adjudication depends." In the years since Frankel broached the contrast, understanding of the relation of judge to lawyers in the inquisitorial mode had hardly penetrated the American legal community. Langbein made the point all over again:

> My theme is that, by assigning judges rather than lawyers to investigate the facts, the Germans avoid the most troublesome aspects of our practice. But I shall emphasize that the familiar contrast between our adversarial procedure and the supposedly

> nonadversarial procedure of the Continental tradition has been grossly overdrawn.
>
> To be sure, since the greater responsibility of the bench for fact-gathering is what distinguishes the Continental tradition, a necessary (and welcome) correlative is that counsel's role in eliciting evidence is greatly restricted.

Who gathers the facts controls the process. "Apart from fact-gathering, however, the lawyers for the parties play major and broadly comparable roles in both the German and American systems. Both are adversary systems. . . ." The German system "combines judicial fact-gathering with vigorous and continuing adversarial efforts in nominating lines of factual inquiry and analyzing factual and legal issues."

A case goes through several stages. Briefly: The aggrieved party's initial complaint lays out facts, presents relevant law, requests a remedy; the defendant's response follows the same pattern. Unlike the American system, the parties also propose means of proof, documents to be examined, others to be sought out, potential witnesses. All this goes into a dossier, which will grow to be the record of the case and is open to either side all the time. The judge will hold an initial hearing with the lawyers present and perhaps the principals themselves, sometimes even witnesses. At this point, the judge will know the fundamentals of the case; if it has not been resolved he can call for more information and begin to schedule witnesses. Lawyers do not call witnesses, nor do they prepare or coach them: Langbein wrote that for a German lawyer to attempt to get in touch with a witness out of court would be treated as highly unethical.

> The judge serves as the examiner-in-chief. At the conclusion of his interrogation of each witness, counsel for either party may

> pose additional questions, but counsel are not prominent as examiners. Witness testimony is seldom recorded verbatim; rather, the judge pauses from time to time to dictate a summary of the testimony into the dossier.

Lawyers can ask for changes in the summaries. If expert witnesses are needed, the judge—in consultation with the two sides—will choose and instruct them. They will not be hired and prepared by one side or the other: the battle of opposing experts cannot take place. "The essential insight of Continental civil procedure is that credible expertise must be neutral expertise."

What Americans expect to be a continuous single trial will be in Germany a series of hearings spaced out over months. The first consequence of judicial control "is that German procedure functions without the sequence rules to which we are accustomed in the Anglo-American procedural world. The implications for procedural economy are large. The very concepts of 'plaintiff's case' and 'defendant's case' are unknown." The judge takes the case in the direction most likely to dig out the real story.

The German system is set about with safeguards. The chief of these are two, and they would be impossible to provide in the Anglo-American legal context. In the United States, the qualifications of judges are varied and often rudimentary; in some states judges are even elected. In Britain, judges are appointed from among the leading barristers, which provides some guarantee of talent, education, and courtroom experience. In Germany, judges do not rise from the ranks of lawyers. "The distinguishing attribute of the bench in Germany (and virtually everywhere else in Europe) is that the profession of judging is separate from the profession of lawyering," Langbein wrote. After legal training alongside would-be lawyers, the would-be judge serves an apprenticeship of several years in the legal system, then sits an examination. Only those with the best examination results are recruited as probationary, entry-level judges—typically in their late twenties. From there the career ladder is clear, and is punctuated with rotations of court assignments and periodic evaluations by senior judges. The second safeguard is simpler to describe, just as difficult to imagine in the American legal landscape: an extremely broad right of appeal.

Langbein's article has been received with great skepticism. The essential objection was posed succinctly in 1990 by John Reitz, a professor at the University of Iowa Law School. "It is also necessary to propose how we are to get there from our present system," he wrote, and went on:

> Unless there is a change in our fundamental cultural definitions, adoption of judicial questioning of witnesses is likely to result in simply engrafting judicial questioning onto our present system of adversarial questioning with relatively unrestricted prehearing witness contact by attorneys and vigorous attempts at cross-examination following judicial questioning.

Perhaps; but Judge Pointer's structuring and disciplining of the use of the expert panel in the breast-implant litigation surely suggests that the worst of such hybridism could be overcome.

"Important changes have occurred in recent years that diminish the contrast between German and American civil procedure," Langbein wrote. "Under the rubric of case management, American trial judges are exercising increasing control of the conduct of fact-gathering." This is called managerial judging, and it developed in response to the coming of the big, complex, technical cases. Langbein took the rise of managerial judging as evidence that, in fact, in certain respects

American practice was converging towards the German. In 1988, in an article defending his earlier presentation against certain ignorant and unpleasant misreadings, he restated this hope for convergence:

> Complex litigation has required us to superimpose upon our lawyer-driven procedure a growing component of judicial management, including judicial involvement in identifying issues, promoting settlement, and sequencing investigation. These techniques are strongly reminiscent of German-style procedure. I observed that it is awkward to reconcile our new practice of managerial judging with our traditional theory of party domination of fact-gathering.

That was five years before *Daubert.*

Since the late nineteenth century, the American and English legal systems have been invaded by administrative law and its accompanying quasi-judicial proceedings. These are neither criminal nor civil; they do not appear in the formal court system. They are internal to the great governmental bureaucracies, and as the bureaucracies have grown so have the reach and importance of administrative law, impinging on most aspects of modern society. The rules are often looser and the safeguards weaker than in the formal court system. The proceedings of the Office of Research Integrity and especially of the Departmental Appeals Board of the Department of Health and Human Services are prime and cautionary examples of the workings of administrative law.

Scientific fraud is an area where the inquisitorial mode ought to be particularly effective, while in the realm of administrative law its approach should be adaptable and adoptable. Recall the hearings before the review board in the matter of Thereza Imanishi-Kari. They were conducted unlike any appeals-court procedure but rather as a *de novo* trial, with witnesses including expert witnesses, exhibits, examination, and cross-examination. They were run by a mixed three-person panel, two lawyers and a scientist, and of course no jury. The bias on the bench was considerable. The hearings were adversarial to an extreme degree. Though Marcus Christ, leading for the Office of Research Integrity—the prosecution—appeared ponderous and, from what I saw at the hearing and read in transcripts, no ornament to his profession, Imanishi-Kari's chief lawyer, Joseph Onek, was the very model of the Washington criminal-defense lawyer, well dressed, blow-dried, aggressive and supremely confident. At one break in the proceedings, I saw him in the corridor pacing rapidly back and forth and muttering, "Pound them into the ground!" The standard of proof was imported from civil law, not "beyond a reasonable doubt" but "by the preponderance of the evidence." On the desk behind Christ was a long row of thick loose-leaf binders containing the product of three years of investigative work, correspondence, documents, photocopies of notebooks, interview transcripts. Not one page of that could be presented to the review-board panel.

Now suppose that this appeal, or more importantly the investigations out of which it arose, had been carried out on the inquisitorial model. Consider other cases in the same light: McCaa, starting with the University of Mississippi's first inquiry; or the Breuning case, starting when the University of Pittsburgh and then the National Institute of Mental Health began investigating Robert Sprague's charges; or Fisher and the falsified cases in the breast-cancer lumpectomy clinical study.

Once again, then: although when fraud is charged the culpability of individuals may need to be ascertained, what the sciences need from the law above all is the protection of the scientific process and of the integrity of the scientific record. Once more, in the Baltimore affair Margot O'Toole was the only one who got it right. For two years, she refused to charge fraud even when told that was the only way forward. She asked that the paper be corrected or retracted.

EPILOGUE

THE TRANSITION TO THE STEADY STATE AND THE ENDS OF THE SCIENCES

> *What we want to know about the science of the future is the content and character of future scientific theories and ideas. Unfortunately, it is impossible to predict new ideas—the ideas people are going to have in ten years' or ten minutes' time—and we are caught in a logical paradox the moment we try to do so. For to predict an idea is to have an idea, and if we have an idea it can no longer be the subject of a prediction.*
>
> —*Sir Peter Medawar*

Late in the nineteenth century, many a scientist thought physics was all done, complete but for a few puzzling, arcane details like the anomalous precession of the perihelion of Mercury or the cause of the dark lines in the spectrum of light from the sun which Newton must have seen but which Joseph Fraunhofer was first to report. In 1894, the physicist Albert Michelson said, in a speech that has achieved an ambiguous celebrity, "While it is never safe to say that the future of Physical Science has no marvels even more astonishing than those of the past, it seems probable that most of the grand underlying principles have been firmly established and that further advances are to be sought chiefly in the rigorous application of these principles to all the phenomena which come under our notice." But predicting the end of science is a fool's conceit. Mind you, some even then were less satisfied than Michelson. But nobody could have predicted the miraculous decade that began just a year later. In 1895 in Würzburg, Wilhelm Conrad Röntgen discovered x rays. In 1896 in Paris, Henri Becquerel discovered radioactivity. In 1897 in Cambridge, Joseph John Thomson discovered the electron. In 1900 in Berlin, Max Planck laid the foundation of quantum theory. In 1905 in Bern, Albert Einstein promulgated the special theory of relativity. Physics was not done.

The conceit recurs. In 1969, for example, Gunther Stent, a molecular biologist and disciple of Max Delbrück's, published *The Coming of the Golden Age: A View of the End of Progress.* Most recently, in 1996 John Horgan, then a writer at *Scientific American,* made a splash with *The End of Science: Facing the Limits of Knowledge in the Twilight of the Scientific Age.* (Ages are popular in this genre and titles run long.) Horgan's message was that *all* the sciences are all but all done. He is a clever man and a graceful, plausible writer, with a vast acquaintance among scientists. He couched an apocalyptic prophecy in sweetly reasonable tones. From scores of interviews and other contacts with scientists of all stripes, he put together a patchwork of quotations, chapter by chapter: "The End of Philosophy," "The End of Cosmology," "The End of Evolutionary Biology," "The End of Neuroscience," and so on. For the End of Physics, he chuckled over Michelson, then went right ahead, assembling a series of short, cute profiles—well, sketches—of luminaries whom he prompted to doubting whether anything is left to do. Sheldon Glashow, Hans Bethe, Steven Weinberg, John Archibald Wheeler, and so forth—all capped off with a gloomy paragraph from Richard Feynman, who wrote, "The age in which we live is the age in which we are discovering the fundamental laws of nature, and that day will never come again. It is very exciting, it is marvelous, but this excitement will have to go." Echoes of Michelson, a century farther on, and Feynman was a very smart man.

Even as Horgan wrote, physics and cosmology were erupting in fresh turmoil—a joyous frustration of new fundamental problems. As for biology, by the turn of the millennium the technologies of molecular genetics, combined with the multiple genome projects and the power of computers, are opening to solution problems that have been intractable since William Harvey turned from heart and blood to embryology—while unveiling new problems, new Everests, undreamed of a decade earlier.

> The bear went over the mountain,
> The bear went over the mountain,
> The bear went over the mountain
> To see what he could see.
>
> And what do you think he saw?
> And what do you think he saw?
>
> He saw another mountain,
> He saw another . . .
> . . . And what do you think he did?
>
> The bear went over the mountain, . . .
>
> —*Boy Scout marching song*

The sciences may not be like *that,* either, though scientists sometimes speak as though they think so. But given the risks, dare we make any predictions at all?

*

So we reach the coming of the steady state, the doctrine of Malthus applied to the sciences. John Ziman is an English physicist turned sociologist of science, a pugnacious little man with a distinctly original slant. For a dozen years and more, Ziman and colleagues he has enlisted have been pursuing the uncomfortable effects of an obvious truth that scientists are reluctant to confront. The reasoning begins with Derek de Solla Price's observation of 1963, which we have visited before, that since the mid-seventeenth century the output of scientific papers has doubled every fifteen years—with, of course, the population of scientists increasing proportionately. At that rate, he said, soon every man, woman, and child in the country would have to be doing scientific research and writing papers. Everyone laughed and said, What a droll idea! Like most assertions of Malthusian limits, this one was taken seriously by few. The rate of growth of funding fluctuated, to be sure, yet the trend has been upwards in constant dollars, from both government and private industry. The fact of exponential growth was confirmed in many ways: the number of doctorates produced, the number of scientists employed, the number of journals and journal pages published, the number of authors per paper, and so on. Exponential growth has some startling aspects: it means as Chapter 1 noted that more than eighty per cent of all scientists who have ever lived are alive today.

Yet the long bull market is over. The transition from exponential growth to the steady state is producing enormous systemic strains, some obvious, some subtle. Continuous advance, Ziman pointed out at a conference I organized in 1996, has been a structural principle of science. He went on:

> Its intellectual dynamism requires it to make itself anew every twenty years or so. All its practices are based upon the assumption that there will be two research jobs soon where there was one job before, that old subjects don't need to be killed off to make room for new ones, that alpha-rated projects will eventually get funded, and so on. Putting a bound to its inputs doesn't just stabilise its

> output. It dislocates the machine inside the black box. It disrupts an established pattern of intellectual debits and credits, social obligations and expectations, moral rights and responsibilities. At every level, the knowledge-producing mechanism is being subjected to stresses that can only be accommodated by deformations, fractures and bodged repairs. In other words, a structural "phase transition" is taking place.

The obvious sign of the transition is the increasing competition for funds to do the work and for priority in publishing it. The individual scientist, even the politically savvy chief of a leading laboratory, sees the grant-review process, in its immediate effects, in terms of success or failure. But the coming of the steady state distorts the funding process systemically. Recall Harold Varmus's characterization of grant review in 1993 as he assumed the directorship of the National Institutes of Health: "the low success rate, the high number of resubmitted applications, the unwillingness of talented people to serve on study sections. They're making distinctions between grants that are equally excellent. It's very, very demoralizing to do that kind of reviewing." With no basis for choice on scientific merit, panelists are inevitably swayed by bias, prejudice, politics. Those applying for grants must fear to propose risky projects, or ones that may take a long time, several funding cycles, to pay out. Chapter 6 looked at these and related problems. Electronic processing of applications and their evaluation, more efficient triage, that is, the early rejection of applications with little merit—such measures, promised and to some extent put in place, are merely palliative.

Publication of research is similarly affected. Rare now is the laboratory that can work for years on a substantial nest of problems, at last issuing the magisterial, definitive report. Most must rush to get even the most minimal step—the least publishable unit—into print.

The coming of the steady state is at the root of these problems, and makes them extremely difficult to fix. Recall, too, the hesitation, in the late nineteen-forties as the Vannevar Bush years were just beginning, of certain great privately financed research institutions to accept federal funding. Perhaps they had justification. Consider the effects of exponential growth on scientific personnel. In the boom years, an increasing proportion of research staff has come to be paid not from the institutions' independent income but out of soft money, that is, out of grants, which can fail to be renewed. This is obvious in biomedical research, but is as true in the physical sciences. An aspect of this is that a great deal of the day-to-day work of the sciences, particularly in the laboratory, is done by gangs of graduate students and postdoctoral fellows—serfs, they are often called. The way the sciences are done now, they are indispensable. Yet some science departments even at elite universities report a falling-off in applications to graduate school by American students, though this is masked by the large numbers of foreign students.

Supply Without Demand. Thus the headline on a lead editorial in *Science,* in February of 2004, by Donald Kennedy and three of his colleagues. It was a reasoned, categorical, urgent attack on the overproduction, in the United States and in Europe as well, of new scientists. The latest version of this drive has been the widely publicized concern that few native-born Americans are entering scientific careers, leading to a call "for an intensified national effort" to increase the supply. "Meanwhile," the editorial said, "unemployment rates for scientists are going up; according to the American Chemical Society, they have doubled among chemists over the past 2 years."

> What is going on here? Why do we keep wishing to expand the supply of scientists even though there is no evidence of imminent shortages, and most jobs are in the private sector, where

> they are immune to management by policy fiat? First, there is a widespread belief that economic progress depends on science and technology; why shouldn't we have more of such a good thing? Second, policies are set mainly by elders, who, like the institutions that employ them, have little incentive to downsize their operations. Instead, academic reward structures and government funding priorities tend to perpetuate the "train more scientists" status quo.
>
> There's one more, uncomfortable, explanation for calls to increase the supply of scientists. The present situation provides real advantages for the science and technology sector and the academic and corporate institutions that depend on it. We've arranged to produce more knowledge workers than we can employ, creating a labor-excess economy that keeps labor costs down and productivity high. . . .
>
> The consequences of this are troubling. To be sure, the best graduates of the most prestigious programs may eventually find good jobs, but only after they are well past the age at which their predecessors were productively established. The rest—scientists of considerable potential who didn't quite make it in a tough market—form an international legion of the discontented.

The transition is already transforming the labor supply. But the analysis must be pushed further. In the fully achieved steady state, each senior scientist would have no more than three or four postdocs—one as a replacement, one or two for industry, one for natural wastage.

Another clear sign of the transition is the ever-increasing pressure from politicians and governmental agencies for directed or targeted research, with slogans like "national needs" and "strategic research." When President Richard Nixon called for a national war on cancer, the community bridled; *The New York Times* reported a "backlash." Cancers, though, are a perversion of normal processes in the cell; scientists soon enough found that they could justify basic cell biology as cancer-related. A useful stratagem, yet increasingly suspect. A while ago, for example, Senator Barbara Mikulski, Democrat of Maryland, who sits on the Senate committee that controls a high proportion of federal research appropriations, told the National Science Foundation to be explicit in its characterization of fields and proposals, to prevent scientists from sneaking basic research in: "The foundation should make clear how it specifically defines each area so as not to shroud curiosity-driven activities under the rubric of strategic activities."

Parsimony and political interference: most scientists read these as characteristic of an abnormal situation, antithetical to the practice of science in the mode that has proved so successful. Scientists long for a return to normality. We shall never see a return to such a nostalgic normality.

The transition has other concomitants. One of these is the internationalization of scientific research. Universities are following corporations towards globalization. Along with that goes the intensifying and increasingly complex linkages between science and industry. This last is especially evident in the biological and biomedical sciences—although many of the traits are found, *mutatis mutandis,* in other sciences, for example, solid-state physics. In 1970, 1972, 1974, the men (and several women) who were developing the methods we call recombinant DNA, or genetic engineering, were pulled by the desire to get at a pure and intractable scientific problem: development and differentiation, or how the fertilized egg, the single cell, becomes the adult organism, multicellular, multifunctional. Since then, many of the men (none of the women, so far as I know) have made fortunes in biotechnology. Naturally, money will influence the direction of research; but the point for our purpose is that the linkages between academic laboratory and industry are changing the career structure of research.

In many sciences, and especially in molecular biology, the membrane between pure and applied research has grown ever more tenuous. It is now vanishingly thin. The trick one laboratory devises to take a tiny step forward in, say, the puzzle of cell development has often turned out to allow manipulations of some other type of cell for the production of something that might be commercially profitable. Indeed, we can no longer predict which kind of advance will come from what sort of laboratory, academic or corporate. These days, too, much of the work done in commercial laboratories is just as publishable in scientific journals; scientists from both sorts of laboratories often collaborate in published research. Multiple examples of such rapid interchange can be cited in any subfield. Consider, for example, the history of the polymerase chain reaction, invented at the biotech company Cetus, or the biosynthesis of human growth hormone, achieved by Genentech.

One pervasive consequence of the transition to the steady state is a change in the patterns of work and advancement in the sciences. The young scientist today sees not one but two career ladders, one academic the other high-tech industrial, and the most talented and highly qualified soon realize that as they ascend they will be able to step from one ladder to the other, either way.

The aspect of the transition to the steady state that is most threatening is the growing emphasis on evaluation and accountability. Peer review, as we have seen, created institutions for evaluating research. The first characteristic of these evaluations is that they are made by scientists. Secondly, evaluation begins with inputs. Targetted research, however, tends to move the choice of problems at least in some degree away from the control of the scientific community. At the other end, judgement of the results of research has also been shifting away from the community. The modern managerial doctrine, in government as in industry, calls for evaluation of outcomes. For a decade at least, government managers of social programs, of industrial policy, and of technological and scientific activity of all sorts have been calling for evaluation of outcomes. For those in science, evaluation of outcomes will mean evaluation of their work by non-scientists, evaluation according to new criteria over which the community will have far less control.

The transition to the steady state underlies the many structural transformations this book has sketched. It unifies, rationalizes, our understanding of them. It is the setting in which the institutions of the sciences have been operating for decades. And since the various patterns of interactions of scientists within those institutions are in turn the settings where scientific fraud takes place or is precluded, the transition to the steady state is fraud's deep context.

To say that the transition to the steady state is inevitable, is taking place now, is not a prediction of the end of science. To be sure, one aspect of what Michelson said more than a century ago will apply: further advances will be found chiefly in the exploitation of discoveries and techniques in hand. But this is trivial, in that it has always been true. Great discoveries remain to be made, and, as has always been true, much of what remains looks intractably difficult. No, the transition to the steady state predicts something else: the end of the growth of science.

At first glance, this is paradoxical: the end of growth while important new work must flourish. Obviously the paradox is resolved by the shifting of resources, a pattern Ziman and his colleagues rather glibly call the dynamic steady state. Yet Malthusian limits are inexorable.

Some fields will wind down. Chemistry is one likely candidate for a Michelsonian reduction. Others will grow. Neurobiology is one where the action is particularly hot now; but neurobiology will sooner or later be subsumed under development and differentiation, which becomes more bewilderingly complex day by day.

Yet in the midst of a great structural transformation its longer-term consequences are not reliably predictable in detail. But keep your eye on such things as changes in the career structure of the sciences including the abandonment of academic tenure, changes in the aims, conditions, and organization of research, and changes in the ways research is evaluated.

*

The golden age, the grand imperium of science, if it ever existed is forever gone. We must still find our way. For science offers values beyond the practical. *Nam et ipsa, scientia potestas est,* wrote Francis Bacon. This is usually translated "Knowledge is power," and taken to proclaim the instrumental utility of knowledge, that science is valuable because of what it enables us to do in the world. By that view, Vannevar Bush was a modern Baconian. But that translation is not correct. The sentence occurs not in one of Bacon's extensive works in philosophy of science later in life but in a slim volume he published in his thirties, a young man on the rise, titled *Religious Meditations.* The first three words, *nam et ipsa,* must be given full weight. The translation is "*In and of itself,* knowledge is power." This view of science is not exclusive, of course. Who could deny that science has had extraordinary instrumental value? But science offers more. What it offers is consonant with Max Weber's valuation of science as a high, austere, disinterested vocation, set off from other vocations by its acceptance—by its embrace—of uncertainty, of provisional, changeable understanding.

Science is the art of our time. Science has several rewards, but the greatest is that it is the most interesting, difficult, pitiless, exhilarating, and beautiful pursuit that we have yet found. We can date a new era to Friday, 30 June 1905, when Albert Einstein submitted a thirty-one-page paper, "Zur Elektrodynamik bewegter Körper," to *Annalen der Physik.* No poem, no play, no piece of music written since then comes near the theory of relativity in its power, as one strains to apprehend it, to make the mind tremble with delight. Or take the molecular structure of the fabric of the gene, the celebrated double helix of deoxyribonucleic acid. Two strands running in opposite directions, hooked together across the space between them by a sequence of pairs of chemical entities—four sorts of entities, making just two kinds of pairs, with exactly ten pairs to a full turn of the helix. It's a sculpture. But observe how form and function are one. That sequence possesses a unique duality: one way, it allows the strands to separate and each to assemble on itself, by the pairing rules, a duplicate of the complementary strand; the other way, the sequence enciphers, as in a four-character alphabet, the specification for the substance of the organism. The structure thus encompasses both heredity and embryological growth, the passing-on of potential and its expression. The structure's elucidation, at the end of February of 1953, was an event of such transcendent explanatory power that it will reverberate through whatever time mankind has got remaining. The structure is also perfectly parsimonious and splendidly elegant. To those who engage with its form and meaning, no sculpture made in the past hundred years is so entrancing.

If to write about science in this way, at the new millennium, seems to understate what science does, at least partly that must be because we now expect art to do so little. The creative faculty, I think, operates in the same manner when the strange-particle physicist is engaged in reasoning or experiment, when the poet is writing a page, or when the playwright and the director collaborate at a rehearsal, and this is by the argument between invention and disposition, between the voice over one shoulder urging *try this, then,* and the voice over the other shoulder whispering *not quite right yet.* "Scientific reasoning," Sir Peter Medawar said in a lecture before the American Philosophical Society some years ago, "is a constant interplay or interaction between hypotheses and the logical implications they give rise to: there is a restless to-and-fro motion of thought, the formulation and rectification of hypotheses, until we arrive at a hypothesis which, to the best of our prevailing knowledge, will satisfactorily meet the case." Thus far, change only the word "hypothesis" and Medawar described well the experience the painter or the poet has of his own work. (He knew this, I suspect, but tactfully left the point for his listeners to fill in.) "Scientific reasoning is a kind of dialogue between the possible and the actual, between what might be and what is in fact the case," he said a moment later—and there the difference lies. The scientist enjoys that harsher discipline of what is and is not the case. It is he, rather than the painter or the poet in the century since Einstein, who pursues in its stringent form the imitation of nature, fundamental to the conception of art since Aristotle told us so. The social system of science, from collaboration at bench or blackboard, at computer, dig, or telescope, to formal publication and response, among its several functions is a means to enlarge the interplay between imagination and judgement from a private into a public activity.

Writing about music is not the same as composing it or performing it—and in just that trivial way writing about science, talking about science, explaining science, is not science. But is the pleasure taken in science by the non-scientist therefore illegitimate? Listening to music with an instructed ear and the aid of a score is certainly a musical activity—and in just that self-evident way, along a continuum from dummy to savant there is a point *p* at which reading about science with an instructed mind and with reference to original research becomes a scientific activity.

I remember the day, 16 January 1957, when a strange report was on the front page of *The New York Times:* two quantum physicists named Chen Ning Yang and Tsung-Dao Lee had overthrown the parity of weak interactions. They got the Nobel prize for that, and in record time. But the front page of the *Times*? Consider the continuing arguments over the Big Bang, the origins of the universe. The news of quasars and black holes. Or of quarks. The accounts of fragments of a comet splashing into Jupiter. Or of still more evidence of the vasty meteorites that we now think caused mass extinctions not once but repeatedly over the billions of years of deep time. Or the excitement generated by the finding of a new protohuman fossil, or of a new Mayan city in Yucatan or pharaonic tomb in Egypt, or the frozen body of a bronze-age hunter shed by an Alpine glacier. Every year, every month, such things are front-page news. Yet what practical result can come from any of this? Why do we pay for the Hubble telescope and its successor now being built? For rockets to Mars, for new archaeological digs?

During the exponential growth of the sciences and now under the strains of transition to the steady state, in pursuit of the instrumental, of the promised payout, I fear we have lost that sharp, esthetic awareness of the end of science—the end in the other and prior sense: purpose, goal. Those great lineages of science, the passing on of mentoring, inculcated that awareness and inoculated against fraud. *Bring me my arrows of desire,* sang William Blake. One of the powers of *scientia* is that it exalts its pursuit.

We want to know *nam et ipsa,* for itself. Since time immemorial, origin stories have been central in every culture. We are all fascinated by origins—of the universe, of the solar system, of life, of species, of language, of civilization. For the general public as for the practicing scientist, the sciences tell better, more comprehensive, more verifiable stories of origins. Again an example from biology, for it has one of the greatest origin stories as yet untold. Two modes of explanation run through all of biology. We have questions of *how,* and these are about physiology, about how creatures grow, eat, reproduce, run around, and die—and so about genetics. Ernst Mayr, the last survivor of the great evolutionary biologists of the 1930s, has called these *proximate* questions. The other mode speaks to questions of *why.* These are about evolution. They are *ultimate* questions. Now recall that to talk of the human-genome project is a gross misnomer, because it is not just the human genome, but bacteria and roundworm and fruit fly and mouse and now chimpanzees and soon more, genomes over the entire range of living creatures. Comparative genomics will give us the fusion of genetics and evolution, of how and why, of proximate and ultimate.

The accounts of origins that the sciences offer are all the more compelling because they are alive, changing, evolving. As scientist or as observer of the sciences, we work on the accounts actively. If we are courageous enough, stoical enough, we refuse to make the intellectual sacrifice, we embrace the knowledge that knowledge is uncertain—and still we want to know. *Bring me my chariot of fire.* Here is the triumph of the scientific world view.

NOTES AND SOURCES

The *Zanes,* at Olympia: Sir Richard Claverhouse Jebb and Ernest Arthur Gardner, "Olympia," in *Encyclopædia Britannica,* vol. XX, 11th edition (Cambridge and New York: Cambridge University Press, 1911), p. 96

Preface

Philip J. Hilts, "Biologist Who Disputed a Study Paid Dearly," *The New York Times,* 22 March 1991

Prologue

William Harvey, *Exercitatio Anatomica de Motu Cordis et Sanguinis in Animalibus,* facsimile of the 1628 Francofurti edition, with translation edited from the first English text, 1653, Geoffrey Keynes (Birmingham, Alabama: Classics of Medicine Library, 1978), pp. 25 (Latin) and 27 (English); the painting of Harvey explaining circulation to Charles I is reproduced in *De Motu Cordis,* Chauncey D. Leake, translator (Springfield, Illinois: Charles C. Thomas, 1941)

Editorial, "War on Cancer," *The New York Times,* 31 May 1971

Chapter 1: A Culture of Fraud

Corporate fraud in the 1920s: John Kenneth Galbraith, *The Great Crash: 1929* (Boston: Houghton Mifflin, 1954), passim

Walter Bagehot: quoted in Galbraith, *The Great Crash,* p. 140

From December 2001 through March 2004, *The New York Times* alone ran upwards of three thousand articles about business frauds; although I employed a variety of other sources as well, including *The Economist* and the *Financial Times,* for final checking and annotating the *Times*'s and *Economist*'s Web sites were invaluable, fast, and comprehensive.

Enron: Richard A. Opel, Jr., and Andrew Ross Sorkin, "Enron Corp. files largest U.S. claim for bankruptcy," *The New York Times,* 3 December 2001; Kurt Eichenwald, "Enron official is reported set to plead guilty," *The New York Times,* 21 August 2002; "Corporate America's woes, continued," and "Investor self-protection," *The Economist,* 28 November 2002; Eichenwald, "Former Enron treasurer enters guilty plea," *The New York Times,* 11 September 2003; Paul Krugman, "Enron and the system," *The New York Times,* 9 January 2004; Eichenwald, "Ex-Chief financial officer of Enron and wife plead guilty," *The New York Times,* 15 January 2004; and Eichenwald, "Enron's Skilling is indicted by U.S. in fraud inquiry," *The New York Times,* 20 February 2004

Global Crossing: Simon Romero, "5 years and $15 billion later, a fiber optic venture fails," *The New York Times,* 29 January 2002; "Survival of the slowest," *The Economist,* 31 January 2001; Romero, "Telecommunications: Global Crossing to Reorganize," *The New York Times,* 29 May 2002; "The firms that can't stop falling," *The Economist,* 5 September 2002

Adelphia: Geraldine Fabrikant, "Market Place; A family affair at Adelphia Communications," *The New York Times,* 4 April 2002

Tyco: Alex Berenson and Carol Vogel, "Ex-Tyco Chief is indicted in tax case," *The New York Times,* 5 June 2002; and Andrew Ross Sorkin, "2 top Tyco executives charged with $600 million fraud scheme," *The New York Times,* 13 September 2002

WorldCom: Seth Schiesel, "WorldCom leader departs company in turbulent time," *The New York Times,* 1 May 2002; Barnaby J. Feder, "Leadership shake-up could spread as WorldCom and its rivals regroup," *The New York Times,* 1 May 2002; [no author listed], "WorldCom bankruptcy filing is said to be set for next week," *The New York Times,* 19 July 2002; "The only way is up, maybe," *The Economist,* 25 July 2002; Kurt Eichenwald, "2 ex-officials at WorldCom are charged in huge fraud," *The New York Times,* 2 August 2002; Eichenwald, "Plea deals are seen for 3 WorldCom executives," *The New York Times,* 29 August 2002; and Feder, "WorldCom's practices enabled huge fraud," *The New York Times,* 10 June 2003

Xerox: Floyd Norris and Claudia H. Deutsch, "Xerox to restate results and pay big fine," *The New York Times,* 2 April 2002; and "Corporate America's woes, continued" and "Investor self-protection," *The Economist,* 28 November 2002

Qwest Communications: Simon Romero, "Echoes of other scandals haunt a chastened Qwest" and "Qwest acknowledges accounting flaws," *The New York Times,* 30 July 2002

Arthur Andersen: Jonathan D. Glater, "Last Task at Andersen: Turning Out the Lights," *The New York Times,* 30 August 2002; and "The charge sheet," *The Economist,* 12 September 2002

Merrill Lynch: Patrick McGeehan, "$100 million fine for Merrill Lynch," *The New York Times,* 22 May 2002

Morgan Stanley Dean Witter & Company: United States of America before the Securities and Exchange Commission . . . Release No. 46578 / 1 October 2002, in the matter of Dean Witter Reynolds Inc. n/k/a Morgan Stanley DW, Inc. . . .

"Corporate America's woes, continued," "Corporate delicti," and "Corporate crookery," *The Economist,* 28 November 2002; the articles also give details about Enron, Adelphia, Tyco, WorldCom, Xerox, Global Crossing, and Arthur Andersen.

United States Olympic Committee: Jere Longman, "U.S. Olympic Committee says it is investigating its leader," *The New York Times,* 31 December 2002; Richard Sandomir, "U.S.O.C. head says he made error in judgment," *The New York Times,* 3 January 2003; Sandomir, "U.S.O.C. chief stays on as latest furor ends," *The New York Times,* 14 January 2003; Sandomir, "U.S.O.C. member resigns," *The New York Times,* 16 January 2003; Sandomir, "A second U.S.O.C. officer resigns," *The New York Times,* 17 January 2003; Sandomir, "3 quitting panel on Olympic ethics," *The New York Times,* 18 January 2003; Sandomir, "President of the U.S.O.C. calls for ethics investigation," *The New York Times,* 21 January 2003; Sandomir, "7 officers of U.S.O.C. press its president to resign," *The New York Times,* 22 January 2003; Sandomir, "Head of inquiry on Olympic ethics has link to Ward," *The New York Times,* 25 January 2003; Sandomir, "Senators favor overhaul of U.S. Olympic group," *The New York Times,* 29 January 2003; Sandomir, "More heat ahead for the feuding Olympic bosses," *The New York Times,* 30 January 2003; Bill Briggs, "Agent waving red flag; USOC president key to allegation," *The Denver Post,* 4 February 2003; Sandomir, "President is quitting U.S. Olympic panel in dispute on ethics," *The New York Times,* 5 February 2003; Sandomir, "Olympic Committee chief loses his bonus," *The New York Times,* 9 February 2003; Sandomir, "Veracity of Olympic Committee's tax returns questioned," *The New York Times,* 14 February 2003; Sandomir, "Effort is turned back to oust U.S.O.C. chief," *The New York Times,* 26 February 2003; Sandomir, "Chief of U.S. Olympic Committee quits amid a furor over ethics," *The New York Times,* 2 March 2003; and Jayson Blair, "After War steps down, there are still concerns," *The New York Times,* 3 March 2003

Royal Ahold: Ian Bickerton, Susanna Voyle, and Neil Buckley, "Europe faces accounting scandal," *Financial Times,* 25 February 2003; Stephanie Kirchgaessner, "Ahold 'knew in 2001 of US problem' says insider," *Financial Times,* 10 March 2003; and Gregory Crouch, "Royal Ahold says fraud was worse than thought," *The New York Times,* 9 May 2003

HealthSouth: Lisa Fingeret Roth and Joshua Chaffin, "SEC alleges HealthSouth carried out $1.4 billion fraud," *Financial Times,* 20 March 2003; Milt Freudenheim, "Hospital chain accused of accounting fraud," *The New York Times,* 20 March 2003; Kurt Eichenwald, "HealthSouth inquiry expands to Medicare," *The New York Times,* 28 March 2003; Gretchen Morgenstern with Milt Freudenheim, "Scrushy ran HealthSouth real estate on the side," *The New York Times,* 14 April 2003; Adrian Michaels, Lisa Fingeret Roth, and Betty Liu,

"Diagnosis of fraud: how employees of HealthSouth fooled colleagues, auditors and investors for 15 years," *Financial Times,* 15 April 2003; and Freudenheim, "HealthSouth audit finds as much as $4.6 billion in fraud," *The New York Times,* 21 January 2004

Mutual funds: Landon Thomas, Jr., "Another fund company let its favored clients time market, memo says," and Riva D. Atlas, "Fund inquiry informant discloses her identity," *The New York Times,* 9 December 2003; "Mutual funds: seller beware," *The Economist,* 17 January 2004, p. 68; and Stephen Labaton, "Fund misconduct is common, panel is told," *The New York Times,* 28 January 2004

Parmalat: John Tagliabue, "Parmalat files for protection from creditors" and "Parmalat said to create ruse for $11 billion," *The New York Times,* 25 December 2003; Tagliabue, "Judge issues arrest order for founder of Parmalat," *The New York Times,* 29 December 2003; Mark Lander and Jason Horowitz, "Investigation of Parmalat widens to include foreign and Italian banks," *The New York Times,* 8 January 2004; Gregory Crouch, "Parmalat units in Netherlands under scrutiny for bond sales," *The New York Times,* 15 January 2004; "Parmasplat," *The Economist,* 17 January 2004; "Italy after Parmalat: Spilt milk," *The Economist,* 7 February 2004; Eric Sylvers, "7 banks said to be added to Parmalat investigation," *The New York Times,* 11 February 2004; Jonathan D. Glater, "Law firm for Parmalat under scrutiny," *The New York Times,* 21 February 2004; and Sylvers, "Indictments are sought in collapse of Parmalat," *The New York Times,* 19 March 2004

First months of 2004: A general roundup of current cases: Alex Berenson, "Prosecutors score white-collar victories," *The New York Times,* 4 April 2004. Enron: Kurt Eichenwald, "Ex-chief financial officer of Enron and wife plead guilty," *The New York Times,* 15 January 2004, and "Enron's Skilling is indicted by U.S. in fraud inquiry," *The New York Times,* 15 January 2004. Adelphia and the Rigas family: Andrew Ross Sorkin, "Adelphia is next in parade of fraud trials," *The New York Times,* 23 February 2004; Barry Meier, "Former insider seen as pivotal in Rigas trial starting today," *The New York Times,* 1 March 2004; Bloomberg News, "Adelphia kept records of deals with Rigases, witness says," *The New York Times,* 28 May 2004. WorldCom: Barnaby J. Feder and Kurt Eichenwald, "Ex-WorldCom chief is indicted by U.S. in securities fraud," *The New York Times,* 3 March 2004; [Business/Financial Desk], "Former chief of WorldCom indicted again," *The New York Times,* 25 May 2004; Kenneth N. Gilpin, "WorldCom restates numbers and settles criminal case," *The New York Times,* 12 March 2004. Tyco: Andrew Ross Sorkin, "Tyco trial ended as juror cites outside pressure," *The New York Times,* 3 April 2004; Jonathan N. Glater, "Latest Tyco trial starts as lawyers' showcase," *The New York Times,* 7 May 2004

Kenneth Lay indicted: United States District Court, Southern District of Texas, Houston Division, United States of America v. . . . Kenneth L. Lay. . . . Filed

July 7 2004; Kurt Eichenwald and Christine Hauser, "Ex-chairman of Enron surrenders and faces 11-charge indictment," *The New York Times,* 8 July 2004

Rigas verdict: Barry Meier, "Founder of Adelphia is found guilty of conspiracy," *The New York Times,* 8 July 2004, and "Michael Rigas is free for now after mistrial is declared," *The New York Times,* 10 July 2004

Scandals in Roman Catholic church: "A tumultuous 2002 in the Boston archdiocese" [a calendar of articles on the church sex-abuse scandal, in the *Boston Globe* in 2002], *Boston Globe,* 13 December 2002; Melinda Henneberger, "Vatican to hold secret trials of priests in pedophilia cases," *The New York Times,* 9 January 2002; Pam Belluck, "Jury finds ex-priest guilty of assaulting boy," *The New York Times,* 19 January 2002; Pam Belluck, "Papers in pedophile case show church effort to avoid scandal," *The New York Times,* 25 January 2002; Daniel J. Wakin, "Brooklyn bishop silent on handling of sex scandal," *The New York Times,* 26 January 2002; "Things it is dangerous to conceal," *The Economist,* 31 January 2002; Pam Belluck, "New Hampshire diocese names 14 priests accused of abuse," *The New York Times,* 16 February 2002; Laurie Goodstein, "Boston priests' sex-abuse scandal has ripple effect; other names are released," *The New York Times,* 17 February 2002; Barbara Whitaker, "Los Angeles cardinal removes priests involved in pedophilia cases," *The New York Times,* 5 March 2002; Pam Belluck, "Maine parish agonizes over a priest's confession," *The New York Times,* 5 March 2002; Laurie Goodstein, "Catholic bishop in Florida quits, admitting sex abuse in the 70's," *The New York Times,* 5 March 2002; Dean E. Murphy and Daniel J. Wakin, "Dioceses, facing great scrutiny, look anew at sex abuse cases," *The New York Times,* 14 March 2002; Anthony DePalma, "Church scandal resurrects old hurts in Louisiana bayou," *The New York Times,* 19 March 2002; "Lexington: The faith of the fathers," *The Economist,* 21 March 2002; Chris Hedges, "Documents allege abuse of nuns by priests," *The New York Times,* 21 March 2002; John Tagliabue, "Polish priests press Vatican on case against bishop," *The New York Times,* 27 March 2002; Anthony DePalma, "[Palm Beach] Diocese reels after losing 2 tainted bishops," *The New York Times,* 27 March 2002; "Sex and the Catholic church: Wolves in the flock," *The Economist,* 28 March 2002; Laurie Goodstein, "St. Louis priest resigns after new accusation," *The New York Times,* 28 March 2002; Brian Lavery, "Irish government plans inquiry into reports of abuse by priests," *The New York Times,* 6 April 2002; James Sterngold, "Los Angeles cardinal at center of strategy to control scandal," *The New York Times,* 7 April 2002; Francis X. Clines, "Cincinnati: Subpoenaed archbishop avoids testifying," *The New York Times,* 19 April 2002; Laurie Goodstein, "Vatican accepts resignation of Milwaukee's archbishop," *The New York Times,* 25 May 2002; Francis X. Clines, "Nearly 100 Kentucky men add to accusations against priests," *The New York Times,* 28 May 2002; "Erring priests: A policy of sorts," *The Economist,* 6 June 2002; Jim Yardley, "Zero tolerance takes big toll in a Texas diocese," *The New*

York Times, 24 August 2002; [Foreign Desk], "Germany: Bishops act on sex abuse," *The New York Times,* 28 September 2002; Warren Hoge, "British cardinal apologizes for ignoring warnings about a pedophile priest," *The New York Times,* 22 November 2002; Michael Rezendes and Matt Carroll, "Documents detail Law's steps on pastor, transferred priest," *Boston Globe,* 12 December 2002; Michael Paulson and Charles M. Sennott, "Cardinal Law resigns," *Boston Globe,* 13 December 2002; Pam Belluck with Frank Bruni, "Law, citing abuse scandal, quits as Boston archbishop and asks for forgiveness," *The New York Times,* 14 December 2002; Michael Paulson, "Diocese gives abuse data," *Boston Globe,* 27 February 2004; "The Catholic bishops: Found wanting: Two studies show the true extent of the paedophilia scandal," *The Economist,* 4 March 2004; Brian MacQuarrie, "Vt. diocese in record settlement over abuse," *Boston Globe,* 7 April 2004; Brian MacQuarrie, "Bishop places priest on leave," *Boston Globe,* 16 April 2004

Robert Bullock on public radio: *All Things Considered,* 27 August 2002

Vannevar Bush, *Science—The Endless Frontier. A Report to the President on a Program for Postwar Scientific Research,* July 1945 (Washington, D.C.: National Science Foundation, reprinted 1960)

Statistics of growth: National Institutes of Health, *NIH Almanac,* http://www.nih.gov/about/almanac; National Science Foundation, *Science & Engineering Indicators 2004*

For classic and recent cases of scientific fraud, see Chapters 2 and 3.

Max Weber, *From Max Weber: Essays in Sociology,* Hans Gerth and C. Wright Mills, editors and translators (New York: Oxford University Press, 1946), pp. 154–55

Robert Merton, "The Normative Structure of Science, 1942" in *The Sociology of Science: Theoretical and Empirical Investigations* (Chicago: University of Chicago Press, 1973) pp. 266–78

Joseph Needham interview with author, Cambridge, 27 October 1975

Max Delbrück, "A physicist's renewed look at biology: twenty years later" (Nobel Lecture, 10 December 1969), text reprinted in *Science* 168 (12 June 1970): 1312–15

Consilience: William Whewell, "*Novum Organum Renovatum,* 1858," in *William Whewell: Theory of Scientific Method,* Robert E. Butts, editor (Indianapolis: Hackett, 1989), p. 139

Francis Crick: personal communication with author

University of California, San Diego: the Spector case is discussed in Chapter 3

Thomas S. Kuhn, *The Structure of Scientific Revolutions,* 2d edition (Chicago: University of Chicago Press, 1970)

Social construction: see, generally, *Handbook of Science and Technology Studies,* Sheila Jasanoff, Gerald E. Markle, James C. Petersen, and Trevor Pinch, editors (Thousand Oaks, California: Sage, 1995); for Marxism in "science studies," see Sal Restivo, "The theory landscape in science studies," in *Handbook,* p. 103; see also Paul R. Gross and Norman Levitt, *Higher Superstition: The Academic Left and Its Quarrels with Science* (Baltimore: Johns Hopkins University Press, 1994), passim; Gerald Holton, *Science and Anti-Science* (Cambridge, Massachusetts: Harvard University Press, 1993), passim; and for comic relief, Alan Sokal and Jean Bricmont, *Fashionable Nonsense: Postmodern Intellectuals' Abuse of Science* (New York: Picador USA, 1998)

Roald Hoffman: quoted in Rudy Baum, "Declining R & D support has ethical implications," *Chemical & Engineering News* (29 April 1996): 66–68

Chapter 2: What's It Like? A Typology of Scientific Fraud

Epigraph: Charles Babbage, *Reflections on the Decline of Science in England, and on Some of Its Causes* (London: B. Fellowes, 1830; facsimile reprint, Shannon, Ireland: Irish University Press, 1971), pp. 167–83

Piltdown man: see extended discussion in Horace Freeland Judson, *The Search for Solutions* (New York: Holt, Rinehart and Winston, 1980), pp. 174–76.

Isaac Newton: Richard S. Westfall, *Never at Rest: A Biography of Isaac Newton* (Cambridge: Cambridge University Press, 1980)

Gregor Mendel, "Experiments on plant hybrids, 1865," in Curt Stern and Eva R. Sherwood, *The Origin of Genetics: A Mendel Source Book,* Eva R. Sherwood, translator (San Francisco: W. H. Freeman, 1966), pp. 1–48; Ronald Aylmer Fisher, "Has Mendel's work been rediscovered?," *Annals of Science* 1 (1936): 115–37, reprinted in Stern and Sherwood, pp. 139–72; Sewall Wright, "Mendel's Ratios," in Stern and Sherwood, pp. 173–75; A. W. F. Edwards, "Are Mendel's results really too close?," *Biological Reviews* 61 (1986): 295–312; Daniel L. Hartl and Vitezslav Orel, "What did Gregor Mendel think he discovered?," *Genetics* 131 (June 1992): 245–53; Franz Weiling, "What about R. A. Fisher's statement of the 'too good' data of J. G. Mendel's Pisum paper?" *Journal of Heredity* 77 (1986): 281–83; and Alain F. Corcos and Floyd V. Monaghan, "Where is the bias in Mendel's experiments?," in Gregor Mendel, *Experiments on Plant Hybrids: A Guided Study* (New Brunswick: Rutgers University Press, 1993), pp. 196–204

Marshall Nirenberg: Heinrich Matthaei interview with author, Göttingen, 22 April 1976

Charles Darwin: Phillip Prodger, "Photography and *The Expression of the Emotions,*" in Charles Darwin, *The Expression of the Emotions in Man and Animals,* 3d edition (New York: Oxford University Press, 1998), pp. 399–415

Louis Pasteur: Gerald L. Geison, *The Private Science of Louis Pasteur* (Princeton: Princeton University Press, 1995), especially Chapters 6 and 7; "Louis Pasteur: Pity a science that needs heroes," *The Economist,* 1 July 1995, pp. 79–80; Peter Monaghan, "Separating fact from legend," *Chronicle of Higher Education* (29 September 1995): A10–11, A18; Max F. Perutz, "The pioneer defended," *The New York Review of Books* (21 December 1995): 54–58, followed by an exchange of letters to the editor by Geison and Perutz, *The New York Review of*

Books (4 April 1996): 68–69; and Adolfo Martínez-Palomo, "The science of Louis Pasteur: a reconsideration," *The Quarterly Review of Biology* 76 (March 2001): 37–45. Other comments by Perutz: personal communications with the author

Robert Millikan: Gerald Holton, "Subelectrons, presuppositions, and the Millikan-Ehrenhaft dispute," in *The Scientific Imagination: Case Studies* (Cambridge: Cambridge University Press, 1978), pp. 25–83; Ullica Segerstråle, "Good to the last drop? Millikan stories as 'canned' pedagogy," *Science and Engineering Ethics* 1 (1995): 197–214; Allan Franklin, "Forging, cooking, trimming and riding on the bandwagon," *American Journal of Physics* 52 (1984): 786–93; and David Goodstein, "Scientific fraud," *American Scholar* 60 (Autumn 1991): 505–15

Frederic Lawrence Holmes, *Meselson, Stahl, and the Replication of DNA: A History of the "Most Beautiful Experiment in Biology"* (New Haven: Yale University Press, 2001)

Ernst Haeckel: M. K. Richardson, J. Hanken, M. J. Gooneratne, C. Pieau, A. Raynaud, L. Selwood, and G. M. Wright, "There is no highly conserved embryonic stage in the vertebrates: implications for current theories of evolution and development," *Anatomy and Embryology* 196 (1997): 91–106; Elizabeth Pennisi, "Hackel's embryos: fraud rediscovered," *Science* 277 (5 September 1997): 1435; M. K. Richardson et al., "Haeckel, embryos, and evolution," *Science* 280 (15 May 1998): 983; Michael K. Richardson and Gerhard Keuck, "A question of intent: when is a 'schematic illustration' a fraud?," *Nature* 410 (8 March 2001): 144, and "Haeckel and the vertebrate archetype," http://zygote.swarthmore.edu/evo5.html

Sigmund Freud: Adolf Grünbaum, *The Foundations of Psychoanalysis: A Philosophical Critique* (Berkeley: University of California Press, 1984); Malcolm Macmillan, *Freud Evaluated: The Completed Arc* (Amsterdam: North-Holland, 1991; reprinted Cambridge, Massachusetts: MIT Press, 1997), p. 625 and passim; Frank Sulloway, *Freud, Biologist of the Mind: Beyond the Psychoanalytic Legend* (London: Burnett Books, 1979), and "Reassessing Freud's case histories: the social construction of psychoanalysis," *Isis* (1991): 245–74; Frederick Crews and His Critics, *The Memory Wars: Freud's Legacy in Dispute* (New York: New York of Review Books, 1995); and Edward Dolnick, *Madness on the Couch: Blaming the Victim in the Heyday of Psychoanalysis* (New York: Simon and Schuster, 1998)

Cyril Burt: Leslie Spencer Hearnshaw, *Cyril Burt: Psychologist* (London: Hodder and Stoughton, 1979), pp. 227–61, and "The Burt affair: a rejoinder," *The Psychologist: Bulletin of the British Psychological Society* 2 (1990): 61–64; Richard Lewontin, "Race and Intelligence," *Bulletin of the Atomic Scientists* (March 1970): 2–8; Leon J. Kamin, "Heredity, intelligence, politics and society," presented to the Thirteenth International Congress of Genetics, 1973, reprinted in *The IQ Controversy: Critical Readings,* Ned Block and Gerald Dworkin, editors (London: Quartet Books, 1977), pp. 242–64; and Oliver Gillie, "Crucial data was faked by eminent psychologist," *The Sunday Times,* 24 October 1976

Chapter 3: Patterns of Complicity: Recent Cases

Epigraph: Philip Handler testimony before the Subcommittee on Investigation of the Committee on Science and Technology, House of Representatives, 31 March 1981

Jayson Blair: Jacques Steinberg, "Times Reporter Resigns after Questions on Article" and "Editors' note," *The New York Times,* 2 May 2003; Dan Barry et al., "Correcting the record; Times reporter who resigned leaves a long trail of deception," "Correcting the record; witnesses and documents unveil deceptions in a reporter's work," and "Editors' note," *The New York Times,* 11 May 2003; Steinberg, "Editor of Times tells staff he accepts blame for fraud," *The New York Times,* 15 May 2003; "Editors' note," *The New York Times,* 12 June 2003; Steinberg, "Changes at the Times: Times's 2 top editors resign after furor on writer's fraud," *The New York Times,* 6 June 2003

William T. Summerlin: Barbara J. Culliton, "The Sloan-Kettering affair: a story without a hero," *Science* (10 May 1974): 644–50, and "The Sloan-Kettering affair (II): an uneasy resolution," *Science* (14 June 1974): 1154–57; and Gail McBride, "The Sloan-Kettering affair: could it have happened anywhere?," *JAMA: The Journal of the American Medical Association* (9 September 1974): 1391–410

John Long: Michael Knight, "Doctor at Harvard quit after faking research data," *The New York Times,* 28 June 1980; and Nicholas Wade, "A diversion of the quest for truth," *Science* 211 (6 March 1981): 1022–25

Elias Alsabti: William J. Broad, "Would-be academician pirates papers," *Science* 208 (27 June 1980): 1438–49; and Patricia Woolf, "Fraud in science: how much, how serious?," *The Hastings Center Report* 11 (October 1981): 91–114

Vijay Soman and Philip Felig: Lawrence K. Altman, "Columbia's medical chief resigns; ex-associate's data fraud at issue," *The New York Times,* 9 August 1980; and William J. Broad, "Imbroglio at Yale (I): emergence of a fraud, *Science* 3 (October 1980): 38–41, and "Imbroglio at Yale (II): atop job lost," *Science* (10 October 1980): 171–73

John Darsee: William J. Broad, "Harvard delays in reporting fraud," *Science* 215 (29 January 1982): 478–82; Daniel C. Tosteson (Dean, Harvard Medical School), "To the Members of the Faculty of Medicine," letter and accompanying "Synopsis," 16 February 1983; Barbara Culliton, "Coping with fraud: the Darsee case" and "Harvard acknowledges it could have done better," *Science* 220 (1 April 1983): 31–35; Richard Knox, "The Harvard fraud case: where does the problem lie?," *JAMA* 249 (8 April 1983): 1797–1807; Knox, "Deeper problems for Darsee: Emory probe," *JAMA* 249 (3 June 1983): 2867–76; Arnold S. Relman, "Editorial: Lessons from the Darsee affair," *The New England Journal of Medicine* 308 (9 June 1983): 1415–17; and Elmer C. Hall et al., "Report of *ad hoc* committee to evaluate research of Dr John R. Darsee at Emory University," *Minerva* 23 (Summer 1985): 276–305

Walter Stewart and Ned Feder: Philip M. Boffey, "Major study points to faulty research at two universities," *The New York Times,* 22 April 1984; Walter W. Stewart and Ned Feder, "The integrity of the scientific literature," John Maddox, "Editorial: Fraud, libel and the literature," and Eugene Braunwald, "On analysing scientific fraud," *Nature* 325 (15 January 1987): 181–82, 207–14, and 215–16; and "Statement of Walter W. Stewart and Ned Feder" before the Subcommittee on Civil and Constitutional Rights of the Committee on the Judiciary, House of Representatives, 26 February 1986

Mark Spector: Gina Bari Kolata, "Reevaluation of cancer data eagerly awaited," *Science* (16 October 1981): 316–18; and Efraim Racker, "A view of misconduct in science," *Nature* (11 May 1989): 91–93

Stephen Breuning and Robert Sprague: Paul W. Valentine, "Drug therapy researcher is indicted," *The Washington Post,* 16 April 1988; Mary Pat Flaherty, "Ex-Western Psych researcher charged in drug-grant fraud," *Pittsburgh Press,* 16 April 1988; Alun Anderson, "Criminal charge in scientific fraud case," *Nature* (21 April 1988): 670; Mark Roman, "When good scientists turn bad," *Discover* (April 1988): 50–58; Robert L. Sprague, "Acceptance of AAAS Scientific Freedom and Responsibility Award," AAAS Convention, San Francisco, 16 January 1989; Daniel S. Greenberg, "Issue of scientific fraud not easily faced," *Chicago Tribune,* 16 August 1989; Christine McGourty, "Scientific misconduct: Threats against witness," *Nature* (24 August 1989): 585; Warren T. Brookes, "The high costs of bad science: Dingell onto something?," *The Washington Times,* 27 September 1989; and Sprague, "Scientific misconduct: recent cases and regulations," presentation at University of Nevada, Reno, 24 April 1992

Robert Slutsky: Robert L. Engler, James W. Covell, Paul J. Friedman, Philip S. Kitcher, and Richard M. Peters, "Misrepresentation and responsibility in medical research," *The New England Journal of Medicine* 317 (26 November 1987): 1383–89

Daniel E. Koshland, Jr, editorial, *Science* 235 (9 January 1987): 41

Viswat Jit Gupta: John A. Talent, "The case of the peripatetic fossils," *Nature* 338 (20 April 1989): 613–15, 604; [unsigned], "News: Gupta takes to the hills," *Nature* 344 (15 March 1990): 187; [unsigned], "Opinion: another rope trick," *Nature* 371 (27 October 1994): 726; and K. S. Jayaraman, "Geologist 'victimized' for role in fraud case," *Nature* 378 (16 November 1995): 227, and "Court allows fossil fraudster to return," *Nature* 380 (18 April 1996): 570

Malcolm Pearce: Owen Dyer, "Consultant struck off for fraudulent claims," *British Medical Journal* 310 (17 June 1995): 1554–55; Stephen Lock, "Editorial: Lessons from the Pearce affair: handling scientific fraud," *British Medical Journal* 310 (17 June 1995): 1547; and Richard Smith, "Editorial: Misconduct in research: editors respond," *British Medical Journal* 315 (26 July 1997): 201–2

Peter Nixon: Clare Dyer, "Cardiologist admits research misconduct," *British Medical Journal* 314 (24 May 1997): 1501; Smith, "Misconduct in research"

John Anderton: Clare Dyer, "Consultant struck off over research fraud," *British Medical Journal* 315 (26 July 1997): 205–10; and Smith, "Misconduct in research"

Mark Williams: Clare Dyer, "Doctor admits research fraud," *BMJ* 316 (25 February 1998): 645; and Smith, "Misconduct in research"

Friedhelm Herrmann and Marion Brach: Alison Abbott, "Forged images lead to German inquiry," *Nature* 387 (29 May 1997): 442, and "Fraud claims shake German complacency," *Nature* 387 (19 June 1997): 750; Quirin Schiermeier, "Gene therapist accused of fraud to seek redress in German court," *Nature* 389 (11 September 1997): 105; Abbott, "Germany tightens grip on misconduct," *Nature* 390 (4 December 1997): 430, and "Researcher sues over 'fraud' sanction," *Nature* 390 (16/25 December 1997): 662; Robert Koenig, "Panel proposes ways to combat fraud," *Science* 278 (19 December 1997): 2049–50; Marion A. Brach, "Correspondence: Scapegoat for fraud in Germany?," *Nature* 392 (2 April 1998): 431; Abbott, "German scientists may escape fraud trial" and "Task force set up to determine the damage," *Nature* 395 (8 October 1998): 532–33 and 533; [unsigned], "Editors debate whether to blow the whistle on suspect papers," *Nature* 398 (4 March 1999): 15; Abbott, "German Science admits to fraud," review of *Der Sündenfall: Betrug und Fälscung in der deutschen Wissenschaft, Nature* 398 (29 April 1999): 765–66; Abbott, "German fraud inquiry casts a wider net of suspicion—as disillusionment reigns in task force," *Nature* 405 (22 June 2000): 871; Michael Hagman, "Panel finds scores of suspect papers in German fraud probe," *Science* 288 (23 June 2000): 2106–7; Andrew Lawler, "Fallout from German fraud case continues," *Science* 291 (9 March 2001): 1876–77; Adam Bostanci, "Germany gets in step with scientific misconduct rules," *Science* 296 (7 June 2002): 1778; and Alison Abbott and Johanna Schwarz, "Dubious data remain in print two years after misconduct inquiry," *Nature* 418 (11 July 2002): 113

Inge Czaja and Richard Walden: Alison Abbott, "Fraud claim puts German rules to test," *Nature* 392 (12 March 1998): 111, "German technician's confession spurs check on suspect data," *Nature* 393 (28 May 1998): 293, and "German institute 'correct' to fire technician," *Nature* 394 (2 July 1998): 6

Peter Seeburg: Alison Abbott, "Seeburg faces misconduct inquiry," *Nature* 399 (10 June 1999): 512; Eliot Marshall, "Startling revelations in UC-Genentech battle," *Science* 284 (7 May 1999): 883; Dennis Henner et al., "UC-Genentech trial" (letter from Genentech and other scientists) and Peter H. Seeburg (letter objecting in part to Marshall's report), *Science* 484 (28 May 1999): 1465; and Michael Hagman, "Researcher rebuked for 20-year-old misdeed," *Science* 286 (17 December 1999): 2249–50

Holger Kiesewetter: Quirin Schiermeier, "German garlic study under scrutiny," *Nature* 401 (14 October 1999): 629

Jan Hendrik Schön: Kenneth Chang, "Bell Labs forms panel to study claims of research misconduct," *The New York Times,* 22 May 2002; Geoff Brumfiel, "Bell Labs launches inquiry into allegations of data duplication," *Nature* (23 May 2002): 367–68; Robert F. Service, "Pioneering physics papers under suspicion for data manipulation," *Science* (24 May 2002): 1376–77; Chang, "A sudden host of questions on Bell Labs breakthroughs," *The New York Times,* 28 May 2002; Service, "Physicists question safeguards, ponder their next moves," *Science* (31 May 2002): 1584–85; Brumfiel, "Bell Labs inquiry spreads to superconductors," *Nature* (4 July 2002): 5–6; David Malakoff, "Winning streak brought awe, and then doubt," *Science* (5 July 2002): 34–35; Chang, "Panel says Bell Labs scientist faked discoveries in physics," *The New York Times,* 26 September 2002; Chang, "On scientific fakery and the systems to catch it," *The New York Times,* 15 October 2002; and Cherry A. Murray and Saswato R. Das (of Bell Labs), "The price of scientific freedom," *Nature Materials* 2 (April 2003): 204–5

Kelley: Blake Morrison and Jacques Steinberg, "A question of credibility: ascent of USA Today reporter stumbled on colleagues' doubts," *The New York Times,* 19 January 2004; Blake Morrison, "Ex-USA Today reporter faked major stories," *USA Today,* 19 March 2004; Rita Rubin, "Material without attribution," *USA Today,* 19 March 2004; Kevin McCoy, "Mileage, expenses and facts don't add up," *USA Today,* 19 March 2004; and Jacques Steinberg, "Writer's work in USA Today called false," *The New York Times,* 20 March 2004; Jacques Steinberg, "Editor of USA Today resigns; cites failure over fabrications," *The New York Times,* 21 April 2004

Blair's book: Jack Shafer, "Dateline: Brooklyn," *New York Times Book Review,* 14 March 2004

Chapter 4: Hard to Measure, Hard to Define: The Incidence of Scientific Fraud and the Struggle Over Its Definition

Committee on Publication Ethics: Richard Smith, "Editorial: Misconduct in research: editors respond," *BMJ* 315 (26 July 1997): 201–2; and Nigel Williams, "Editors call for misconduct watchdog," *Science* 280 (12 June 1998): 1685–86

Fraud in mathematics: Benoit Mandelbrot, personal communication with author

Patricia Woolf, "Deception in scientific research," *Jurimetrics Journal* 29 (Fall

1988): 67–95, and "Fraud in science: how much, how serious?," *The Hastings Center Report* 11 (October 1981): 91–114

Richard Smith, "Editorial: The need for a national body for research misconduct," *BMJ* 316 (6 June 1998): 1686–87

Ian St James-Roberts, "Are researchers trustworthy?," *New Scientist* (2 September 1976): 481–83, and "Cheating in science," *New Scientist* (25 November 1976): 466–69

Judith P. Swazey, Melissa S. Anderson, and Karen Seashore Louis, "Ethical problems in academic research," *American Scientist* 81 (November–December 1993): 542–53; Philip J. Hilts, "Misconduct in science is not rare, a survey finds," *The New York Times,* 12 November 1993; and Lawrence K. Altman, "Her study shattered the myth that fraud in science is a rarity," *The New York Times,* 23 November 1993

Nicholas H. Steneck, "Assessing the integrity of publicly funded research," in *Proceedings: Investigating Research Integrity* (2001)

National Academy of Sciences, *Responsible Science: Ensuring the Integrity of the Research Process* (Washington, D.C.: National Academy Press, 1992); Warren E. Leary, "Panel opposes government policing of scientific misconduct cases," *The New York Times,* 23 April 1992; editorial, "Defining misconduct: US Academy offers strict definitions but ducks other urgent issues," *Nature* 356 (30 April 1992): 730–31; Daniel S. Greenberg, "Academy Misconduct study calls for more study," *Science & Government Report,* 1 May 1992; and David P. Hamilton, "A shaky consensus on misconduct," *Science* 256 (1 May 1992): 604–5

Howard K. Schachman, "What is misconduct in science?," *Science* 261 (9 July 1993): 148–49, 183

Karen A. Goldman and Montgomery K. Fisher, "The constitutionality of the 'other serious deviation from accepted practices' clause," *Jurimetrics* 37 (Winter 1997): 149–66

Kenneth J. Ryan, MD, Chair, Commission on Research Integrity, *Integrity and Misconduct in Research: Report of the Commission on Research Integrity* (Washington, D.C.: U.S. Department of Health and Human Services, Public Health Service, 1995)

Opposition to Ryan Commission report (Bruce Alberts, Ralph Bradshaw): Bruce Alberts, President and Chairman of the Executive Committee, National Academy of Sciences, et al., letter to Dr William Raub, Science Advisor, Department of Health and Human Services, 15 March 1996; Jocelyn Kaiser, "Commission proposes new definition of misconduct," *Science* 269 (29 September 1995): 1811; and Meredith Wadman, "'Unrealistic' misconduct plans under fire" and "Hostile reception to US misconduct report," *Nature* 381 (23 May 1996): 299 and (20 June 1996): 639

Office of Scientific and Technology Policy: The White House: Office of Science and Technology Policy, "Research misconduct: a new definition and new procedures for federal research agencies," *The Federal Register,* 14 October 1999; "Research misconduct: a new definition and guidelines for federal research agencies," *The Federal Register,* 6 December 2000, pp. 76260–76264; Jocelyn Kaiser, "A misconduct definition that finally sticks?," *Science* 286 (15 October 1999): 391

National Academy Town Meeting on Research Misconduct, National Academy of Sciences, 17 November 1999

Chapter 5: The Baltimore Affair

Epigraph: Howard Temin interview with author, University of Wisconsin, 16 March 1993

David Weaver, Moema H. Reis, Christopher Albanese, Frank Costantini, David Baltimore, and Thereza Imanishi-Kari, "Altered Repertoire of Endogenous Immunoglobulin Gene Expression in Transgenic Mice Containing a Rearranged Mu Heavy Chain Gene," *Cell* (25 April 1986): 247–59

Margot O'Toole, 7 May 1986: O'Toole interview with author, Cambridge, Massachusetts, 13 April 1991; response of O'Toole to the Draft Investigative Report from the Office of Scientific Integrity, 13 May 1989; statement of O'Toole to the Subcommittee on Oversight and Investigations of the Committee on Energy and Commerce, House of Representatives, One Hundred First Congress, First Session, 9 May 1989; and O'Toole interview with Walter Stewart, 26 August 1986

Network theory in immunology: Niels Kai Jerne, "Toward a network theory of the immune system," *Annals of Immunology* 125C (1974): 373–89

David Baltimore's background: Baltimore interview with author, Rockefeller University, 31 July 1991; Baltimore dissertation at Rockefeller, *The Diversion of Macromolecular Synthesis in L-Cells Towards Ends Dictated by Mengovirus,* draft lightly annotated by Norton Zinder; Anthony Cerami interview with author, New York, 14 May 1996; and Norton Zinder interview with author, Rockefeller University, 18 May 1995

Margot O'Toole's background: O'Toole interviews with author, Cambridge, Massachusetts, 13 April 1991 and 1 March 1992; O'Toole *curriculum vitae,* undated but after mid-1992; Philip J. Hilts, "Biologist who disputed a study paid dearly," *The New York Times,* 22 March 1991

Thereza Imanishi-Kari's background and education: Japanese in Brazil: "Eine kleine samba," *The Economist,* 4 November 1995; "Biographical sketch: Thereza Imanishi-Kari," in Grant Application No. 2 P01 CA28900-06, received 1 February 1985; letter to author from Darci Pareja de Almeida of the University of São Paulo, 13 April 1993; letter, Laboratory of Developmental Biology, Biological Institute, College of Science, University of Kyoto, 26 August 1970, "To whom it may concern," signed Mikita Kato and Atsuyoshi Hagiwara, with their titles; and letter to Dr John W. Krueger, Office of Research Integrity, from Faculty of Science, Kyoto University, 18 May 1995, signed Yoshio Miyabe, with title

University of Helsinki rules: letter to John Krueger from Maarit Alaluusua, Study Secretary, University of Helsinki, Faculty of Science, 19 June 1995, and e-mail message, 20 June 1995, from Timo-Jussi Hamalainen, Study Counselor

Valto Eero Olavi Mäkelä interview with author, World Congress of Immunology, Budapest, 24 August 1992

Klaus Rajewsky interviews with author, University of Cologne Institute of Genetics, 2 April 1992, and Waldorf-Astoria Hotel, New York, 18 June 1992

Claudia Berek interview with author, Charité, Berlin, 1 April 1992

Robert Jack interview with author, University of Cologne Institute of Genetics, 2 April 1992

Gertud Giels interview with author, near Lake Shasta, California, 26 April 1992

Herman Eisen interview with author, Massachusetts Institute of Technology, 12 February 1993

David Baltimore's view of the importance of Weaver et al.: Baltimore interviews with author, Rockefeller University, 31 July 1991 and 10 September 1991; and Baltimore, "Open letter to Paul Doty," *Nature* 353 (5 September 1991): 9

Nancy Hopkins interview with author, Cambridge, Massachusetts, 29 February 1992

Susumu Tonegawa interview with author, Massachusetts Institute of Technology, 7 August 1992

Philip Sharp, who was associate director of the Center for Cancer Research at Massachusetts Institute of Technology at the time, also remembered that the work that led to Weaver et al. related to the Jerne network hypothesis: "As Weaver was talking about this, that was clearly the interesting issue in the air, that by expressing one idiotype you affected the idiotype response of another"; Sharp interview with author, 17 September 1991.

Alfred Nisonoff interview with author, Brandeis University, 7 August 1992

Morale in Thereza Imanishi-Kari's laboratory and her behavior: O'Toole interviews with author, cited above; Charles Maplethorpe interview with author, Washington, D.C., 19 April 1991; Philip Cohen interview with author, University of California, San Francisco, 26 March 1993; Giels interview with author; Mary White-Scharff interview with author, Boston, 7 August 1992; Gene Brown interview with author, Massachusetts Institute of Technology, 18 September 1991

Bernard Davis interview with author, Cambridge, Massachusetts, 14 April 1991

Grant application: "Control of antigen-specific T cell responses," Department of Health and Human Services, Public Health Service, National Institutes of

Health; Grant Application No. 2 P01 CA28900-06, Principal Investigator Herman N. Eisen; Biographical Sketch, Thereza Imanishi-Kari, page 00012

Seventeen pages in Reis's studbook: handwritten data, in various forms; for O'Toole's response and analysis, see interviews and testimony cited above.

Response to O'Toole: for example, "Memorandum, December 30, 1986 / To: Maury Fox / From: Herman N. Eisen / Re: Allegations of misconduct in a research study entitled, 'Altered Repertoire . . . ,'" last two paragraphs, exhibit at the Subcommittee on Oversight and Investigations of the Committee on Energy and Commerce hearings, pp. 312–13

Benjamin Lewin's price for *Cell*: according to Vitek Tracz, the chairman of Current Science Group, who said in conversation in London, 28 August 2002, that he knew the figure because he had been interested in buying the journal

Benjamin Lewin interview with author, Cambridge, Massachusetts, 16 April 1991; criticism of some of his editorial policies: for example, James Darnell interview with author, Rockefeller University, 20 June 1991, and Paul Berg conversation with author, Stanford University Medical School, 28 July 1998

Klaus Rajewsky's review: anonymous, but acknowledged by him, "Re: Baltimore M1236"

Margot O'Toole to Brigitte Huber to Henry Wortis; meeting with Wortis and Huber: O'Toole interview with author, 13 April 1991; statement of O'Toole to the Subcommittee on Oversight and Investigations, pp. 4–6; Wortis interview with author, Tufts University, 25 June 1991; the Wortis memorandum, "Minutes of Ad-Hoc Committee Meeting / Brigitte Huber, PhD, Robert Woodland, PhD, Henry Wortis, MD (chair) / June 4, 1986," an exhibit at the Subcommittee on Oversight and Investigations hearings of May 4 and 9, 1989, cited above, pp. 303–5; questioned by Dingell at the hearing on May 9, Wortis said that he had drafted it in May of 1987, the Subcommittee on Oversight and Investigations hearing pp. 245–46

Margot O'Toole to Gene Brown to Herman Eisen: Brown interview with author; O'Toole interview with author, 13 April 1991; meeting with Eisen, that interview, and handwritten letter to Gene Brown, 9 June 1986, transmitting her "Memorandum / June 6, 1986 / To: Dr. Herman Eisen / From: Dr. Margot O'Toole [signed] / Subject: the *Cell* 45:247–259 paper," 5 pages. The letter and memorandum were exhibits at the Subcommittee on Oversight and Investigations hearings of May 4 and 9, 1989, pp. 307–10; "Memorandum, June 17, 1986 / To: Gene Brown, Maury Fox, Phil Sharp, and Mary Rowe / From: Herman N. Eisen / Re: Margot O'Toole and the paper by D. Weaver, M. Ries, C. Albanese, F. Costantini, D. Baltimore, and T. Imanishi-Kari," 2 pages, annotation "never sent to anyone"; that Eisen did not transmit this memorandum he explained in an 11 May 1987 letter to O'Toole, an exhibit at the Subcommittee on Oversight and Investigations hearings of May 4 and 9, 1989, p. 321;

"Memorandum," 30 December 1986; and statement of O'Toole to the Subcommittee on Oversight and Investigations hearing, pp. 7–9

Henry Wortis and Herman Eisen efforts not an investigation: Brown interview with author; Eisen interview with author; memorandums written by Eisen; that the Wortis group had no standing, his replies to Dingell at the Subcommittee on Oversight and Investigations hearing, 9 May 1989

Philip M. Boffey, "Major study points to faulty research at two universities," *The New York Times,* 22 April 1986

Charles Maplethorpe interviews with author, Washington, D.C., 19 April 1991 and 8 September 1997, and by telephone 18 May 1998; and testimony of Maplethorpe before the Subcommittee on Oversight and Investigations of the Committee on Energy and Commerce, House of Representatives, One Hundredth Congress, Second Session, 12 April 1988, pp. 106 *et seq.*

Subcommittee on Oversight and Investigations hearings: 12 April 1988 and May 4 and 9, 1989, cited above

James Wyngaarden interview with author, Washington, D.C., 19 April 1991

Davie panel report: an exhibit at Subcommittee on Oversight and Investigations hearings, May 4 and 9, 1989, pp. 324–34; O'Toole's response: O'Toole interviews with author, 13 April 1991, by telephone, 4 December 1991, and Cambridge, 1 February 1992; and her testimony at Subcommittee on Oversight and Investigations hearing, 9 May 1989

David Baltimore's explosion: Subcommittee on Oversight and Investigations hearing, 4 May 1989, pp. 172–74

Philip Sharp's letter: Philip Sharp interview with author, 17 September 1991

David Baltimore's move to Rockefeller University: William K. Stevens, "Nobel Prize winner asked to head Rockefeller U.: Dr Baltimore figured in a study that is being investigated," *The New York Times,* 4 October 1989, "Dispute on new president shatters the tranquillity at Rockefeller U.: Choice of new head splits a citadel of research," *The New York Times,* 10 October 1989, "The ruckus at Rockefeller," *The New York Times,* 12 October 1989, and "Baltimore accepts post despite faculty outcry," *The New York Times,* 18 October 1989; Gunther Blobel interview with author, Rockefeller University, 22 February 1992; Anthony Cerami interview with author; James Darnell interviews with author, Rockefeller University, 20 June 1991 and 7 January 1992; and Norton Zinder interview with author, Rockefeller University, 18 June 1991

The scandal raged: John Maddox, "The end of the Baltimore saga," and David Baltimore, "Dr Baltimore says 'sorry,'" *Nature* 351 (9 May 1991): 85 and 94–95; Margot O'Toole, "Margot O'Toole's record of events," *Nature* 351 (16 May 1991): 180–83; Herman Eisen, "Origins of MIT inquiry," *Nature* 351 (30 May 1991): 343–44; Brigitte T. Huber, Henry H. Wortis, and Robert T. Woodland, "Opinions from an inquiry panel," *Nature* 351 (13 June 1991): 514; Walter W.

Stewart and Ned Feder, "Analysis of a whistle-blowing," *Nature* 351 (27 June 1991): 687–90; Thereza Imanishi-Kari, "Imanishi-Kari's riposte," *Nature* 351 (27 June 1991): 691; O'Toole, "O'Toole re-challenges," *Nature* 351 (27 June 1991): 691–93; letters from Mark Ptashne, John Cairns, and Herman Eisen, "More views on Imanishi-Kari," *Nature* 352 (11 July 1991): 101–2; Barbara J. Culliton, "NIH need clear definition of fraud," *Nature* 352 (15 August 1991): 563; O'Toole, "Imanishi-Kari (continued)," *Nature* 352 (15 August 1991): 560; Eric Selsing, "Response to Ptashne," *Nature* 352 (22 August 1991): 657; Ptashne, "Baltimore's unanswered questions," *Nature* 353 (10 October 1991): 495 (published with Paul Doty's rejoinder to Baltimore, cited below); and Maddox, "Baltimore's defence," *Nature* 353 (10 October 1991): 484

Paul Doty's letter, David Baltimore's angry response, Doty's rejoinder: Paul Doty, "Responsibility and Weaver et al.," *Nature* 352 (18 July 1991): 183–84; David Baltimore, "Open letter to Paul Doty"; Doty, "Baltimore's unanswered questions," *Nature* 353 (10 October 1991): 495

Urging David Baltimore to ignore Paul Doty: Maxine Singer interview with author, National Institutes of Health, 4 March 1992

Maddox, "Baltimore's defence"

Cerami interview with author, 14 May 1996

Gerald Edelman interview with author, Rockefeller University, 21 October 1991

David Rockefeller interview with author, Rockefeller Center, 27 February 1992

Norton Zinder: earlier views, Zinder interview with author, Rockefeller University, 18 June 1991; on the trustees' meeting, Zinder telephone interview with author, 3 December 1991

Paul Berg interview with author, Stanford, 7 February 1992

Richard Furlaud interview with author, Furlaud's office, 26 February 1992

David Baltimore's resignation: Furlaud interview with author; Baltimore interview with author, Rockefeller University, 28 February 1992

Office of Scientific Integrity report: Office of Research Integrity / Investigation Report / The Massachusetts Institute of Technology / ORI 072; letter to Imanishi-Kari by her attorney, Bruce A. Singal, Office of the Secretary, Department of Health and Human Services, 26 October 1994, signed Terrence [*sic*] J. Tychan, Deputy Assistant Secretary for Grants and Acquisition Management, and Lyle W. Bivens, Director, Office of Research Integrity

Appeals-board decision: Department of Health and Human Services / Departmental Appeals Board / Research Integrity Adjudications Panel / June 21, 1996, Subject: Thereza Imanishi-Kari, PhD: Decision

Donald Kennedy on financial responsibility at universities: Subcommittee on Oversight and Investigations of the Committee on Energy and Commerce, House of Representatives, One Hundred and Second Congress, First Session, March 13 and May 9, 1991; Kennedy's testimony, March 13, p. 138 *et seq.*

Bernadine Healey: Subcommittee on Oversight and Investigations of the Committee on Energy and Commerce, House of Representatives hearings on Apparent Wrongdoing at the NIH Laboratory of Tumor Cell Biology and Actions Involving the Office of Scientific Integrity, March 6 and August 1, 1991; Healey's testimony, August 1, p. 169 *et seq.*

Joseph Onek's cross-examination begins, in the Transcript of Proceedings, p. 504, line 13 (of course Judith Ballard's silent gesture is not recorded); the passage from the cross-examination of Baltimore is from the hearing of 26 June 1995, pp. 2092–93

Joseph Davie: Subcommittee on Oversight and Investigations, May 4 and 9, 1998; earlier draft of first Davie panel report, pp. 14–17; Davie on misconduct, p. 16

Howard Temin interview with author, University of Wisconsin, March 15 and 16, 1993

Klaus Rajewsky on fate of network theory: Rajewsky telephone interview with author, 22 December 1996

Chapter 6: The Problems of Peer Review

Epigraph: Philip Kitcher, *Science, Truth, and Democracy* (Oxford University Press, 2001), p. 3; and Richard Smith, "The future of peer review," in *Peer Review in Health Sciences,* 2d edition, Fiona Godlee and Tom Jefferson, editors (London: BMJ Books, 2003)

Among the important general treatments of peer review: Stephen Lock, *A Difficult Balance: Editorial Peer Review in Medicine* (London: Nuffield Provincial Hospitals Trust, 1985); the four international congresses on peer review in biomedical publication, organized by *JAMA* and led by Drummond Rennie, deputy editor (west) and which were subsequently published in *JAMA*: congress in Chicago, May 1989: *JAMA* 263 (9 March 1990), congress in Chicago, September 9–11, 1993: *JAMA* 272 (13 July 1994), congress in Prague, September 1997: *JAMA* 280 (15 July 1997), congress in Barcelona, September 14–16, 2001: *JAMA* 287 (5 June 2002); and *Peer Review in Health Sciences*

Horace Freeland Judson, "Structural transformations of the sciences and the end of peer review," *JAMA* 272 (13 July 1994): 92–94

Elizabeth Knoll, "The communities of scientists and journal peer review," delivered at First International Congress, *JAMA* 263 (9 March 1990): 1330–32

Andrew Carnegie and John D. Rockefeller, Sr: Gerald Jonas, *The Circuit Riders: Rockefeller Money and the Rise of Modern Science* (New York: W. W. Norton, 1989); Robert E. Kohler, *Partners in Science: Foundations and Natural Scientists 1900–1945* (Chicago: University of Chicago Press, 1991), pp. 15–105; and Andrew Carnegie, "Wealth," *North American Review* 148 (June 1889): 663–64, and *North American Review* 149 (December 1889): 682–98, reprinted in

The Gospel of Wealth and Other Timely Essays (Cambridge, Massachusetts: Harvard University Press, 1962)

Warren Weaver: Jonas, *The Circuit Riders,* passim, especially pp. 177–89 and 201–10; Robert E. Kohler, "The management of science: the experience of Warren Weaver and the Rockefeller Foundation programme in molecular biology," *Minerva* 14 (Autumn 1976): 279–306

Boris Ephrussi and Jacques Monod: Horace Freeland Judson, *The Eighth Day of Creation: Makers of the Revolution in Biology,* expanded edition (Cold Spring Harbor, New York: Cold Spring Harbor Laboratory Press, 1996), pp. 350–52

Plate tectonics: Bill Glen, *The Road to Jaramillo* (Palo Alto, California: Stanford University Press, 1981)

Andrew Wiles: Gina Kolata, "At last, shouts of 'Eureka!' in an age-old math mystery," 24 June 1993, *The New York Times,* A1; in fact, a flaw was found in the proof, but fixed: Gina Kolata, "How a gap in the Fermat proof was bridged," *The New York Times,* 31 January 1995

Vannevar Bush, *Science: The Endless Frontier: A Report to the President on a Program for Postwar Scientific Research* (Washington, D.C., 5 July 1945, reprinted, Washington, D.C.: National Science Foundation, July 1960)

Robert Pollack: personal communication with author

National Science Foundation Web site: http://www.nsf.gov/; Stephen Cole, Leonard Rubin, and Jonathan R. Cole, *Peer Review in the National Science Foundation: Phase One of a Study* (Washington, D.C.: National Academy of Sciences, 1978); Jonathan R. Cole and Stephen Cole, *Peer Review in the National Science Foundation: Phase Two of a Study* (Washington, D.C.: National Academy Press, 1981)

Christine Wennerås and Agnes Wold, "Nepotism and sexism in peer-review," *Nature* 387 (22 May 1997): 341–43

Elvera Ehrenfeld: telephone conversation with author, 14 June 2004

Howard Hughes Medical Institute: telephone check with public information office, 16 July 2004

Richard Smith interview with author, BMJ Publishing Group, 29 August 2002

Royal Society, *Philosophical Transactions*: Robert William Frederick Harrison (assistant secretary of the Royal Society, London), "The Royal Society," *Encyclopædia Britannica,* vol. XXIII, 11th edition (Cambridge, England: Cambridge University Press, 1911), pp. 791–93

History of journals and refereeing: David A. Kronick, "Peer review in 18th-century scientific journalism," *JAMA* 263 (9 March 1990): 1321–22; John C. Burnham, "The evolution of editorial peer review," *JAMA* 263 (9 March 1990): 1323–29; and Burnham, "How journal editors came to develop and critique peer review procedures," in *Research Ethics, Manuscript Review, and Journal*

Quality, Mayland and Sojka, editors (Madison, Wisconsin: American Society of Agronomy, 1992), pp. 55–62

Ernest Hart: quoted and cited in Burnham, "The evolution of editorial peer review"

Donald Kennedy: Subcommittee on Oversight and Investigations of the Committee on Energy and Commerce, House of Representatives, One Hundred Second Congress, First Session, hearings on Indirect Cost Recovery Practices at U.S. Universities for Federal Research Grants and Contracts, March 13 and May 9, 1991; the Stanford group's testimony begins p. 138

David Botstein interview with author, Stanford Medical School, 14 May 1991

Richard Horton, "The hidden research paper," delivered at Fourth International Congress, *JAMA* 287 (5 June 2002): 2775–78

Tom Jefferson, Philip Alderson, Elizabeth Wager, and Frank Davidoff, "Effects of editorial peer review: a systematic review," delivered at Fourth International Congress, *JAMA* 287 (5 June 2002): 2784–86

Knoll, "The communities of scientists and journal peer review," cited above

Richard Smith: interview

Chapter 7: Authorship, Ownership: Problems of Credit, Plagiarism, and Intellectual Property

Epigraph: Stephen Lock, "Fraud in medicine," *BMJ* 296 (6 February 1998): 376–77

Raymond V. Damadian: full-page advertisement, "The Shameful Wrong That Must Be Righted . . . Paid for by the friends of Raymond Damadian," *The New York Times,* 10 October 2003; Horace Freeland Judson, "No Nobel prize for whining," *The New York Times,* 20 October 2003, op-ed page; two-page advertisement, "This is the great voyage of scientific discovery . . . ," *The New York Times,* 9 December 2003; [unsigned], "Campaigns: the price of honor," *Science* 302 (19 December 2003): 2065; and John Gore, "Out of the shadows—MRI and the Nobel prize," *The New England Journal of Medicine* 349 (11 December 2003): 2290–91

Barbara McClintock's biography: Nathaniel C. Comfort, *The Tangled Field: Barbara McClintock's Search for the Patterns of Genetic Control* (Cambridge, Massachusetts: Harvard University Press, 2001)

LaDeana W. Hillier et al., "The DNA sequence of human chromosome 7," *Nature* 424 (10 July 2003): 157–64

Genetic code: Horace Freeland Judson, *The Eighth Day of Creation,* expanded edition (Cold Spring Harbor, Long Island: Cold Spring Harbor Laboratory Press, 1996), pp. 468–71; and F. H. C. Crick, "The recent excitement in the coding problem," in *Progress in Nucleic Acid Research* 1 (New York: Academic Press, 1963), pp. 163–217

John Crewdson, "Fraud in Breast Cancer Study: Doctor Lied on Data for Decade," *Chicago Tribune,* 13 March 1994

Bernard Fisher case: Lawrence K. Altman, "Researcher falsified data in breast cancer study" and "Flawed study raises questions on U.S. research: fall of a man pivotal in breast cancer research," *The New York Times,* 4 April 1994.

Genome project: International Human Genome Sequencing Consortium, "Initial sequencing and analysis of the human genome," *Nature* 409 (15 February 2001): 860–921, and associated articles, pp. 813–59 and 922–41; Horace Freeland Judson, "Talking about the genome," *Nature* 409 (15 February 2001): 769; and J. Craig Venter et al., "The sequence of the human genome," *Science* 291 (16 February 2001): 1304–51

Fraud in Francis Collins's laboratory: John Crewdson, "Disclosures of fraud rock gene project; leukemia research set back," *Chicago Tribune,* 29 October 1996; Lawrence K. Altman, "Falsified data found in gene studies," *The New York Times,* 30 October 1996; Eliot Marshall, "Fraud strikes top genome lab," *Science* 274 (8 November 1996): 908–10; Francis Collins e-mail to author, 8 February 2004; and letter by Charles F. Wooley and letter by Paul de Sa and Ambuj Sagar, *Science* 274 (6 December 1996): 1593

Donald Kennedy, "Next steps in the Schön affair," *Science* 298 (18 October 2002): 495

Editorial, "Ghost with a chance in publishing undergrowth," *The Lancet* 342 (18–25 December 1993): 1498–99

Drummond Rennie at Nottingham meeting and Vancouver Group: Fiona Godlee, "Definition of 'authorship' may be changed," *Nature* 312 (15 June 1996): 1501–02

Drummond Rennie, Linda Emanuel, and Veronica Yank, "When authorship fails: a proposal to make contributors accountable," *JAMA* 278 (1997): 579–85

Drummond Rennie letter, *Science* 298 (22 November 2002): 1554

Gordon Research Conference: Web site, http://www.grc.uri.edu

Committee on Publication Ethics: Nigel Williams, "Editors seek ways to cope with fraud," *Science* 278 (14 November 1997): 1221; Web site: http://www.publicationethics.org.uk/

Marek Wroński: e-mail to Discussion of Fraud in Science (SCIFRAUD), 3 December 1997

Eliot Marshall, "Medline searches turn up cases of suspected plagiarism" and "The Internet a powerful tool for plagiarism sleuths," *Science* 279 (23 January 1998): 274–76

Tadeusz Wilczok, "An open letter to Scifraud," 9 March 1998, distributed 11 March 1998

Fintan Culwin: telephone interview with author and e-mail, 9 February 2004

Caroline Phinney: Philip J. Hilts, "Scholar who sued wins $1.2 million: University of Michigan cited in theft-of-research case," *The New York Times,* 22 September 1993, and "University forced to pay $1.6 million to researcher," *The New York Times,* 10 August 1997; [unsigned], *Science* 277 (12 September 1997): 1611; and e-mail exchange, Phinney with author, March 2004

Heidi Weissmann: Daniel S. Greenberg, "Plagiarist gets promoted; his victim loses her job," *The Medical Post,* 25 September 1990): 11 (reprinted from *Science and Government Report* [1 May 1990]); Daniel S. Greenberg, "Copyright-suit winner faces misconduct charge," *Science and Government* Report (1 November 1990); and "Case No. 7: plagiarism: taking sides at Montefiore Medical Center," United States House of Representatives, *Nineteenth Report by the Committee on Government Operations,* One Hundred and First Congress, 10 September 1990, pp. 39–43 and 63

Robert Gallo affair: Mikulas Popovic et al., "Detection, isolation, and continuous production of cytopathic retroviruses (HTLV-III) from patients with AIDS and pre-AIDS," *Science* 224 (4 May 1984): 497–500, followed by three more reports on research with HTLV-III, with Popovic and Gallo and various other authors, pp. 500–8; John Crewdson, "Special report: the great AIDS quest: science under the microscope," *Chicago Tribune,* 19 November 1989; Crewdson, "U.S. Probe cites lies, errors, in AIDS article," *Chicago Tribune,* 15 September 1991; Philip J. Hilts, "Censure is urged for AIDS scientist: panel finds no wrongdoing by Gallo but faults him for lapses by others," *The New York Times,* 16 September 1991; Joseph Palca, "Draft of Gallo report sees the light of day," *Science* 253 (20 September 1991): 1347–48; Barbara J. Culliton, "Popovic rebuts," *Nature* 353 (26 September 1991): 288–89; Crewdson, *Science Fictions: A Scientific Mystery, a Massive Cover-up, and the Dark Legacy of Robert Gallo* (Boston: Little, Brown, 2002); and Jaap Goudsmit, "Book review: Lots of peanut shells but no elephant," *Nature* 416 (14 March 2002): 125–26

Joint history: Robert C. Gallo, M.D., and Luc Montaigner, M.D., "The discovery of HIV as the cause of AIDS." *The New England Journal of Medicine* 349 (11 December 2003): 2283–85

Chapter 8: The Rise of Open Publication on the Internet

Epigraph: Steven Harnad, "Open Archiving for an Open Society: Freeing the Scholarly and Scientific Research Literature Online Through Public Self-archiving," July 2000, http://www.cogsci.soton.ac.uk/~harnad/

Richard Smith: David Nicholson, "The controversial editor of the BMJ discusses how he sees the future of electronic biomedical publishing," *The Scientist* (7 September 2001)

Global knowledge network: Paul Ginsparg, "Creating a global knowledge network," invited contribution for conference held at UNESCO headquarters, Paris, France, 19–23 February 2001, Second Joint ICSU Press—UNESCO

Expert Conference on Electronic Publishing in Science, during session responses from the scientific community, 20 February 2001

Joshua Lederberg: "Communication as the root of scientific progress," lecture to the Sixth International Conference of the International Federation of Science Editors, Woods Hole, Massachusetts; edited version in *The Scientist* (8 February 1993): 10, 11, 14

Paul Ginsparg interviews with author, Los Alamos National Laboratory, 2 July 1998 and 29 June 2001

Gary Taubes, "Publication by electronic mail takes physics by storm," *Science* 259 (26 February 1993): 1246–48; and "E-mail withdrawal prompts spasm," *Science* 262 (8 October 1993): 173–74

Ginsparg, UNESCO: cited above

Andrew Odlyzko, "Tragic loss or good riddance? The impending demise of traditional scholarly journal," condensed version, 26 September 1994, in *Notices of the American Mathematical Society* (January 1995); full version, 6 November 1994, in *Electronic Publishing Confronts Academia: The Agenda for the Year 2000* (Cambridge, Massachusetts: MIT Press/ASIS monograph, 1995)

Derek J. de Solla Price, *Little Science, Big Science* (New York: Columbia University Press, 1963), pp. 5–27

Steven Harnad interview with author, University of Southampton, 16 June 2000, and "Open archiving"

Ginsparg's response to Harnad's numbers: interviews cited above; e-mail commentary on a draft of this chapter, 15 January 2004

Craig Bingham, "The future of medical publishing," HMS Beagle.com, issue 46, posted 22 January 1999, http://gateways.bmn.com/hmsbeagle

George Lundberg, "Re: electronic archiving and IIS talk," e-mails, 8 September 2000, 16:03:04 and 14:25:35 BST, in "September 1998 *American Scientist* Forum: Problems with the Ginsparg Model: Debate 2000–2003," archive generated 16 October 2003

Querulous correspondent: Chris Armstrong, "Re: electronic archiving," 8 September 2000, 10:46:36 BST; Harnad's reply, 8 September 2000, 12:35:24 BST

"artefact of print": Steve Hitchcock, "Re: electronic archiving," 8 September 2000, 13:51:04 BST

Odlyzko, "Tragic loss or good riddance": cited above

David Goodman, "Re: librarians and the Ginsparg model," e-mail, 8 September 2000, 17:54:45 BST in *American Scientist* Forum (September 1998)

Martin Rees and Alvaro de Rújula: A. de Rújula, "The new paradigm for gamma-ray bursts: a case of unethical behaviour?," arXiv:physics/0310134 v[ersion] 2, 2 November 2003; and Jim Giles, "Critical comments threaten to open libel floodgate for physics archive," *Nature* 426 (6 November 2003): 7

"no global court": Paul Ginsparg, e-mail, 15 January 2004

John Maddox, "*Nature* on the Internet (at last)!," *Nature* 377 (12 October 1995): 475

Journals on the World Wide Web: Gary Taubes, "Science journals go wired" and "Electronic preprints point the way to 'author empowerment,'" *Science* 271 (9 February 1996): 764–66, 767–68

Language gene: Cecillia S. L. Lai et al., "A forkhead-domain gene is mutated in a severe speech and language disorder," *Nature* 413 (4 October 2001): 519–23; and Wolfgang Enard et al., "Molecular evolution of *FOXP2*, a gene involved in speech and language," *Nature* 418 (22 August 2002): 869–72

Harnad, "Open archiving": cited above

Reed Elsevier rapacious: for example, Brock Read, "Editorial board of scientific journal quits, accusing Elsevier of price-gouging," *Chronicle of Higher Education* (9 February 2004)

Harold Varmus, "E-BIOMED: a proposal for electronic publications in the biomedical sciences," May 5, 1999 (draft), from NIH Web site, 6 May 1999; Eliot Marshall, "NIH weighs bold plan for online preprint publishing," *Science* 283 (12 March 1999): 1610–11; Declan Butler, "NIH plan brings global electronic journal a step nearer reality," *Nature* 398 (29 April 1999): 735; and [News note], "Varmus circulates proposal for NIH-backed online venture," *Science* 284 (30 April 1999): 718

Objections to E-biomed: Robert Pear, "N.I.H. plan for journal on the web draws fire," *The New York Times,* 8 June 1999; Meredith Wadman, "Varmus defends plan for global environmental e-journal," *Nature* 399 (24 June 1999): 720; and Eliot Marshall, "Varmus defends E-biomed proposal, prepares to push ahead," *Science* 284 (25 June 1999): 2062–63

Vitek Tracz and BioMed Central: Tracz interview with author, London, 28 August 2002; and Kate Galbraith, "British researchers get to publish in 90 online medical journals free," *Chronicle of Higher Education* (11 July 2003)

Public Library of Science: "PloS history," at PloS Web site; and Pat Hagan, "PLoS plans to publish its own journals," *The Scientist* (7 September 2001)

Conferences: meeting at Howard Hughes Medical Institute, 11 April 2003, "Bethesda statement on open access publishing," released 20 June 2003; and "Berliner Erklärung über offenen Zugang zu wissenschaftlichen Wissen," published simultaneously as "Berlin declaration on open access to knowledge in the sciences and the humanities," 22 October 2003

Cornell library: Jonathan Knight, "Cornell axes Elsevier journals as prices rise," *Nature* 426 (20 November 2003): 217

Chapter 9: Laboratory to Law: The Problems of Institutions When Misconduct Is Charged

Epigraph: Stephen Breyer, "Introduction," Reference Manual on Scientific Evidence, second edition (Washington, D.C.: Federal Judicial Center, 2000), p. 2

Jan Hendrik Schön at Koblenz: Cherry A. Murray and Saswato R. Das (Bell Labs), "The price of scientific freedom," *Nature Materials* 2 (April 2003): 204–5; and Kenneth Chang, "Panel says Bell Labs scientist faked discoveries in physics," *The New York Times,* 26 September 2002

Courses in ethics: Gina Kolata, "Ethics 101: A course about the pitfalls," *The New York Times,* 21 October 2003

Morale in Thereza Imanishi-Kari's lab: author's interviews with Margot O'Toole, Charles Maplethorpe, Gertrud Giels, Philip Cohen, and others cited in Chapter 5

Interview with David Baltimore: at the Carnegie Institution dinner on 21 May 1991, Baltimore lectured on "Biological diversity: the immune system"; interview with author, set up by Ray O'Rourke of Burson-Marsteller, at Rockefeller University, 31 July 1991

C. K. Gunsalus: *Responding to Allegations of Search Misconduct: A Practicum,* sponsored by the American Association for the Advancement of Science and the Association of American Medical Colleges, San Francisco, 14 December 1992; Gunsalus interview with author that same day; *Investigating Allegations of Research Misconduct: A Practicum,* sponsored by the AAAS/ABA National Conference of Lawyers and Scientists, Association of American Medical Colleges, Association of American Universities, and the National Association of State Universities and Land-Grant Colleges, 12–13 December 1994, San Francisco; C. K. Gunsalus, JD, "Institutional structure to ensure research integrity," *Academic Medicine* 68 (September Supplement, 1983): S33–S38

Robert McCaa: Susan Lauscher, "Scientific misconduct: a case study," *Grants Magazine* 10 (September 1987): 143–49

Sloppiness in tax law: I owe the point to George Thomsen, Esq.

Stephen Breyer, "The interdependence of science and law," *Science* 280 (1998): 537

Frye v. United States, 54 App.D.C. 46, 47, 293 F. 1013, 1014 (1923); see also Federal Rules of Evidence, Rule 702, 28 U.S.C.A.

William Daubert, Joyce Daubert, individually and as Guardians Ad Litem for Jason Daubert, a minor; Anita Young, individually and as Guardian Ad Litem for Eric Schuller, v. Merrell Dow Pharmaceuticals, Inc., 509 U.S. 579, 113 S. Ct. 2786, 125 L.Ed.2d 469

Wolfgang Pauli: Breyer and elsewhere

Joe Cecil: e-mail attachment, 8 March 2004

General Electric Company v. Robert K. Joiner, 522 U.S. 136, 118 S. Ct. 512

ipse dixit: see Joe S. Cecil, "Construing science in the quest for '*ipse dixit*': a comment on Sanders and Cohen," *Seton Hall Law Review* 33 (2003): 967–86

Scientific community's response: I owe the point to Joe Cecil; Breyer mentions the American Association for the Advancement of Science pilot program

Kumho Tire Company, Ltd. v. Carmichael, 526 U.S. 137, 119 S. Ct. 1167

Observer of questioning: Olivia Judson, covering the hearing for *The Economist,* personal comunication with author

Laural L. Hooper, Joe S. Cecil, and Thomas E. Willging, *Neutral Science Panels: Two Examples of Panels of Court-Appointed Experts in the Breast Implants Product Liability Litigation* (Washington, D.C.: Federal Judicial Center, 19 November 1999)

Response to Pointer panel: editorial, "An Unnatural Disaster," *The Wall Street Journal,* 10 December 1998

Marvin E. Frankel, "The search for truth: an umpireal [*sic*] view," Benjamin N. Cardozo lecture delivered to the Association of the Bar of the City of New York in 1974, *University of Pennsylvania Law Review* (May 1975): 1031–67

John H. Langbein, "The German advantage in civil procedure," *The University of Chicago Law Review* (Fall 1985): 823–82

John C. Reitz, "Why we probably cannot adopt the German advantage in civil procedure," *Iowa Law Review* (May 1990): 987–1014

John H. Langbein, "Trashing the German advantage," *Northwestern University Law Review* (Spring 1988): 763–90

Administrative law: A. V. Dicey, *Lectures on the Relationship between Law and Public Opinion in England in the Nineteenth Century,* 2d edition (London: Macmillan, 1914)

Epilogue: The Transition to the Steady State and the Ends of the Sciences

Epigraph: Peter Medawar, "A biological retrospect," in *Pluto's Republic* (Oxford University Press, 1982), p. 287

Albert Michelson, *Physics Today* (April 1968): 9; quoted and cited in John Horgan, *The End of Science: Facing the Limits of Knowledge in the Twilight of the Scientific Age* (Reading, Massachusetts: Addison-Wesley, 1996), pp. 19 and 270.

X rays, etc.: more detail in Horace Freeland Judson, *The Search for Solutions* (New York: Holt, Rinehart and Winston, 1980), passim

Gunther S. Stent, *The Coming of the Golden Age: A View of the End of Progress* (Garden City, New York: Natural History Press, 1969)

Horgan, *The End of Science*: passim, especially pp. 60–91

Richard Feynman, *The Character of Physical Law* (lectures delivered at Cornell University in 1965; reprinted New York: Modern Library, 1994), p. 166

John Ziman, "The transition to the steady state," paper read at the Symposium on Science in Crisis at the Millennium, George Washington University, 19 September 1996; see also *The Research System in Transition,* Susan E. Cozzens, Peter Healey, Arie Rip, and John Ziman, editors (Dordrecht: Kluwer Academic Publishers, 1990), and John Ziman, *Prometheus Bound: Science in a Dynamic Steady State* (Cambridge University Press, 1994)

Harold Varmus on peer review: Eliot Marshall, "Varmus: the view from Bethesda," *Science* 262 (26 November 1993): 1364–66

Donald Kennedy, Jim Austin, Kirstie Urquhart, and Crispin Taylor, "Supply Without Demand," *Science* 303 (20 February 2004): 1105

Editorial, "War on Cancer," *The New York Times,* 31 May 1971; and Harold M. Schmeck, Jr., "None of the enemies retreat," *The New York Times,* 5 August 1973

The art of our time: portions adapted from a longer, earlier article, Horace Freeland Judson, "Has science won the argument?," *The Spectator,* 4 June 1977, pp. 24–25

Peter Medawar on scientific reasoning: "Induction and intuition in scientific thought," in *Pluto's Republic,* p. 103

INDEX